Climate Change Management

Series Editor

The aim of this book series is to provide an authoritative source of information on climate change management, with an emphasis on projects, case studies and practical initiatives – all of which may help to address a problem with a global scope, but the impacts of which are mostly local. As the world actively seeks ways to cope with the effects of climate change and global warming, such as floods, droughts, rising sea levels and landscape changes, there is a vital need for reliable information and data to support the efforts pursued by local governments, NGOs and other organizations to address the problems associated with climate change. This series welcomes monographs and contributed volumes written for an academic and professional audience, as well as peer-reviewed conference proceedings. Relevant topics include but are not limited to water conservation, disaster prevention and management, and agriculture, as well as regional studies and documentation of trends. Thanks to its interdisciplinary focus, the series aims to concretely contribute to a better understanding of the state-of-the-art of climate change adaptation, and of the tools with which it can be implemented on the ground.

Notes on the quality assurance and peer review of this publication

Prior to publication, the quality of the works published in this series is double blind reviewed by external referees appointed by the editor. The referees are not aware of the author's name when performing the review; the referees' names are not disclosed.

More information about this series at http://www.springer.com/series/8740

Johannes M. Luetz · Patrick D. Nunn
Editors

Beyond Belief

Opportunities for Faith-Engaged Approaches to Climate-Change Adaptation in the Pacific Islands

Springer

Editors
Johannes M. Luetz
School of Social Sciences
Christian Heritage College (CHC)
Brisbane, QLD, Australia

School of Law and Society
University of the Sunshine Coast (USC)
Maroochydore, QLD, Australia

School of Social Sciences
University of New South Wales (UNSW)
Sydney, NSW, Australia

Patrick D. Nunn
School of Law and Society/Australian
Centre for Pacific Islands Research
University of the Sunshine Coast (USC)
Sippy Downs, QLD, Australia

ISSN 1610-2002 ISSN 1610-2010 (electronic)
Climate Change Management
ISBN 978-3-030-67601-8 ISBN 978-3-030-67602-5 (eBook)
https://doi.org/10.1007/978-3-030-67602-5

This Springer imprint is published by the registered company Springer Nature Switzerland AG
The registered company address is: Gewerbestrasse 11, 6330 Cham, Switzerland

Preface—Authors' Note

The American academic John Richardson Jr. (1992) once said, "When it comes to the future, there are three kinds of people: those who let it happen, those who make it happen, and those who wonder what happened" (p. 2).[1] This book is not about religion, nor is it about climate change. It is about survival. This book is intended to help inform and map effective and sustainable adaptive pathways that will allow people in the Pacific Islands to weather the challenges of future climate change so that, two hundred years hence, these people will still be able to identify as unique. Today we acknowledge the uniqueness of Pacific Island belief systems and their manifold expressions, and we acknowledge the uniqueness of Pacific Island places and that of the cultures and livelihoods they support. All this singularity is threatened by climate change and how Pacific Island peoples confront these threats, especially within the next twenty years, will largely determine whether or not these peoples and the places they aspire to continue occupying are still recognizable centuries from now. As the editors of *Beyond Belief: Opportunities for Faith-Engaged Approaches to Climate-Change Adaptation in the Pacific Islands*, we propose that incorporating religious perspectives into climate-change adaptation discourse and practice is the critical "missing link" to more effective and sustainable adaptation to climate change in the Pacific Island region.

The editors hope that the chapters in this volume will appeal to a broad readership of researchers, practitioners, religious leaders, policymakers, and other stakeholders who will derive encouragement and orientation from its manifold perspectives and prospects. Inviting and leveraging spiritual leadership to bring their influence to bear in public policy fora will help shape and support policy development and enable regulatory change to be more holistic and effective. The editors also hope that the book may inspire others to venture forth courageously into many new research spaces where faith-informed climate-change adaptation still remains in its infancy. As such, the book also embodies an invitation to other colleagues to build this nascent field of faith-engaged climate-change adaptation in the Pacific Island region and beyond.

[1] Richardson JM (1992) Making it happen: A positive guide to the future. U.S. Association for the Club of Rome, Washington, D.C.

We wish to thank the diverse range of chapter authors from various disciplines for their willingness to share their expertise as well as the anonymous peer reviewers who gave generously of their time and expertise to ensure each chapter is well aligned with contemporary understandings and the most recent research. With a chapter line-up as rich in multidisciplinary diversity, conceptual originality and practice-informed cogency as you are now holding in your hands, it is our joy and privilege to present to you *Beyond Belief: Opportunities for Faith-Engaged Approaches to Climate-Change Adaptation in the Pacific Islands*. We trust that you will enjoy a thought-provoking time of contemplation, inspiration and imagination towards the *implementation* and *normalisation* of faith-informed climate-change adaptation narratives and practices.

Brisbane, Australia
Sippy Downs, Australia
March 2021

Johannes M. Luetz
Patrick D. Nunn

Acknowledgements

Johannes Luetz wishes to thank Wendy for her selfless support of this book project. Having lived in countries around the world, she exemplifies and embodies the truism by William Langewiesche, "So much of who we are is where we have been". Her ceaseless reminding to view the world and people through the lens of their socio-historical context continues to be a source of ongoing encouragement to him.

Patrick Nunn wishes to thank all the Pacific Islanders whom over the course of some thirty years have educated him about the way they understand the world and especially Rosie and Petra, the two Pacific Islanders closest to him, who have selflessly supported him to work on this book.

Contents

Contributors

Jalasayi Atkinson-Nolte School of Health and Behavioural Sciences, University of the Sunshine Coast, Maroochydore, QLD, Australia

Clive W. Ayre University of Queensland, Brisbane, QLD, Australia

Kurt Bangert Bad Nauheim, Germany

Zina Bird Pacific Centre for Environment and Sustainable Development (PaCE-SD), The University of the South Pacific, Suva, Fiji

Shannon Brincat School of Law and Society, University of the Sunshine Coast (USC), Maroochydore, QLD, Australia

Helene Jacot Des Combes Pacific Centre for Environment and Sustainable Development (PaCE-SD), The University of the South Pacific, Suva, Fiji;
National Disaster Management Office, Majuro, Republic of the Marshall Islands

Richard A. Davis Wesley House, Cambridge, UK

A. De Ramon N'Yeurt Pacific Centre for Environmental and Sustainable Development, The University of the South Pacific, Suva, Fiji

Rowan Gard University of St. Andrews, St. Andrews, Scotland

Francisco Gelves-Gómez School of Law and Society, University of the Sunshine Coast (USC), Maroochydore, QLD, Australia;
Environmental Social Sciences Research Group, School of Ecosystem and Forest Sciences, University of Melbourne, Melbourne, VIC, Australia

John D. Griffiths Alphacrucis College, Parramatta, NSW, Australia

P. H. Havea Live and Learn Tonga, Tongatapu, Tonga

Karenne Hills Australian Institute of Professional Counsellors, Brisbane, QLD, Australia;
Christian Heritage College, Brisbane, QLD, Australia

Viliamu Iese Pacific Centre for Environment and Sustainable Development (PaCE-SD), The University of the South Pacific, Suva, Fiji

Guy Jackson School of Earth and Environmental Sciences, The University of Queensland, St Lucia, QLD, Australia;
Lund University Center for Sustainability Studies, Lund University, Lund, Sweden

Claudia Ledderucci Dipartimento di Culture, Politica e Società, Università degli Studi di Torino, Turin, Italy

Richard G. Leo School of Education, Humanities and Business, Christian Heritage College (CHC), Brisbane, QLD, Australia

Johannes M. Luetz School of Social Sciences, Christian Heritage College (CHC), Brisbane, QLD, Australia;
School of Law and Society, University of the Sunshine Coast (USC), Maroochydore, QLD, Australia;
School of Social Sciences, University of New South Wales (UNSW), Sydney, NSW, Australia

Kylie McKenna Divine Word University, Madang, Papua New Guinea

Prudence Millear School of Health and Behavioural Sciences, University of the Sunshine Coast, Maroochydore, QLD, Australia

Patrick D. Nunn School of Law and Society, University of the Sunshine Coast, Maroochydore, QLD, Australia

Ian N. Parker Department of Anthropology, University of California San Diego, La Jolla, CA, USA

Clive Pearson Charles Sturt University, Parramatta, NSW, Australia

Annah Piggott-McKellar School of Geography, University of Melbourne, Melbourne, VIC, Australia

Thelma A. Raman Macquarie University, Sydney, NSW, Australia

Jacqueline Ryle School of Law and Social Sciences, The University of the South Pacific, Suva, Fiji

Shirleen Sahai School of Law and Social Sciences, The University of the South Pacific, Suva, Fiji

Tammy Tabe Pacific Centre for Environment and Sustainable Development, The University of the South Pacific, Suva, Fiji

Morgan Wairiu Pacific Centre for Environment and Sustainable Development (PaCE-SD), The University of the South Pacific, Suva, Fiji

Lorelle Tekopiri Yakam Divine Word University, Madang, Papua New Guinea

Chapter 1
Beyond Belief

Patrick D. Nunn and Johannes M. Luetz

Abstract Given that almost every Pacific Island resident is spiritually engaged to a degree that is uncommon in western secular societies, an evaluation of the role of religion in climate-change adaptation in this exposed region is overdue. This chapter explains the nature of Pacific Island people's religious engagement, its undoubted links with culturally-grounded tradition, and its importance in rationalizing extraneous phenomena like climate change and determining communal action in the face of its impacts. Each chapter of this book is summarized. The need to reassess the roles of different agencies in climate-change action in the Pacific is explained, as is the importance of effective action for survival.

Keywords Belief · Coping · Religion · Pacific · Survival

Preamble

It was a balmy evening. The sun was setting majestically, throwing a roseate glow up behind the western mountains, slowly darkening the coast of the island lying in their shadow. You could hear children playing on the beach, splashing in the shallows, dogs barking in the distance, small boats creaking as they tugged at their anchors, the gentle swash of the waves against the beach. All these sounds framed a group

[1] When Fiji chaired COP-23 in 2017, it introduced the concept of *talanoa* or constructive storytelling, a Pacific way of discussing issues to reach a shared understanding and move forward (Winkler and Depledge 2018).

P. D. Nunn (✉) · J. M. Luetz
School of Law and Society, University of the Sunshine Coast, Maroochydore, QLD, Australia
e-mail: pnunn@usc.edu.au

J. M. Luetz
e-mail: jluetz@chc.edu.au

J. M. Luetz
School of Social Sciences, Christian Heritage College (CHC), Brisbane, QLD, Australia

School of Social Sciences, University of New South Wales (UNSW), Sydney, NSW, Australia

J. M. Luetz and P. D. Nunn (eds.), *Beyond Belief*, Climate Change Management,
https://doi.org/10.1007/978-3-030-67602-5_1

of people—Pacific Islanders—that came together here, as they often did in the late afternoon, to sit on the sand or upturned boats, to yarn, to *talanoa.*[1] They joshed with one another, some smoked, some were occasionally voluble, many more silent, contemplative.

This particular evening, they eventually came around to discussing their dilemma—one which countless coastal communities across the vast Pacific Islands region are increasingly acknowledging to be theirs as well. The climate is changing. The air is becoming warmer, night and day; the crops are struggling to grow as well as they once did; the fish are less plentiful around the colourless corals you now commonly find offshore. Food is becoming more difficult to acquire but thankfully, one person said, we can always buy rice and noodles, even tinned fish, from the village store. But it's not the same as before.

And then there is the sea, once our friend who succoured us, gave us food, especially in times of exigency. Today its waters are not just getting warmer but it is encroaching on the land more and more. *Ua mai*, 'the water comes', these people say. It is eating away the land, the land in which our ancestors lie; some villages have already started shifting inland, up the slope, but what do we do? How can we know the future?

Our nation's leaders say the climate is changing but they don't live here, so what do they actually know of our dilemma? Our elderly island chief says don't worry because the chances are that sea level will start falling one day very soon and everything will return to normal. But we worry anyway because this is our ancestral land, given to us by God. Why would He have done that if one day He was going to take it away from us and force us to move elsewhere? Maybe God is punishing us, testing our resolve like He tested that of Job.[2] Maybe we have not been sufficiently devout, maybe some of us have strayed? After all, how many different churches do people in this village now attend? What if some of them are not the right ones? What if Satan has gained a foothold in our community? What should we do? What does God want us to do?

The Dilemma of the Pacific

For many thousands of coastal communities across the vast Pacific Islands region, which spans almost one third of the earth's surface, the dilemma described above is very real, its immediacy having increased sharply over the past decade or so as the pace of climate change has demonstrably accelerated (WMO 2020). Unprecedentedly strong tropical cyclones have lashed parts of the region—Tropical Cyclone Pam (2015) was the strongest ever to hit Vanuatu; Tropical Cyclone Winston (2016) packed the strongest winds ever recorded in the southern hemisphere (Le De et al. 2018; Terry and Lau 2018). Rates of sea-level rise in the western Pacific are among

[2]In the Christian bible, it is said that God allowed Satan to test the faith of the 'blameless upright' man named Job and see whether, in the face of repeated hardships, he would ever renounce his faith—which he did not.

the highest in the world; over the past five decades, the ocean surface in Micronesia and Solomon Islands has risen 4–5 times faster than the global average (currently 3.2 mm/year) causing massive impacts along island coasts and even the disappearance of entire (low sandy) islands (Albert et al. 2016; Becker et al. 2012; Nunn et al. 2017). And in the background, the steady rise in temperature is impacting crop growth, stressing familiar ecosystems, and forcing changes to farming practices (McGree et al. 2019; Taylor et al. 2016; Luetz and Havea 2018).

While global onlookers understandably rationalize such occurrences through a secular scientific lens, most people living on Pacific Islands view them quite differently, privileging Pacific worldviews and seeking spiritual-religious explanations for the causes of these occurrences and, most importantly, spiritually-framed guidance as to how best to respond to them (Cox et al. 2018; Luetz and Nunn 2020).

This is not a trivial issue. For if you seek to bring about behaviour change (like appropriate climate-change adaptation) in communities whose members have a different worldview to your own, then you will not achieve your goal by simply insisting that you are correct—and they, by implication, are wrong. Of course, you may be cut a bit of slack if your insistence that you are correct is accompanied by a sizeable wad of cash but eventually such communities will fall back on their own beliefs, their time-honoured ways of rationalizing and coping with change. For such reasons, huge amounts of adaptation funding given by well-intentioned outside agencies to Pacific Island nations over the past few decades have failed to bring about any effective or sustainable adaptation, especially at community level (Nunn and McNamara 2019).

It is often claimed that Pacific Island peoples and communities lack resilience to external livelihood stressors like climate change but that is a superficial judgement which disregards the fact that people have lived o n Pacific Islands for hundreds if not thousands of years.[3] How could this have happened if they (and the societies of which they were part) were not extraordinarily resilient? The issue is that 'resilience' as defined at a global level for the specific purpose of identifying climate-change adaptation needs has, when scaled down, little relevance in many non-western contexts since it sidelines their culturally-grounded resilience.

For outsiders to help achieve effective and sustainable adaptation to climate change in non-western contexts, it is essential to meaningfully engage with the worldviews of target communities. This is a preliminary to having them 'own' the issue of climate-change adaptation rather then consider it "a foreign construct, a preoccupation of developed nations" (Nunn 2009, p. 213). But it is now clear that outsiders need to go further than has been usual in the past and allow these communities to design and drive adaptive solutions along pathways that are familiar to them (particularly in terms of languages and cultural protocols). Into these pathways, the insights from western science needed to fine-tune these solutions should be inserted in as ethical

[3]The earliest-known colonization of oceanic islands in the Pacific is that of the Mariana Islands (northwest Pacific), plausibly from Luzon (Philippines), about 3500 years ago, a journey that involved an open-ocean crossing of at least 2300 km (almost 1500 miles) (Hung et al. 2011).

and non-partisan fashion as possible (McNamara et al. 2020; Nalau et al. 2018; Luetz 2019).

Almost everyone living in the Pacific Islands region has a religious affiliation expressed through regular practice and which influences everyday decision-making to a degree that many people from outside this region sometimes consider astonishing (Luetz and Nunn 2020; Nunn et al. 2016). While the roots of such beliefs and practice are clearly grounded in culture history, they underwent a massive transformation during the nineteenth century when Christian missionaries arrived o n most inhabited Pacific Islands (McNeill 1994; Nunn 2007). Given that this was also a time when diseases (like influenza and measles), against which Pacific Island peoples had no genetic resistance, arrived in the region, huge numbers of them died in a short period of time post-contact, upending traditional societies and power relations, readying the ground for missionization (Cliff and Haggett 1985). And while the Christianization of the Pacific Islands region was indeed comparatively rapid and comprehensive, the particular brand of Christian belief which evolved in most situations was one that acknowledged and incorporated pre-Christian practice, as it commonly still does (Munro and Thornley 1996; Winch-Dummett 2010).

Spiritual and religious beliefs, Christianity in particular, are today part of the worldview of most Pacific Island people. And where religious beliefs come into conflict with secular ones, the former generally prevail although it may seem otherwise when it suits political or pragmatic agendas. But at a community level, where few decision-makers have western-scientific training but where religious practice is a daily routine, there is no contest. Spiritual beliefs influence almost every type of decision-making in Pacific Island communities, including that governing how a community should respond to the effects of climate change.

This point is also important because most Pacific Island governments, especially those of archipelagic nations, have insufficient capacity (and possibly always will) to enable effective sustainable adaptation in each of their constituent rural communities (Nunn 2009). Recent research, produced mostly by Pacific Islanders, shows that the degree to which a community is distant from national developmental cores (its peripherality) largely measures its ability to cope autonomously with environmental adversity. Peripheral communities tend to have a greater stock of traditional knowledge for coping, while near-core communities have more global knowledge (Korovulavula et al. 2019; Nunn and Kumar 2019b).

An understanding of this situation, that which exists around the end of the second decade of the twenty-first century, is key to the future. For it can be argued that current trends of growing dependency by Pacific Island nations on outside (donor) funding for climate-change adaptation are not sustainable. Over the next decade or so, as the costs of domestic adaptation in donor countries soar, they will inevitably become far less generous towards their Pacific Island neighbours (Nunn and Kumar 2019a). So the time is ripe to conserve traditional knowledge, to disseminate and privilege at least its most useful aspects,[4] in order that Pacific Island countries can better prepare for a future in which they will be forced to stand alone far more than they do at present.

Religion will play a huge role in this, so we consider that the timing of this book—which captures the gamut of faith-grounded understandings of the natural world across the Pacific Islands region—to be especially opportune. Together with traditional knowledge, the spiritual and religious beliefs of Pacific Island peoples will form the foundation of their survival into the future. The time is thus opportune to elevate the role of religion in climate-change adaptation discourse from footnote to foreground, welcoming it at the table of interlocking climate change adaptation concerns.

The Pacific Islands region has suffered for more than a hundred years from the consequences of outsiders' superficial judgements about what is right and what is deviant. For example, when James Cook landed on the Pacific Island of Niue on 2nd June 1774, he encountered islanders whose faces were coloured with the juice of red bananas; quick to judge, he wrote "*The Conduct and aspect of these Islanders occasioned my giving it the Name of Savage Island*", an uninformed interpretation that coloured outsiders' views of Niuean society for centuries thereafter. Today, the 'suffering' continues; outsiders with little experience of them adjudge 'small' islands to be inherently 'vulnerable', their peoples clearly lacking 'resilience', unquestionably 'exposed'; millions of dollars are poured into the region to try and reduce this abominable problem … and most Pacific Islanders wonder why.

In this book, we have strived not to exhibit or indulge such biases. There is nothing wrong with having religious beliefs, in believing in a God, following His tenets, living His values. Neither faith nor secularism are the human norm; one is not better than the other, although people on either side often argue otherwise. And climate change, while certainly a wicked problem that is robustly explainable by western science, does not require in any particular place on earth either a secular or a faith-grounded solution; such choices refer to process and pathways, not to outcomes.

Yet one of the key audiences for this book are climate-change influencers—like donor representatives, scientists, politicians and members of international agencies—who develop and implement policy to enable climate-change adaptation in places like the Pacific Islands region that are assumed to be unable to do this alone. And the key message for you is that, not only do the spiritual and religious beliefs held by almost all Pacific Islanders contribute hugely to resilience, but that your help can be overdone, almost to a point of being maladaptive. There is now ample scientific research supporting the points made above, all of which have led to the conclusion that "the role that implementers and donors should and can play is to become 'facilitators' of the desired adaptation aspirations for communities, rather than 'doing' adaptation 'to communities'" (McNamara et al. 2020, p. 638).

---

[4]This extends beyond traditional ways of coping with environmental adversity into things like the use of traditional pharmacopoeias for primary health care and the attainment of food and nutritional security through a return to more traditional diets.

This book will also be published during, possibly in the aftermath of, the worst global pandemic for a century. And while to date (November 2020), Pacific Island countries have not had to bear the direct effects of COVID-19 to the degree that most continental countries have, the economic impact has been severe, and its outfall will endure many years (Leal Filho et al. 2020). Such global events give us pause to reflect on what is important in our lives; for many people, it has been an opportunity to be re-acquainted with a spiritual self they lost touch with some time ago. For such people, this book may be revelatory. They will discover that Pacific Island peoples are not merely religious but also commonly more spiritually engaged to the extent that routine decisions are guided by faith rather than (western) science. Has this affected the quality of their lives? Will it diminish the quality of their lives a hundred years from now? Probably neither.

Pacific Islanders who read this book will inevitably be less surprised by the degree and the diversity of spiritual-religious beliefs that exist in their region and the extent to which these beliefs influence decision-making, especially for issues like those of climate change. But there are nevertheless points of relevance, not least (a) the need for the co-production of knowledge (the need to combine traditional and spiritual-religious understandings) to allow optimal decision-making, (b) the need for appropriate action, and (c) the benefits of leveraging religion and spirituality for adaptation.

Several chapters in this book note the role of spiritually-informed conviction as a motivational force for environmentally sustainable decision-making that is unparalleled by either rational scientific explanations or economic pursuits. Fair (2018) makes the point that spiritually-shaped local worldviews should be proactively engaged as "a motivational force not mirrored by economics or science" (p. 4). Building on the premise that "men and women at their deepest level are motivated by metaphysical beliefs" (Knight 2006, p. 19), we argue that, rather than constituting a barrier, religion and spirituality should be leveraged as a "cultural resource" (Hulme 2017, p. 15) that is grounded in the understanding that the most influential messages in such contexts are those that engage with people's spiritual beliefs. This holds particular relevance in the 'liquid continent' given the widespread prevalence of Christianity across this region (Ernst 2012; Rubow and Bird 2016; Ernst and Anisi 2016). There is therefore an argument that climate-change adaptation efforts should seek to tap into religious worldviews as an auspicious 'thick' cultural resource that has so far remained essentially eclipsed, misunderstood, misrepresented and/or grossly under-appreciated (Hulme 2017; Luetz et al. 2018; Luetz and Nunn 2020).

Book Focus and Contribution

Exploring the science-spirituality nexus through a multidisciplinary lens enables unique and timely perspectives to emerge on effective and sustainable climate-change adaptation in the Pacific Islands region and beyond. This volume features case studies, literary analyses, field projects and empirical research, and it deliberates faith-engaged approaches through context, theory, practice, doctrine and engagement.

Comprising peer-reviewed works by scholars, professionals and practitioners from across the Pacific region, the book closes a critical gap in the literature as it aims for scientifically sound, spiritually attuned, locally meaningful, and contextually compelling analysis and synthesis. It seeks to create consilience by integrating interdisciplinary contributions from the sciences and humanities, including case studies, literary analyses, field projects and empirical research (Luetz and Nunn 2020); it inquires how positionality affects insider–outsider adaptation discourse, and probes limitations, opportunities and future prospects; it presents a strong practice-oriented focus that comprises peer-reviewed works by scholars, professionals and practitioners; it investigates influences on decision-making of tradition, local precedent, spirituality, indigeneity and connectedness to Nature; it endeavours to spawn faith-engaged approaches to climate change adaptation that are "more-than-scientific yet not anti-scientific" (Fair 2018, p. 11); it probes locally relevant theological, epistemological, hermeneutical and eschatological perspectives on ecological meaning-making—or, expressed in simpler language, it seeks to understand how different interpretations of religious teachings can affect 'buy-in' and environmental outcomes.

Looking Back to the Future: The Clash of Religion and Secularism

Owing to its geography—what many perceive as a vast empty space rather than a 'sea of islands' in Epeli Hau'ofa's memorable phrase[5]—the Pacific has often been a zone of conflict, a place across which continental powers battle (or trial future battles) with little recognition of the effects on island peoples and ecologies. Yet as explained in Chapter 2 by **Clive Ayre** and colleagues, people of faith in the Pacific and beyond can contribute to a more peaceful, reconciliatory and sustainable future through their engagement on two simultaneous existential dangers—nuclear war and climate change—that threaten to undermine society's ability to endure.

A grounded case study from Solomon Islands comprises Chapter 3 in which **Zina Bird** and others discuss what changes in the environment and its climatic controls the people living in subsistence communities in northern Malaita have noticed in recent years. This chapter gives insights into how many Pacific Island people in such contexts rationalize and respond to changes that can be attributed to climate change.

[5]This phrase was introduced in Epeli's 1993 book (Hau'ofa 1993) and was conceived to emphasize the contrast between the common outsider impression of the Pacific as a vast ocean peppered with small islands and the common insider view of the Pacific as a 'sea filled with islands'. It is an issue of perspective that underpins contrasting worldviews. An excellent discussion of the evolution of the Pacific as 'Earth's empty quarter' was given by Gerard Ward (1989).

The overwhelming importance of religious and cultural-spiritual beliefs is clearly enunciated.

The Roots of Spirituality and Religion in the Pacific Islands Region

Pacific people's beliefs in a spiritual dimension to the tangible world are deep-rooted in Pacific pasts, a notion that is explored in Chapter 4 by **Francisco Gelves-Gómez** and **Shannon Brincat** in reference to Fiji. Their soundly-researched and compellingly-argued study seeks to understand how the harmonization of the relationship between humans and Nature can be used to confront the challenges of future climate change. They ask to what degree such culturally-grounded understandings can become incorporated into policy, something that may optimize future adaptation processes.

Many of these points are reinforced in Chapter 5 in which **Guy Jackson** and **Annah Piggott-McKellar** discuss Pacific people's understandings of natural hazards, including those otherwise attributable to climate change, in contrasting contexts from the highlands of Papua New Guinea and the low atolls of Kiribati. The fact that these understandings have been largely ignored in most externally-sponsored efforts to achieve effective and sustainable adaptation here accounts for their commonly negligible impact and points the way forward.

The degree to which religious interpretations of disasters in the Pacific, as well as more insidious effects of climate change (like sea-level rise), permeate contemporary understandings is explained in Chapter 6 by **Richard Davis**. He argues not just for the mandatory recognition by outsiders of the extent of these interpretations but also, using insights from Martin Luther's reformism, how climate change can be perceived as an outcome of human misuse of the natural world and that Pacific Island peoples, while having contributed little directly to this situation, have a role to play in rectifying it.

Religion as Culture, Religion for Wellbeing

The next group of chapters focuses on the practical expressions of religion in the Pacific Islands, especially as it influences behaviour, decision-making and the prospects for the future. In Chapter 7, a study of the role of religious belief during and after an unprecedently strong tropical cyclone in a rural Fiji community is presented by **Shirleen Sahai** and two colleagues from the University of the South Pacific, the premier tertiary teaching and research institution in the Pacific Islands region. In this chapter, we learn about the power of prayer and the calming effects of belief and hope that facilitated recovery in the aftermath of this event.

Written by **Kylie McKenna** and **Lorelle Yakam**, Chapter 8 reports the results of a survey of market vendors in Papua New Guinea who demonstrate a clear understanding of the effects of recent climate changes on food supply. While recognizing that their understanding would benefit from additional information, their consensus is that climate change is a sign of the 'end times', a fatalistic interpretation that clearly constrains the range of adaptive responses.

Following from this, Chapter 9 by **Kurt Bangert** looks in more detail at such apocalyptic beliefs in the Pacific (and elsewhere) and discusses three distinct variations of 'end-times' beliefs. The underlying theme is that while God is merciful and compassionate, He is also in control and responsive to human action, which reinforces humanity's responsibility in the age of climate changes.

For **Clive Pearson**, the profound and multifarious challenges of climate change in the Pacific Islands region represent a *kairos* moment, a time requiring an urgent reappraisal of biblical and theological traditions in the context of the Anthropocene. Chapter 10 argues that for future policies to be well-grounded in communal life, there is an opportunity for scientific, political and sociological discourses to be matched with contemporary theological thinking. The importance of blending religious insights with traditional knowledges and values is highlighted.

Weaving Worldviews: Science and Spirituality Together

Leading on from the blended approaches discussed in Chapter 10, the next four chapters deal with the ways in which spiritual-religious and scientific-secular worldviews can meaningfully co-exist in Pacific societies, as complementary rather than contradictory. In Chapter 11, **Rowan Gard** characterizes this as 'paddling on both sides of the canoe', something that propels you forward fastest. She argues that optimal adaptation to climate change will be gained this way, through the co-production of knowledge that is at once scientifically rigorous but also incorporates Pacific Islanders' spiritual beliefs and values.

Chapter 12 presents a similar model, 'the cord of three strands', that is argued by **Peni Havea** and **Antoine N'Yeurt** to characterize an explanatory framework for climate-change adaptation in five representative coastal communities in Tonga. The three strands—the state, the people, the church—should be treated equally and woven together to produce culturally-acceptable frameworks for adaptation and for the pursuit of sustainability.

A case study from a remote island in Fiji by **Jalasayi Atkinson-Nolte** and two colleagues comprises Chapter 13 and focuses on how people's spiritual-religious beliefs influence their interpretation of observed climate change. But this study also found that these people had developed a system of 'flexible attribution' of particular phenomena to particular causes, either science-based or religious-based, and that this proves n o hindrance to effective action and planning.

In a similar vein, Chapter 14 discusses the complementary approaches to conservation practices adopted by island communities in Western New Guinea. As **Ian**

Parker explains, in response to climate-linked deterioration in ecosystem health, communities initiated different conservation strategies that either privileged traditional spiritual understandings or Christian ones. But both overlapped, revealing the deep cultural roots of the former, and dualistic beliefs in the efficacy of both.

Belief and Action: A Way Forward?

Action consistent with belief is the theme of the next two chapters. In Chapter 15, **Claudia Ledderucci** discusses the Pacific Climate Warriors, a popular culture-based activist group of Pacific Islanders who represent the fight against climate change both in terms of social justice and religious advocacy. This demonstrates the growing belief in the Pacific that effective and sustained action in the face of climate change has both practical and spiritual dimensions—and that one makes little sense without the other.

The nexus between religion and pro-environmental behaviours, which feed into appropriate climate-change adaptation, is the subject of Chapter 16. In this, **Thelma Raman** and **Johannes Luetz** report how the influence of pro-environmental education in Fiji's high schools led to better behavioural outcomes in religious-based rather than secular institutions, demonstrating how religion and environmental education often reinforce one another.

Faith for Survival

The last three chapters in this book focus on the future, the ways in which it is viewed through various religious and cultural lenses, and how this could affect the ways in which particular groups of people behave in response to the unfolding challenges of climate change. The worldview of Pentecostalism is the topic of Chapter 17 by **John Griffiths** who attributes much of its success to the alignment of its basic tenets with those of traditional spiritual beliefs in the Pacific. While privileging the power of prayer and repentance, Pentecostal beliefs also espouse an ecology that is aligned with science and, through its restoration and conservation, gives hope for the future.

This theme is discussed more broadly by **Johannes Luetz** and **Richard Leo** in Chapter 18 which looks at ecotheologies, namely the ways in which Christian (and other) beliefs can positively influence issues of sustainability, including through faith-based ecosystem management and environmental stewardship. The chapter concludes that certain religious beliefs may both enable and/or disable environmental conservation in different contexts. Conceptualising religion as an effective cultural adaptation resource whose potential in the Anthropocene remains under-appreciated, the discussion points the way to promoting faith-informed climate change adaptation prospects in the future.

In the final chapter of this book, Chapter 19, **Clive Ayre** argues that Christian responses to climate change should not be merely pragmatic and reactive but should embrace biblical values of restoring ecosystems and sustaining livelihoods. Casting theology, culture and ecology as essential aspects of sustainability, the chapter offers a summary of faith-based opportunities for a more sustainable Pacific.

Afterword

This book is not about religion, nor is it about climate change. It is about survival. This book is intended to help inform and map effective and sustainable adaptive pathways that will allow people in the Pacific Islands to weather the challenges of future climate change so that, two hundred years hence, these people will still be able to identify as unique. Today we acknowledge the uniqueness of Pacific Island belief systems and their manifold expressions, and we acknowledge the uniqueness of Pacific Island places and that of the cultures and livelihoods they support. All this singularity is threatened by climate change and how Pacific Island peoples confront these threats, especially within the next twenty years, will largely determine whether or not these peoples and the places they aspire to continue occupying are still recognizable centuries from now.

References

Albert S, Leon JX, Grinham AR, Church JA, Gibbes BR, Woodroffe CD (2016) Interactions between sea-level rise and wave exposure on reef island dynamics in the Solomon Islands. Environ Res Lett 11(5):054011

Becker M, Meyssignac B, Letetrel C, Llovel W, Cazenave A, Delcroix T (2012) Sea level variations at tropical Pacific Islands since 1950. Glob Planet Change 80–81(1):85–98

Cliff AD, Haggett P (1985) The spread of Measles in Fiji and the Pacific: Spatial components in the transmission of epidemic waves through island communities. Australian National University, Canberra

Cox J, Finau G, Kant R, Tarai J, Titifanue J (2018) Disaster, divine judgment, and original sin: Christian interpretations of tropical cyclone Winston and climate change in Fiji. Contemp Pac 30(2):380–411

Ernst M (2012) Changing Christianity in Oceania: a regional overview. Archives De Sciences Sociales Des Religions. Archives de sciences sociales des religions. https://doi.org/10.4000/assr.23613

Ernst M, Anisi A (2016) the historical development of Christianity in Oceania. In: Sanneh L, McClymond MJ (eds) The Wiley Blackwell companion to world Christianity. Wiley, pp 588–604. https://doi.org/10.1002/9781118556115.ch44

Fair H (2018) Three stories of Noah: navigating religious climate change narratives in the Pacific Island region. Geo Geogr Environ 5(2). https://doi.org/10.1002/geo2.68

Hau'ofa E (ed) (1993) Our sea of islands. University of the South Pacific, Suva, Fiji

Hulme M (2017) Climate change and the significance of religion. Econ Political Wkly 52(28):14–17

Hung H-C, Carson MT, Bellwood P, Campos FZ, Piper PJ, Dixon E, Bolunia MJLA, Oxenham M, Chi Z (2011) The first settlement of Remote Oceania: the Philippines to the Marianas. Antiquity 85:909–926

Knight GR (2006) Philosophy & education: an introduction in Christian perspective, 4th edn. Andrews University Press.

Korovulavula I, Nunn PD, Kumar R, Fong T (2019) Peripherality as key to understanding opportunities and needs for effective and sustainable climate-change adaptation: a case study from Viti Levu Island, Fiji. Climate and Development

Le De L, Rey T, Leone F, David G (2018) Sustainable livelihoods and effectiveness of disaster responses: a case study of tropical cyclone Pam in Vanuatu. Nat Hazards 91(3):1203–1221

Leal Filho W, Luetz JM, Sattler DN, Nunn PD (2020) Coronavirus: COVID-19 transmission in Pacific small island developing states. Int J Environ Res Public Health 17(15):5409

Luetz JM (2019) Over-researching migration 'hotspots'? Ethical issues from the Carteret Islands. Forced Migr Rev (61): 20–22. The ethics issue (June 2019), Refugee Studies Centre, University of Oxford. https://www.fmreview.org/ethics/luetz

Luetz JM, Havea PH (2018) "We're not refugees, we'll stay here until we die!"—climate change adaptation and migration experiences gathered from the Tulun and Nissan Atolls of Bougainville, Papua New Guinea. In: Leal Filho W (ed) Climate change impacts and adaptation strategies for coastal communities. Springer Nature, Switzerland, pp 3–29. https://doi.org/10.1007/978-3-319-70703-7_1

Luetz JM, Nunn PD (2020) Climate change adaptation in the Pacific Islands: a review of faith-engaged approaches and opportunities. In: Leal Filho W (ed) Managing climate change adaptation in the Pacific region. Springer Nature, Cham, Switzerland, pp 293–311.https://doi.org/10.1007/978-3-030-40552-6_15

Luetz JM, Buxton G, Bangert K (2018) Christian theological, hermeneutical and eschatological perspectives on environmental sustainability and creation care—the role of Holistic Education. In: Luetz JM, Dowden T, Norsworthy B (eds) Reimagining Christian education—cultivating transformative approaches. Springer Nature, Singapore, pp 51–73. https://doi.org/10.1007/978-981-13-0851-2_4

McGree S, Herold N, Alexander L, Schreider S, Kuleshov Y, Ene E, Finaulahi S, Inape K, Mackenzie B, Malala H, Ngari A, Prakash B, Tahani L (2019) Recent changes in mean and extreme temperature and precipitation in the western Pacific Islands. J Clim 32(16):4919–4941

McNamara KE, Clissold R, Westoby R, Piggott-McKellar A, Kumar R, Clarke T, Namoumou F, Areki F, Joseph E, Warrick O, Nunn PD (2020) An assessment of community-based adaptation initiatives in the Pacific Islands. Nat Clim Chang 10:628–639

McNeill JR (1994) Of rats and men: a synoptic environmental history of the island Pacific. J World Hist 5(2):299–349

Munro D, Thornley A (eds) (1996) The Covenant makers: islander missionaries in the Pacific. Institute of Pacific Studies, Suva

Nalau J, Becken S, Schliephack J, Parsons M, Brown C, Mackey B (2018) The role of indigenous and traditional knowledge in ecosystem-based adaptation: a review of the literature and case studies from the Pacific Islands. Weather, Clim, Soc 10(4):851–865

Nunn PD (2007) Climate, environment and society in the Pacific during the last millennium. Elsevier, Amsterdam

Nunn PD (2009) Responding to the challenges of climate change in the Pacific Islands: management and technological imperatives. Climate Res 40(2–3):211–231

Nunn PD (2017) Sidelining god: why secular climate projects in the Pacific Islands are failing. The Conversation, 17 May. https://theconversation.com/sidelining-god-why-secular-climate-projects-in-the-pacific-islands-are-failing-77623

Nunn PD, Kumar R (2019a) Cashless adaptation to climate change in developing countries: unwelcome yet unavoidable? One Earth 1:31–34

Nunn PD, Kumar R (2019b) Measuring peripherality as a proxy for autonomous community coping capacity: a case study from Bua Province, Fiji Islands, for improving climate change adaptation. Soc Sci 8:225

Nunn PD, McNamara KE (2019) Failing adaptation in island contexts: the growing need for transformational change. In: Klock C, Fink M (eds) Dealing with climate change on small islands: towards effective and sustainable adaptation. Gottingen University Press, Gottingen, pp 19–44

Nunn PD, Mulgrew K, Scott-Parker B, Hine DW, Marks ADG, Mahar D, Maebuta J (2016) Spirituality and attitudes towards nature in the Pacific Islands: insights for enabling climate-change adaptation. Clim Change 136(3–4):477–493. https://doi.org/10.1007/s10584-016-1646-9

Nunn PD, Kohler A, Kumar R (2017) Identifying and assessing evidence for recent shoreline change attributable to uncommonly rapid sea-level rise in Pohnpei, Federated States of Micronesia, Northwest Pacific Ocean. J Coast Conserv 21(6):719–730

Rubow C, Bird C (2016) Eco-theological Responses to climate change in Oceania. Worldviews 20:150–168

Taylor M, McGregor A, Dawson B (2016) Vulnerability of Pacific Island agriculture and forestry to climate change. Pacific Community, Noumea

Terry JP, Lau AYA (2018) Magnitudes of nearshore waves generated by tropical cyclone Winston, the strongest landfalling cyclone in South Pacific records. Unprecedented or unremarkable? Sed Geol 364:276–285

Ward RG (1989) Earth's empty quarter—the Pacific Islands in a Pacific century. Geogr J 155:235–246

Winch-Dummett C (2010) Christianity and cultural reconfigurations in South West Pentecost. Oceania 80(1):78–101

Winkler H, Depledge J (2018) Fiji-in-Bonn: will the 'Talanoa spirit' prevail? Clim Policy 18(2):141–145

WMO (2020) WMO statement on the state of the global climate in 2019. World Meteorological Organization, Geneva

Patrick D. Nunn is Professor of Geography at the University of the Sunshine Coast (Queensland, Australia). For more than thirty years, his research has focused on climate change issues in the Pacific Islands, understanding past and (likely) future human-climate interactions and their implications for coastal livelihoods. This work has seen the publication of several books including Oceanic Islands (Blackwell, 1994) and Climate, Environment and Society in the Pacific (Elsevier, 2007) and more than 320 peer-reviewed publications. A long involvement with the Intergovernmental Panel on Climate Change (IPCC) led to Patrick sharing its 2007 Nobel Peace Prize; Patrick was a lead author on the chapter about 'Sea Level Change' in the IPCC's 5th Assessment Report and is currently a lead author on the chapter about 'Small Islands' for its 6th Assessment. Another of Patrick's research interests focuses on ancient understandings of coastal-environmental change and how these have been culturally filtered and encoded in narrative and myth. This research was laid out in his books, Vanished Islands and Hidden Continents of the Pacific (University of Hawaii, 2009) and The Edge of Memory (Bloomsbury, 2018), and led to the award (with linguist Nick Reid) by the Geographical Society of New South Wales of the Best Paper 2016 in the journal Australian Geographer and in 2018 the Gold Medal of the Royal Geographical Society of Queensland. His new book, due out from Bloomsbury in August 2021, is Worlds in Shadow: Submerged Lands in Science, Memory and Myth. He has an active website at www.patricknunn.org.

Johannes M.Luetz (BA/USA, MBA/Germany, Ph.D./Australia) is Adjunct Senior Lecturer at the University of New South Wales (Sydney) in the School of Social Sciences, where he also completed his Ph.D. (Environmental Policy and Management) into human climate migration. His Ph.D. research has featured in national news and documentaries [Bolivia; Bangladesh; Maldives]. Dr. Luetz has consulted extensively for World Vision International on research projects raising awareness of the growing effects of climate change on vulnerable communities in Asia Pacific,

Africa and Latin America. He was raised in Sierra Leone and Switzerland, cares deeply about poverty reduction, holistic education, justice and equity, and has researched and published conceptually and empirically in these areas. He is a big picture thinker and has interdisciplinary research interests at the science-policy interface. He has spoken at the UNSW Town & Gown annual signature dinner and has been a guest on ABC Radio National The Science Show. His research has attracted grants, prizes, scholarships, awards and distinctions, including best thesis prize (2006), UNSW UIP Award (2009), UNSW ASPIRE Award (2012) and best paper awards (2017, 2018). He coordinates the Bachelor of Human Services at Christian Heritage College (CHC Brisbane) and also heads the Master of Social Science Leadership program, which through its research and teaching activities on change leadership seeks to give rise to equitable social transformation. Dr. Luetz is the Research Chair at CHC, Adjunct Fellow at the University of the Sunshine Coast (USC), Deputy Editor at Emerald journal IJCCSM and Editorial Board member at Springer Nature journal Discover Sustainability. He has an active website at www.luetz.com.

Part I
Looking Back to the Future: The Clash of Religion and Secularism

Chapter 2
Ecology, War, and Nuclear Testing in the Pacific: The Path of Reconciliation

Clive W. Ayre, Karenne Hills, and Johannes M. Luetz

Abstract Armed conflict has been part of the human experience for all recorded history, and together with peoples and cultures, the environment has been a major casualty of this. The causes and costs of war are numerous, complex and invariably expensive, leaving a toll that is measured not only in financial terms, but also in terms of the impact on the natural ecology, communities and wildlife, along with the wastage of finite resources. The ecological footprint from war and nuclear testing in the Pacific is noted particularly. Faith communities have a role to play in reaching for a solution to global conflict, and scriptural records begin to point the way towards a wholistic attitude towards reconciliation. This chapter identifies ten aspects of that charter, which taken together, form a significant and challenging opportunity for people of faith to contribute to a more peaceful and sustainable future. Armed conflict rarely has any long-term winners, wherefore the way forward both for the environment and indeed all life, will be found in the more unifying goal of reconciliation.

C. W. Ayre (✉)
University of Queensland, Brisbane, QLD, Australia
e-mail: c.ayre@bigpond.com

K. Hills
Australian Institute of Professional Counsellors, Brisbane, QLD, Australia

Christian Heritage College, Brisbane, QLD, Australia

J. M. Luetz
School of Social Sciences, Christian Heritage College (CHC), Brisbane, QLD, Australia
e-mail: jluetz@chc.edu.au

School of Law and Society, University of the Sunshine Coast (USC), Maroochydore, QLD, Australia

School of Social Sciences, University of New South Wales (UNSW), Sydney, NSW, Australia

J. M. Luetz and P. D. Nunn (eds.), *Beyond Belief*, Climate Change Management,
https://doi.org/10.1007/978-3-030-67602-5_2

Introduction

Over recent decades it has become increasingly apparent that not only is the global climate changing (IPCC 2014), but also that humans have often not exercised responsible stewardship of the earth's resources. In short, there is a tendency to take the natural environment for granted, although public concern is increasing (IPCC 2018; Corner et al. 2018). The causes of ecological damage around the globe have been discussed and analysed in numerous official reports, conferences and books (White 2008), and the causes are complex and varied. However, the scientific literature presents a high degree of consistency and scientific unanimity concerning the conclusion that humans are largely responsible (IPCC 2018). The new reality of the Anthropocene (the geological age of humans) is becoming uncomfortably clear, both scientifically (Steffen et al. 2018) and theologically (Deane-Drummond et al. 2017).

One of the more serious and arguably least studied causes of ecological damage is the impact of modern warfare on the biosphere, and especially nuclear warfare (Westing 1981; Caldicott 2017; Brewer et al. 2020). From a global perspective, in 2014 the UN Secretary General Ban Ki-moon said:

> The environment has long been a silent casualty of war and armed conflict. From the contamination of land and the destruction of forests to the plunder of natural resources and the collapse of management systems, the environmental consequences of war are often widespread and devastating. (Mathieson 2014, para. 2)

Although the primary focus of this volume is the situation faced by the Pacific Islands, it must be emphasised that this is essentially a global and whole-of-life issue. To a greater or lesser extent, all humans are complicit in the causes of degradation. Certainly, current and future generations will experience the outcomes. It is therefore imperative that the whole human community understand as clearly as possible the ecological impacts of warfare so that more support for a peaceful and sustainable future can be raised (Westing 1981; ICAN 2014, 2017).

This chapter focuses primarily on the destruction caused by nuclear testing in the Pacific Islands, a region that is characterised by a high degree of Christianisation (Nunn et al. 2016; Fair 2018; Luetz and Nunn 2020). In developing its contribution, this discussion is written from a specifically Christian perspective in an attitude of respect for the population living in this region. This involves the discussion of the biblical concept of Shalom in the latter parts of this chapter and also involves the occasional citation of scripture from the New International Version (NIV) of the Bible. Even so, the authors wish to reflect a deep respect for those who hold other points of view (Sire 2009). Religion and religious language have been noted as key themes in international politics and reconciliation efforts (Siddiqui 2005). It is from such a perspective that this chapter recognises the underpinning efforts of religious groups who purport reconciliation initiatives as a solution to ecological destruction. The discourse will use the term "Earthcare" to describe the need for consistent environmental care on the part of all people (Ayre 2014). Even though the significance of Christianity is relatively diminished in contemporary western society, from a global perspective, faith communities are nevertheless well-placed to contribute towards a more peaceful and sustainable future (Bhagwan 2020). In particular, any such contribution towards peace will (ideally) involve embracing a theology that includes practical and earth-friendly ways of living.

The primary purpose of this chapter is to explore ways in which faith-based reconciliation could contribute to a viable pathway to peace and thus a more positive future for both human and non-human life. This prospect will be provided against a backdrop that outlines the impact of war on the ecology of the planet. At the heart of this goal for a more positive and peaceful future is the scriptural call to a ministry of reconciliation, which is without doubt one of the biggest challenges facing the world today (Harper 2016; Moltman 2012; Moore 2014). The authors of this chapter posit that according to scripture, God seeks to work in and through human beings towards an outworking of reconciliation. We understand from the Christian scriptures that God was working to reconcile the world to himself through the life of Jesus Christ, and then committed all his followers to propagate this message of reconciliation (2 Corinthians 5:19). Accordingly, this discussion seeks to contribute to scholarship by attempting to synthesise the key linchpins of this message of reconciliation. More specifically, by grappling with questions involving war, ecological damage and reconciliation, this discourse seeks to create consilience by conjoining contributions from both science and faith. This has been previously synthesised as a fertile space for scholarly engagement:

> climate change adaptation initiatives in Oceania will have a better chance of being effective and sustainable if they are scientifically sound [and] spiritually attuned … The study recommends fostering consilience through propagating locally sensitive science-faith integration. (Luetz and Nunn 2020, p. 305)

By venturing into this little explored territory, this chapter responds to such previous calls for "more-than-scientific yet not anti-scientific responses to climate change that are locally meaningful and morally compelling" (Fair 2018, p. 11). Accordingly, it invites its readers to view reconciliation with Nature through the prism of scriptural reconciliation narratives, perspectives and representations (Hoffman and Sandelands 2005). By looking at the ecological destruction brought on by nuclear testing i n the Pacific retrospectively, the discourse seeks to advance prospects for peace and better disaster preparedness prospectively. In doing so, this chapter builds support for spiritual leadership in Oceania as it engages proactively in areas of policy development, regulatory change and peace research (Bhagwan 2020).

The discussion is organised as follows: after briefly exploring the two concepts of ecology and war, the section "The Causes and Impacts of War" will set the scene globally by considering the causes and impacts of conflict. The section "Nuclear Testing in the Pacific" will then discuss the significant impacts of nuclear testing in the Pacific before moving to consider a solution based on a theology of reconciliation in the section "Towards a Faith-Informed Solution". Thereafter, the section "Ten Steps Toward a Practical Ministry of Reconciliation" will propose ten steps towards a practical ministry of reconciliation. Finally, the "Conclusion" will close the discussion with a concluding synthesis.

The Causes and Impacts of War

War

Historically, it appears that war has always been part of the human experience, yet it seems that humanity does not readily learn the lessons history has to teach. Given the devastating human, ecological and financial implications of war, it is difficult to understand why alternative solutions to political unrest have not been more intentionally pursued and implemented. For nations and peoples who regard themselves as "civilized", war seems a particularly barbaric way of settling differences. The causes of wars, both in a general and a particular sense, have long been debated in a variety of fora, and at the risk of over-simplification, a significant factor would certainly relate to variations in political relationships and the power of larger nations (Vasquez 2012). That concentration of power is in turn fed by attitudes apparent in contemporary society, including extreme nationalism, along with the lure of greed or the desire for power and domination. This appears especially so, when resources are unexpectedly depleted by climate change (Nunn 2007; Zhang et al. 2007).

Wars have been fought over access to important minerals, including fossil fuels such as oil. Furthermore, there is a very real possibility that in the foreseeable future this could be extended to include conflict over essential resources such as water (Farinosi et al. 2018). Another contributing factor is the propensity for intolerance of those who may be perceived as different (Swinton 2012). There appears to be little logic behind many of the "isms" prevalent in contemporary discourse, such as racism, which brings the focus to ethnic diversity, and sexism, which encompasses gender issues—or classism, ageism, ableism, etc. These "isms" are in direct opposition to the spiritual principle of *Imago Dei*, which "supports the inherent worth and value of *all* human beings as created in the image of God" (Hills et al. 2019, p. 12). The desperate need for a change of course away from the potentially destructive path humanity finds itself walking is highlighted (Karkkainen 2013; Volf 1996; Bhagwan 2020). The causes of any given war may be defined by the unique historical circumstances. Whether the rise of ancient empires or the world wars of the twentieth century are being examined, underlying factors such as political control and territorial expansion remain largely constant. Even so, one major change, especially since the second half of the twentieth century, is that wars may nowadays threaten far-flung nuclear annihilation (Caldicott 2017; Bhagwan 2020; Brewer et al. 2020). It must therefore be questioned whether wars really resolve anything, and pondered if they create at least as many problems as they might solve (BAS 2020).

Ecology

Most people have some understanding of ecology. At first consideration, it may seem little different from a more general concept such as nature, with available definitions certainly varying in their scope. From a narrowly scientific point of view, ecology may refer simply to non-human organisms in relationship with their environment.

For the purposes of this discussion, the term ecology is used in the wider sense of a network of relationships between different life forms, or in Santmire's (1985) words, "a network of interrelationships between God, humanity, and nature" (p. 9). This concept can be understood at a global, regional, or local level. For example, every rainforest has its own unique ecosystem, set within the larger network of global ecology. Specific to this discussion, Machlis and Hanson (2008) noted that "environmental research related to warfare is limited in depth and fragmented by discipline" (p. 729), wherefore they have proposed the term "warfare ecology" as a means of more appropriately focusing specific research in this area of concern.

The nature of the relationship between humans and the rest of creation is an important issue, and it is clear that in spite of differing emphases, a wide umbrella of philosophical or spiritual approaches would agree that humankind is part of what may be called "the web of life" (Capra 1996). That does not mean that all life, human and non-human, is necessarily of equal value, but it remains true that humankind and ecology are indelibly intertwined.

Inevitably, war leads to the enormous waste of finite resources as well as a host of other destructive consequences such as the contribution to global warming caused by additional greenhouse gas emissions. Each of these issues in turn exacerbates other related consequences, compounding and complicating the impacts to humans and the environment. While the purpose of this discourse is not to document in fine detail the ecological damage that war has caused and continues to cause, it is nevertheless necessary to provide at least a sense of how this damage often presents. Historically, the impact has always been significant although present day approaches including "scorched earth" policies, the use of chemical and other endlessly destructive ('sophisticated') weapons create ever-increasing stress on ecosystems.

Opportunities for Future Research

The recent War in Iraq is illustrative of these kinds of issues, with three areas of impact noted, although it is acknowledged that there is significant overlap between these. Firstly, the wastage of non-renewable resources represents an enormous cost. For example, between 1975 and 2017, the U.S. military alone consumed 181,436 million gallons of jet fuel, in addition to other forms of fuel (Crawford 2019, p. 10). That translates into 3,629 million metric tons of CO_2 emissions (p. 32). Crawford (2019) provides further details:

> In the 2003 invasion of Iraq, oil wells were set alight and burned for several months. Oil infrastructure was targeted again in 2015, when the US bombed oil infrastructure in Iraq and Syria as a means of reducing ISIS revenue. And when ISIS retreated, it set oil wells and pipelines on fire in Iraq and Syria. In many cases, these fires burned for several months. (p. 16)

These damages are just the tip of a very large iceberg of wastage and ecological bereavement, and reflect but one commodity. Analysis of debates on "peak oil" conclude that the notion mainly lost currency because it clashed with "the commonly held views that see economic growth as always necessary and desirable and depletion/pollution as marginal phenomena that can be overcome by means of technological progress" (Bardi 2019, para. 22).

Furthermore, when the impacts of war on the human community are observed, invariably those impacts are greatest in war zones. News bulletins regularly depict and report the devastation to buildings, historical artefacts, and of course people living in such communities (BAS 2020). The impacts on community health are overwhelming in their own right, offering vast opportunities for further research. Among some of the studies that have been conducted, Savabieasfahani et al. (2015) concluded that the "anthropogenic release of pollutants into the environment is especially harmful to growing foetuses and young children" (p. 1). Data for their study included distressing photographs of newborn babies with severe congenital abnormalities or medical issues including cancer, brain damage, and other associated dysmorphia. The troops themselves are also victims of respiratory disorders and other effects of air pollution, while psychological consequences and mental health concerns amongst returned veterans are becoming ever more apparent. To consider the devastation caused by landmines or soil contamination from chemical toxicity adds to the ghastly array of issues and stresses. Furthermore, and by no means the least significant, is the impact on the natural ecology and wildlife (Westing 1981; Gwynne 2012). Machlis and Hanson (2008) present the case for more research in this field:

> The development and advance of warfare ecology is both a scientific and a moral necessity. Scientific reasons include the widespread ecological consequences of warfare, the complexity of warfare's interactions with natural and social systems, and the distinctive characteristics of war – the deliberateness, destructiveness, and the intensity of its ecological impacts (p. 734)

Research into the effects of war on biodiversity and the environment has generally found that those impacts are overwhelmingly deleterious (Lawrence et al. 2015). Research also suggests that future studies may have a growing role to play in averting resource conflicts and thereby strengthening peace and security (Machlis and Hanson 2008, p. 734).

Nuclear Testing in the Pacific

In terms of the Pacific Islands, many of the more significant impacts of war have been imposed on them, in part by past colonial practice. To date, many visible reminders of war can be found throughout the Pacific (Fig. 2.1). Such relics yield a grim glimpse into the past and may remain capricious, unreckonable and treacherous today. More specifically, it is worth reflecting on the more sustained issue of nuclear testing over a period of 50 years where the Pacific Islands region lends itself as a useful cautionary

Fig. 2.1 Islanders viewing unexploded WWII ordnance in Waor Village, Nissan Atoll, Bougainville/PNG (Photo taken during fieldwork by J. M. Luetz)

tale against the effects of war and the need to work towards peace (Barker 2013, 2017; Krieger 2017; Moruroa Files 2021).

Following the end of World War II, the United States commenced sustained testing of nuclear weapons in the so-called "Pacific Proving Grounds" (Fig. 2.2). According to Martin (2018),

> As the rest of the world worried about the possibility of atomic violence, Marshall Islanders lived it. Between 1946 and 1962, the [United States Atomic Energy Commission (AEC)] conducted 105 atmospheric and underwater nuclear weapons tests at the Pacific Proving Grounds, releasing the equivalent power of more than seven thousand Hiroshima bombs. (p. 581, attributed to Jackson 1993)

Accounted differently, the Marshall Islands "experienced the equivalent of 1.6 Hiroshima-sized bombs every day for the twelve-year period between 1946 and 1958" (Barker 2017, p. 193). The long-term, devastating effects of these nuclear tests in the Pacific on both U.S. veterans as well as on the natural environment have been amply documented in the literature (Minutaglio 1994; Gwynne 2012; Martin 2018). Furthermore, these nuclear tests have imposed a profound impact on the region and its people (Davis 2005; Barker 2013; ICAN 2014, 2017). Notably, the land became contaminated by high levels of radiation, which meant that it was no longer safe for habitation or food production (Barker 2017). Davis (2005) notes how involuntary islander relocations were justified on the grounds of both the presumed U.S. ethnocentric superiority and the region's ascribed geographic peripherality:

Fig. 2.2 Bikini Atoll, Micronesia: Nuclear test "Baker", conducted by the United States military on 25 July 1946 (*Source* U.S. Department of Defense; Library of Congress: https://memory.loc.gov/master/pnp/cph/3b10000/3b13000/3b13500/3b13566u.tif)

> The imagined cultural inferiority of the Bikinians and their supposed weak attachment to Bikini were used by the U.S. military as legitimization for their removal from their atoll. [With] a focus on its nonplace attributes, Bikini was represented as a marginalized place. (pp. 614–615)

According to Martin (2018), "Marshall Islanders suffered forced relocations, destruction of ancestral lands, and radiation sickness" (p. 581). The long-term adverse consequences of forced Bikinian islander relocations reverberate through the decades to this day (Dibblin 1998; Barker 2013, 2017), as exemplified by statements made by a Bikini councillor; "As a result of being displaced we've lost our cultural heritage—our traditional customs and skills, which for thousands of years were passed down from generation to generation" (AFP 2014, para. 8). Research elsewhere in the Pacific has documented the strong attachment of atoll islanders to their ancestral lands (Luetz and Havea 2018; Farbotko 2019).

With each nuclear blast, "paradise" (Davis 2005) progressively morphed and mutated into what Gwynne (2012) has termed "Paradise With an Asterisk". "Today, three-quarters of a century after the Bikini Atoll test by the United States, the people of that once beautiful coral paradise remain displaced or physically scarred" (Bhagwan 2020, para. 12). Importantly, it seems that what was primarily 'proven' in the so-called 'Pacific Proving Grounds' was that the Pacific Islands were considered 'expendable' and that their primary raison d'être existed (in the eyes of their nuclear colonialists) in 'validating' the efficacy of modern weaponry. This utilitarian appropriation of the Pacific is reminiscent of what Farbotko (2010) has argued in relation to,

> [t]he emergence of Tuvalu as a climate change 'canary' … a location where developed world anxieties about global climate change are articulated … Expressions of 'wishful sinking' create a problematic moral geography … only after they disappear are the islands useful as

> an absolute truth of the urgency of climate change, and thus a prompt to save the rest of the planet. (p. 47)

According to Farbotko's logic, rather than embodying inherent value in and of itself, 'sinking' Tuvalu may seemingly serve a better (and more enduring) purpose by 'going under' and thereby serving humanity through its bigger utilitarian function of epitomising a warning to the rest of the world that climate change is real and needs urgent attention. Relatedly, only as the utter destruction, enduring contamination and long-term uninhabitability of the Bikini and Enewetak atolls in the Marshall Islands were laid bare during (and following) nuclear testing—along with the destruction brought on in Christmas (Kiritimati) Island, French Polynesia and Johnston (Kalama) Atoll—did these unspoilt ecosystems serve a meaningful utilitarian purpose by embodying a warning to humanity to guard against the risks of nuclear apocalypse (cf. ICAN 2014, 2017; Krieger 2017; Moruroa Files 2021). The inhabitants of these impacted island regions could be forgiven for concluding that, despite some expressions of regret, in the eyes of the great powers their lives and homes were ultimately regarded as being of little consequence as hundreds of nuclear tests were conducted in the Pacific (Barker 2017) (Fig. 2.3).

According to the Australian Living Peace Museum (ALPM, n.d.), testing of nuclear weapons was widespread in the region:

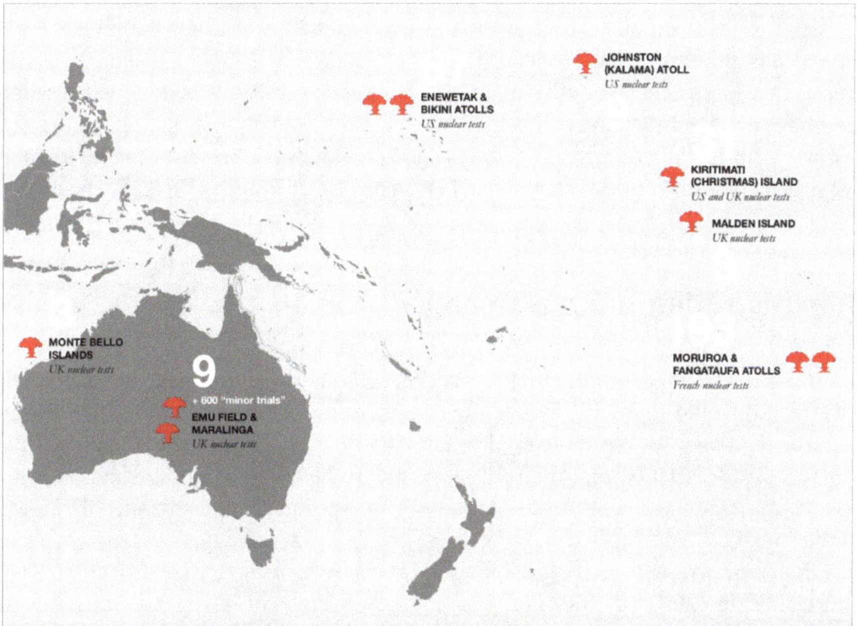

Fig. 2.3 Map of Pacific exhibiting US, UK and French nuclear test sites (Reproduced with permission from ICAN [2017, p. 4])

> Between 1946 and 1996, the United States, Britain and France conducted Cold War programs of nuclear testing in the deserts of Australia and the atolls of the central and south Pacific. Over five decades, more than 315 nuclear tests were held across the region. (para. 2)

There were many both in and beyond the Pacific who opposed the tests. Such opposition in the 1980s led to the declaration of a nuclear-free Pacific by Vanuatu, Palau, and New Zealand, while members of the Pacific Islands Forum signed the Rarotonga Treaty for a South Pacific Nuclear Free Zone (ICAN 2014, p. 18; 2017). Today "Pacific nations are enthusiastic supporters of the United Nations Treaty on the Prohibition of Nuclear Weapons" (Bhagwan 2020, para. 17). The legacies of testing have exacted a heavy and enduring toll on the health of both environments, veterans and local islanders alike (Gwynne 2012; Minutaglio 1994; Barker 2017; Martin 2018; Bhagwan 2020; Moruroa Files 2021).

This outline has been no more than a fragment of the terrible story still waiting to be told. Martin (2018) notes that nuclear testing in the Pacific also contributed to a shift in scientific understanding. Local-level empirical research conducted in the wake of the nuclear blasts highlighted the interconnectedness of life and demonstrated that 'environment' should be more appropriately conceived as 'ecosystem' where lifeforms co-exist together within the web of life (Martin 2018; Capra 1996). Relatedly, the loss of biodiversity also impacts on the resilience of ecosystems and the capacity of life to recover from natural or human-related disasters.

Of course, this background reality of the interconnectedness of life also raises issues of the human toll of war, of lives lost or maimed, as well as the incalculable mental and psychological costs of war. The wastage of life and resources on an epic scale causes needless suffering, which has been aptly described in the Christian scriptures as a kind of 'universal' anguish that invites a redemptive and reconciliatory response: "It is plain to anyone with eyes to see that at the present time all created life groans in a sort of universal travail" (Romans 8:22; Phillips Bible Translation).

Towards a Faith-Informed Solution

Solutions to the impact of conflict on Earth's ecosystems are needed to ensure their sustainability, and that will involve the consideration of many factors. In one way or another all people contribute to planetary health and/or suffer from the ecological impacts of war, so it is important for all people to be involved in the solution. This applies at a personal level in terms of relationships, political engagement, solicitation of the so-called 'poor' as subject matter experts of sustainability, and certainly from a faith perspective (Luetz et al. 2019).

Machlis and Hanson (2008) propose that a warfare ecology agenda may lead to at least three outcomes. First, it may encourage the inclusion of ecological science in military planning. Second, it may encourage the rehabilitation of former damaged military sites to areas of conservation. Third, it can promote policies that support peace and security.

At a deeper level, ecology can inform the way we live and interact at a global level. Ecosystems have evolved in such a way that they contain both a wide variety and essential unity in a fine balance (Miller 1982). Difficulties ensue when that balance is disturbed (McKibben 2010). Even seemingly violent aspects of such systems, such as the way in which wild animals co-exist, nevertheless have their place in the wider ecological balance.

When searching for answers in the sacred writings, one can find a mixed result. From a religious perspective, ancient codes such as the Ten Commandments attempted to give societies a basis on which they could live together in just peace. To this end, the Hebrew Bible introduced the concept of Shalom, which expands on the understanding of peace by speaking to wholeness, well-being and human flourishing. The term Shalom was often used to comfort and encourage and is still a traditional greeting in many Jewish communities. Shalom does not simply refer to peace as in the absence of war, but of peace in the awareness of God's presence in whatever circumstance. This teaching is far-reaching and inclusive in its sweep and was expanded through the nonviolent teachings of Jesus Christ. The love on which the concept of Shalom is based should not be sentimentalised. The various facets of both the Old Testament concepts and New Testament transformations of these concepts are all intertwined; this is a love that, for example, shows no hostility, is not predatory and does not retaliate (Boring 1995).

A prime example of that is Christ's treatment of the biblical commandment "you shall not murder" (Exodus 20:13), in which he goes beyond observable behaviours and delves into internal attitudes that propagate such behaviours. Relatedly, Jesus adds to his teaching the invitation to "love your enemies and pray for those who persecute you" (Matthew 5:43). The Greek word for 'love' in this text is 'agape', which is not about emotional attachment, but can be defined as an "intentional response that promotes well-being" (Oord 2005, p. 935). It is an approach which seeks the highest good of others. Such an approach certainly applies at the level of interpersonal relationships and can make a positive difference. In the larger realm of international affairs, the situation is obviously far more complex, and the risk of destructive warfare is never far away (BAS 2020; Mecklin 2020). Nevertheless, even in this arena, Christ's teaching concerning attitudes toward a perceived enemy offers hope.

The world's major religions promulgate common concepts regarding peace, brotherhood / sisterhood, tolerance and reconciliation, which promise rewards in the hereafter (Anjum 2017). Boring (1995) identified that this type of peacemaking "does not connote a passive attitude, but positive actions for reconciliation" (p. 180). From a Christian perspective Parkinson (2011) argues that the Beatitudes[1] are about "building the best possible world" (p. 65), so that when Jesus spoke of "peacemakers" he had in mind those who had learned to forgive and to let go of old hurts.

After the birth of the early Christian Church, the Apostle Paul purported that true Shalom can come only from God as a by-product of salvation (Romans 5:1–3). As

[1]The Beatitudes of Jesus are a list of eight blessings recorded in the Gospel of Matthew (Ch 5) and describe essential life attitudes.

such, the Church is called to recognise her God-given responsibility as the messenger of Shalom and sustainable peace. It is clear that the success of this charter has been limited.

Shalom implies a state of mind that is satisfied and has relationships which are characterised by harmony. Shalom speaks to the "quality of a person's life and quality of their relationships with God, with one another and with the rest of creation" (Swinton 2000, p. 58; see also Harper 2016). The call of contemporary Christian faith communities is to translate the concept of Shalom into present day contexts.

The beginning of the answer must lie in the inner recesses of individuals and communities, rather than in external factors. Deeper exploration into the human condition is essential, with Jesus' teachings in the Beatitudes (Matthew 5:3–12) offering a helpful starting point. Central to all faith and mission is a ministry of reconciliation, however the term "reconciliation" must be clearly understood. In popular understanding within Australia, reconciliation is typically associated with Indigenous issues. Even so, no one issue, however important, can own this term. In a religious sense it reflects a full-orbed and holistic concern that can speak to a multiplicity of dimensions of human life and society. As such, consensus on a concrete understanding or definition of the term 'reconciliation' can seem complicated. Perhaps it is time to re-define the term to include contemporary voices. Katongole and Rice (2008) offer the following characterisation: "reconciliation is an ongoing spiritual process involving forgiveness, repentance and justice that restores broken relationships and systems to reflect God's original intention for all creation to flourish" (p. 23). This understanding of reconciliation is relevant to the Pacific Islands context (Rubow and Bird 2016).

Therefore, from a Christian point of view, the very heart of the ministry of reconciliation is demonstrated in the person of the Christ. Regardless of where it leads, the process begins with one's own reconciliation with God, which Christians identify as made possible through Jesus Christ (Harper 2016). For example, the Apostle Paul exhorts the early Church: "since we have been justified through faith, we have peace with God through our Lord Jesus Christ, … through whom we have now received reconciliation." (Romans 5:1, 11). If reconciliation is birthed from a spiritual mindset, there is a need, as Hall (1996) posited, to "shift the need for reconciliation from heaven to earth" (p. 595). Thus, "[w]e need to be reconciled with other human beings, not only personally but socially and politically as well. There is also a vital need for reconciliation with nonhuman forms of life and to the earth itself" (Hall 1996, p. 595; see also Harper 2016). An eschatological promise of ultimate reconciliation carries little weight and meaning if it is unable to impact on sustainable and peaceful human coexistence in the present (Luetz et al. 2018).

Ten Steps Toward a Practical Ministry of Reconciliation

It is within that very maelstrom of activity and the more immediate context of ecology and war that we consider a multi-faceted Christian response and contribution that will have much in common with a more general humanitarian position and may resonate with other faith traditions. Such a response should note Archbishop Desmond Tutu's perspectives as Chair of the Truth and Reconciliation Commission in South Africa, in which the primary objective was to preach forgiveness as a way of healing the wounds of hatred and anger. According to The Forgiveness Project (TFP, n.d.), Tutu describes forgiveness as "the best form of self-interest" (para. 2) as it allows one to become "a better person than the one being consumed by anger and hatred" (para. 3). Tutu argues that "[i]f you can find it in yourself to forgive then you are no longer chained to the perpetrator" (para. 4). Importantly, Tutu asserts that "the process of forgiveness also requires acknowledgment on the part of the perpetrator that they have committed an offence" (para. 5). This in turn suggests that a practical ministry of reconciliation needs to move beyond a 'do no harm' approach, to restorative action. Ultimately, success of such an action will depend on many others walking the same path of reconciliation. It will not be easy, but even if one initially walks alone, one must take the first step. In that reconciliatory spirit a 10-point plan will now be proposed to assist with this process.

First, the path to a more peaceful and ecologically friendly future begins with a vision. This may engender a charge of idealism, but the alternative of realism is clearly not working. Such vision has been leveraged before when 193 countries ratified the UN Sustainable Development Goals (SDGs) that were quasi-universally adopted by and for the world in 2015. Rather than simply lurching from one crisis to the next, the leaders and citizens of nations should be encouraged to envisage the kind of world they would like to see and to share that insight with others. From that standpoint, it is at least theoretically possible for a reconciliatory plan of action to be formulated.

Second, the path of reconciliation and peace begins within each person, as an individual. If one is not at peace within one's own self, with life in this universe and with God (however one may comprehend God), then sooner or later that incongruency will expose itself (Harper 2016). At this point the pleading of the Apostle Paul that humans be reconciled to God may well be heeded (2 Corinthians 5:20). But the issue is greater still. The first step on this path is a deep conviction that humanity must work together with God to break down the many dividing walls of hostility that persist in this modern world (Moltman 2012).

Third, peace and reconciliation may be supported through prayer (Andrews 2017). Christians are surely called to pray for justice and peace, which in turn demands engagement at multiple levels. This is a fundamental message of Shalom. In effect, Christians become part of the answer to their own prayers when they pray (and live out) the prayer Christ taught believers everywhere to pray: "your kingdom come, your will be done, on earth as it is in heaven" (Matthew 6:10). Lukas Vischer (1980) noted:

> Prayer for all the churches who confess Christ is therefore a standing protest against the *status quo* of our divisions. God's gift of reconciliation can and should find full expression… Its aim is that the Church should become a real sign of reconciliation in the world. (p. 3)

By engaging prayerfully with pressing contemporary issues such as climate change, Christians may faithfully move towards peace and reconciliation, even if that is not always easy.

Fourth, reconciliation will often need the advocatory input of the 'prophets', who are described in the scriptures as single-minded individuals who are uncompromisingly outspoken against injustice and structural violence (Andrews 2021). Inevitably, there comes a time when calls for justice cannot be silenced, when platitudes are ineffectual, and when the people of God must take a stand against perceived evil (Bonhoeffer 2015). Such a stand is not the sole preserve of those with a faith perspective; popular movements such as Black Lives Matter, climate change protests, or even local community development projects have the capacity to bring people together for a cause. Offering an example from the Pacific, Barker (2017) has observed how nuclear testing has subjected the Marshallese people to forms of "structural violence";

> [I]n the Marshall Islands, the United States detonated its largest weapon ever tested, the Bravo shot of March 1, 1954, the equivalent of one thousand Hiroshima-sized bombs. Bravo exposed the crew of a Japanese fishing boat near Bikini, Marshallese residents downwind from Bikini, and U.S. servicemen to levels of radiation that caused death and lifelong illness. Following Bravo, U.S. government researchers evacuated some of the islanders and enrolled them in a secret medical experiment, called Project 4.1, to study the effects of radiation on human beings. Later, the U.S. government resettled the unwitting participants in this program on an island highly contaminated with radiation to learn firsthand how human beings ingest and absorb radiation from their environment. [...] More than a decade ago, the Marshall Islands submitted a petition to the U.S. Congress for additional assistance, primarily to create the capacity to respond to the health care burdens resulting from the U.S. nuclear weapons testing program. The U.S. government has not responded to the Marshall Island government's request for assistance. (pp. 195, 199; see also Barker 2013)

Barker (2017) further notes that "[t]he people of the Marshall Islands—people who are in the United States legally, people whose homelands are irradiated, people whose bodies carry the radiogenic burdens of the Cold War—are not eligible for Medicaid. The failure to extend Medicaid to the Marshallese is another form of structural violence" (pp. 197–198). This example underscores the enduring need for 'prophetic' protestation today. A timely message, coupled with an unwavering spirit, can take humanity along the path of justice, peace and reconciliation. These two prophetic characteristics are also illustrated by Martin Luther King Jr., who spoke out against racism, but whose famous speech "I have a dream" also presented a compelling picture of a preferential future.

Fifth, and following from the previous point, advocacy speaks to more than human need, but also to other voiceless victims of the impact of war. As Sampley (2000) proposed, "reconciliation is at the heart of life's business" (p. 98). This implies courageous engagement with political processes. More specifically, voiceless forests and wildlife need those who will speak on their behalf, and it is in such a context that raising an advocatory voice on behalf of the natural world becomes truly significant. This is reminiscent of scripture, which expressly calls for such advocatory engagement: "Speak out for those who cannot speak, for the rights of all the destitute. Speak out, judge righteously, defend the rights of the poor and needy" (Proverbs 31:8–9). In doing so, the end purpose is not condemnation but rather reconciliation—the renewal and healing of relationships and of creation itself (Harper 2016; Moltman 2012).

Sixth, it has never been more important than now to work at countering any hostility between world religions (Andrews 2017, 2021). Interfaith dialogue has long been part of the global ecumenical scene and more localised interfaith gatherings are playing their part. Yet far from being a 'lowest common denominator' approach, interfaith dialogue has always begun with a clear faith identity on the part of each participating community of faith. Writing from a Christian perspective, Migliore (2004) advocates that "[t]he encounter of Christians and non-Christians requires genuine dialogue, yet without relinquishing the responsibility to communicate the gospel as faithfully and compellingly as possible" (p. 328; see also Andrews 2017, 2021). Genuine dialogue has had many practical applications in modern times, with an increasing trend to practical cooperation. Examples range from interfaith action groups such as Australian Religious Response to Climate Change (ARRCC) to the book *Earth and Faith* (UNEP 2000), published by the Interfaith Partnership for the Environment as part of the United Nations Environment Programme. Furthermore, as noted above, there are examples from the Pacific as well (Nunn et al. 2016; Rubow and Bird 2016).

Seventh, prospects for reconciliation may be advanced by recognising that there are no significant differences between major world religions in the area of Earthcare. People of different world faiths can be more proactive in collaborating with non-faith environmental groups in the interests of advancing the sustainable stewardship of the Earth. Not only will all lifeforms and the environment itself benefit, but new areas of trust between different groups may foster greater chances of success.

Eighth, social and economic justice is key to peace. When the gap between North and South, rich and poor or the haves and have-nots becomes too wide, the possibility of conflict is greatly increased. Quite apart from the moral and ethical dimensions, survival in the poorest regions of the world is challenging and people are often living unsustainably with respect to the ways they interact with the environment, yet not by choice (Luetz et al. 2019). Therefore, working to redress inequality in the world will heighten prospects for peace and reconciliation.

Ninth, while it can never be a substitute for more substantial systemic change, a spirit of practical compassion can help reduce the burden of suffering on the part of millions of the world's poorest people (Andrews 2017). It is important that this is enacted in terms of a partnership and a shared search for justice and peace. Peace and reconciliation can flourish more readily whenever justice and compassion abound.

Tenth, questions of identity can shift attention onto unifying elements shared by all people. For example, in the Preface to his book *Rainbow of Mysteries*, Norman Habel (2012) writes: "I identify myself now as an Earth being, and not simply as a human being. Earth is my Mother, my primal parent, and the Christian church my adoptive parent" (p. 9). Christian readers who may feel uncomfortable with the idea of Earth as 'Mother' may perhaps take comfort in Chesterton's (1908) characterisation of Earth as 'Sister': "Nature is not our mother: Nature is our sister. We can be proud of her beauty, since we have the same father; … Nature is a sister, and even a younger sister: a little, dancing sister, to be laughed at as well as loved" (pp. 115–116). Irrespective of which analogy of the Earth is preferred, it can be argued that reconciliation is best achieved through a more familial identity with the Earth where the fate of all earthly creatures is viewed as indelibly intertwined.

Conclusion

This discourse has argued that true reconciliation extends far beyond human reconnection. In the context of the significant impact of war on human life and ecology, there is arguably no more important role for faith communities than to urge all parties to seek reconciliation. This imperative to advance peace and reconciliation has also been noted implicitly in the latest Doomsday Clock[2] assessment:

> [T]he Bulletin of the Atomic Scientists Science and Security Board today moves the Doomsday Clock 20 seconds closer to midnight—closer to apocalypse than ever. In so doing, board members are explicitly warning leaders and citizens around the world that the international security situation is now more dangerous than it has ever been, even at the height of the Cold War … Humanity continues to face two simultaneous existential dangers—nuclear war and climate change—that are compounded by a threat multiplier, cyber-enabled information warfare, that undercuts society's ability to respond. (Mecklin 2020, p. 3)

This vulnerability assessment shows that contemporary security threats are multiplex and need to be tackled in concert. Furthermore, it suggests that people of faith can play a special role to promote peace and reconciliation, both in the Pacific Islands region and beyond. One of the best strategies is to work with God to generate a genuine spirit of community as the best antidote to alienation (Peck 1998). When this is based on core values such as acceptance and respect for difference, it will give substance to the deep need that all humans have to belong to something larger than themselves (Swinton 2012). As the net is cast wider, we also negate those negative issues that lead to division and conflict. On 3rd April 1968, the day before his assassination, Martin Luther King Jr gave a speech which included this important observation:

> Men for years have been talking about war and peace. Now we can no longer just talk about it. It is no longer a choice between violence and non-violence in this world, it is non-violence or non-existence. That is where we are today. (Burrowes 2018, para. 3)

King, speaking in the tradition of Gandhi, is correct. There is a choice between non-violence and non-existence, or in other words, between one world and none. Or in the words of Rev James Bhagwan (2020), General Secretary of the Pacific Conference of Churches, "Let's end the nuclear age, before it ends us" (para. 1). Against that backdrop, humans are bound to become part of a movement toward a more peaceful co-existence with all living beings. Walker (2009) has demonstrated how peace with

---

[2]The so-called Doomsday Clock is a globally recognised indicator that reflects human vulnerability to disasters. It was created in 1947 by scientists who had been involved in developing the first atomic weapons as part of the Manhattan Project. The Doomsday Clock uses "imagery of apocalypse (midnight) and the contemporary idiom of nuclear explosion (countdown to zero) to convey threats to humanity and the planet" (Mecklin 2020, p. 3).

God is intertwined with all other aspects of life (Shalom)—peace within oneself, peace with others in society, and peace with the environment (see Harper 2016). Thus, the path of reconciliation is also the way of justice and peace for both humans and all living beings on the planet.

References

AFP - Agence France Presse (2014) Bikini Atoll nuclear test: 60 years later and islands still unliveable, The Guardian, 2 March, https://www.theguardian.com/world/2014/mar/02/bikini-atoll-nuclear-test-60-years

ALPM—Australian Living Peace Museum (n.d.) Nuclear testing in the Pacific 1950s–80s. https://www.livingpeacemuseum.org.au/omeka/exhibits/show/nuclear-weapons-in-aus-pacific/nuclear-testing-pacific

Andrews D (2017) Building a better world: developing communities of hope in times of despair, Anniversary edn. Australia, Albatross Books

Andrews D (2021) To right every wrong: the making and unmaking of an improbable prophet. Wipf & Stock

Anjum M (2017) Concept of peace in world's major religions: an analysis. Int J Sci Res Publ 7(4):248–259

Ayre C (2014) Earth, faith and mission: the theology and practice of earthcare. Morning Star Publishing, Northcote, VIC

BAS—Bulletin of the Atomic Scientists (2020) A statement on the bombings of Hiroshima and Nagasaki, Bulletin Science and Security Board, 6 August. https://thebulletin.org/2020/08/a-statement-on-the-bombings-of-hiroshima-and-nagasaki/

Bhagwan J (2020) Let's end the nuclear age, before it ends us. Assembly, 6 August. Uniting Church in Australia. https://assembly.uca.org.au/news/item/3213-let-s-end-the-nuclear-age-before-it-ends-us

Bardi U (2019) Peak oil, 20 years later: failed prediction or useful insight? Energy Res & Soc Sci 48(February):257–261. https://doi.org/10.1016/j.erss.2018.09.022

Barker HM (2013) Bravo for the Marshallese: regaining control in a post-nuclear, post-colonial world, 2nd edn. Wadsworth/Thomson

Barker HM (2017) Persistent violence and silent suffering: Marshallese migrants in Washington state. In: Caldicott H (ed) Sleepwalking to Armageddon: the threat of nuclear annihilation. The New Press, Ch 20

Boring E (1995) "Matthew". In: The new interpreters Bible, Vol. VIII. Abingdon Press, Nashville

Bonhoeffer D (2015) The cost of discipleship, Anniversary edn. SCM Press, London

Brewer E, Goldenberg I, Rodgers J, Simon M, Thomas K (2020) Toward a more proliferated world? The geopolitical forces that will shape the spread of nuclear weapons. Center for New American Security (CNAS), 51 pp

Burrowes RJ (2018) Nonviolence or Nonexistence? The legacy of Martin Luther King Jr. Counter Punch, 3 April. https://www.counterpunch.org/2018/04/03/nonviolence-or-nonexistence-the-legacy-of-martin-luther-king-jr/

Caldicott H (2017) Sleepwalking to Armageddon: the threat of nuclear annihilation. The New Press

Capra F (1996) The web of life: a new scientific understanding of living systems. Anchor Books.

Chesterton GK (1908) Orthodoxy. Dodd, Mead, New York

Corner A, Shaw C, Clarke J (2018) Principles for effective communication and public engagement on climate change: a handbook for IPCC authors. Climate Outreach. https://climateoutreach.org/resources/ipcc-communications-handbook ; https://climateoutreach.org/download/13492/

Crawford NC (2019). Pentagon fuel use, climate change, and the costs of war. https://watson.brown.edu/costsofwar/files/cow/imce/papers/2018/crawford

Davis JS (2005) Representing place: 'deserted Isles' and the reproduction of Bikini Atoll. Ann Assoc Am Geogr 95(3):607–625. https://doi.org/10.1111/j.1467-8306.2005.00477.x
Deane-Drummond C, Bergmann S, Vogt M (2017) Religion in the anthropocene. Cascade Books, Eugene
Dibblin J (1998) Day of two suns: US nuclear testing and the Pacific Islanders. New Amsterdam Books.
Fair H (2018) Three stories of Noah: navigating religious climate change narratives in the Pacific Island region. Geo Geogr Environ 5(2). https://doi.org/10.1002/geo2.68
Farbotko C (2010) Wishful sinking: disappearing islands, climate refugees and cosmopolitan experimentation. Asia Pac Viewp 51(1):47–60. https://doi.org/10.1111/j.1467-8373.2010.001413.x
Farbotko C (2019) Climate change displacement: towards ontological security. In: Klöck C, Fink M (eds) Dealing with climate change on small islands: towards effective and sustainable adaptation? Gottingen University Press, Gottingen, pp 251–266
Farinosi F, Giupponi C, Reynaud A, Ceccherine G, Carmona-Moreno C, De Roo A, Gonzalez-Sanchez D, Bidogilo G (2018) An innovative approach to the assessment of hydro-political risk: a spatially explicit, data driven indicator of hydro-political issues. Glob Environ Chang 52(September):286–313. https://doi.org/101016?j.gloenvcha.2018.07.001
Karkkainen V (2013) Christ and reconciliation: a constructive Christian theology for the pluralistic world. William B. Eerdmans Publishing Company, Grand Rapids, MI
Krieger D (2017) The Heroic Marshall Islanders: nuclear zero lawsuits. In: Caldicott H (ed) Sleepwalking to Armageddon: the threat of nuclear annihilation. The New Press, Ch 19
Gwynne SC (2012) Paradise with an asterisk. Outside Magazine, 5 October. https://www.outsideonline.com/1905076/paradise-asterisk?page=all
Habel N (2012) Rainbow of mysteries: meeting the sacred in nature. Cooper House, Kelowna, BC, Canada
Hall DJ (1996) Professing the faith: Christian theology in a North American context. Fortress Press, Minneapolis
Harper L (2016) The very good gospel: how everything wrong can be made right. Waterbrook, New York
Hills K, Clapton J, Dorsett P (2019) Spirituality in the context of nonverbal autism: practical and theological considerations. Pract Theol 12:186–197. https://doi.org/10.1080/1756073X.2019.1592927
Hoffman AJ, Sandelands LE (2005) Getting right with nature: anthropocentrism, ecocentrism, and theocentrism. Organ & Environ 18(2):141–162. https://doi.org/10.1177/1086026605276197
ICAN—International Campaign to Abolish Nuclear Weapons (2014) Banning nuclear weapons: a Pacific Islands perspective. https://uk.icanw.org/wp-content/uploads/2014/01/ICAN-PacificReport-FINAL-email.pdf
ICAN—International Campaign to Abolish Nuclear Weapons (2017) Prohibiting nuclear weapons: a Pacific Islands priority. https://icanw.org.au/wp-content/uploads/Pacific-Report-2017.pdf
IPCC (2014) Climate change 2014: synthesis report. Contribution of working groups I, II and III to the fifth assessment report of the intergovernmental panel on climate change [Core Writing Team, Pachauri RK and Meyer LA (eds)]. IPCC, Geneva, Switzerland, 151 pp
IPCC (2018) Summary for policymakers. In: Masson-Delmotte V, Zhai P, Pörtner H-O, Roberts D, Skea J, Shukla PR, Pirani A, Moufouma-Okia W, Péan C, Pidcock R, Connors S, Matthews JBR, Chen Y, Zhou X, Gomis MI, Lonnoy E, Maycock T, Tignor M, Waterfield T (eds) Global Warming of 1.5°C. An IPCC Special Report on the impacts of global warming of 1.5°C above pre-industrial levels and related global greenhouse gas emission pathways, in the context of strengthening the global response to the threat of climate change, sustainable development, and efforts to eradicate poverty. World Meteorological Organization, Geneva, Switzerland, 32 pp
Jackson R (1993) Guide to U.S. atmospheric nuclear weapons effects data. Defense Nuclear Agency, Alexandria. https://apps.dtic.mil/dtic/tr/fulltext/u2/b178624.pdf

Australia, and also operates a private disability, counselling and health services practice. Karenne is passionate about inclusion and the provision of equal opportunities for all people to enjoy societal benefits. She provides professional development workshops and consultation to disability organisations and community groups and is currently working on an innovative program designed to support students with specialized learning abilities to achieve success in post school educational and employment opportunities. Karenne's research has gained international attention, and she regularly speaks at local, national and international conferences on various topics pertaining to people with disabilities, autism, theology, spirituality and inclusive education. Karenne has published in several academic journals and books.

Johannes M. Luetz (BA/USA, MBA/Germany, PhD/Australia) is Research Chair at Christian Heritage College (CHC) in Brisbane, Australia. He has consulted extensively for World Vision International on research projects raising awareness of the growing effects of demographic, environmental and climate change on vulnerable communities in Asia, Africa and Latin America. He was raised in Sierra Leone and Switzerland, cares deeply about poverty reduction, holistic education, justice and equity, and has researched and published conceptually and empirically in these areas. He is a big picture thinker and has interdisciplinary research interests at the science-policy interface. His research has attracted grants, prizes, scholarships, awards and distinctions, including best thesis prize (2006), UNSW UIP Award (2009), UNSW ASPIRE Award (2012) and best paper awards (2017, 2018). Dr. Luetz is an Adjunct Fellow at the University of the Sunshine Coast (USC), Adjunct Senior Lecturer at the University of New South Wales (UNSW) Sydney, Deputy Editor at Emerald journal *International Journal of Climate Change Strategies and Management (IJCCSM)*, and he sits on the Editorial Board of Springer Nature journal *Discover Sustainability*.

Chapter 3
Religious and Cultural-Spiritual Attributions of Climate-Driven Changes on Food Production: A Case Study from North Malaita, Solomon Islands

Zina Bird, Morgan Wairiu, Helene Jacot Des Combes, and Viliamu Iese

Abstract Increasing concern exists over the impacts of climate change on food production amongst the Pacific Island Countries (PICs). Changing weather patterns are already having effects on soil fertility, pests and diseases, heat stress in plants, soil moisture, saltwater incursion from rising sea levels, and increased damage to agriculture and crops from climate-related extreme events. Although climate-related extreme events such as cyclones, floods, droughts and storm surges are causing damages on food production, farmers in PICs often attribute the occurrences and frequency of climatic hazards to God, spiritual beliefs and cultural practices. This paper evaluates farmers' perceptions of how their food production systems and livelihoods are influenced by religious and cultural-spiritual beliefs, as a proxy for understanding climate impacts. We conducted 98 in-depth, semi-structured one-on-one interviews, a social-demographic survey (n = 98) and personal observations on farmers' behaviours using the snowballing method in the Malu'u and Sikwafata communities on Malaita in the Solomon Islands. We thematically analysed the transcripts placing emphasis on understanding commonalities and differences within and between the study sites. We found that climate impacts on food production are one of the major problems identified by 92.3% of the respondents (29.0% of youths, 25.8% of middle-age, and 37.5% of elderlies) who attribute climate impacts to God's punishment or the fulfilment of a prophecy. Human factors including folklore (cultural behaviours/sorcery) were also factors affecting food production for 76% of the respondents (19.3% of youths, 31.6% of middle-age, and 25.1% of elderlies). We also found that 5.4% of farmers had no knowledge or awareness about climate change and 21.7% claimed that lack of farming equipment (tilling tools and planting materials) was a major problem. More community awareness on climate change and its impacts and further research is needed to enable farmers to make informed decisions about improving their food production systems.

Z. Bird (✉) · M. Wairiu · H. J. D. Combes · V. Iese
Pacific Centre for Environment and Sustainable Development (PaCE-SD),
The University of the South Pacific, Suva, Fiji

H. J. D. Combes
National Disaster Management Office, Majuro, Republic of the Marshall Islands

J. M. Luetz and P. D. Nunn (eds.), *Beyond Belief*, Climate Change Management,
https://doi.org/10.1007/978-3-030-67602-5_3

Introduction

In many parts of the world, the impact of climate change on agriculture is evident, having effects on average temperature, evapotranspiration, soil stability and structure, reduced nutrient absorption, and increased erosion (Karmakar et al. 2016). In addition, the spread of pests and diseases, and increased heat stress on plants are increasingly damaging to agricultural crops (Goklany 2015; Rauff and Bello 2015). If not effectively managed, they will have serious implications for the future of global food security.

While farmers in the Pacific Island Countries (PICs) have been coping with changing weather patterns and climate-related extreme events for millennia, the observed increase in frequency and intensity of extreme events over the last two decades has caused widespread loss of crops, affecting their livelihoods (Wairiu et al. 2011). For instance, periods of drought associated with the 1997–1998 and 2015–2016 El Niño events resulted in famine in some PICs, including Fiji, Samoa, Tonga and Papua New Guinea (PNG) (Kuleshov et al. 2020; Wairiu et al. 2011). Recent tropical cyclones in the Pacific have also increased in intensity and destructiveness (Walsh et al. 2016).

Farmers are becoming increasingly aware that their limited understanding of appropriate adaptive strategies to cope with the long-term impacts of climatic changes is a major challenge in effectively sustaining their livelihoods (Le Dang et al. 2014; Wang et al. 2020). Farmers have different perceptions of the factors affecting their farming activities (Morton et al. 2017; Arbuckle et al. 2014). In an African case study of a predominantly Christian community, the attitude towards climate change was found to be divided between two main views: one group seeing climate change as a natural process caused by God, and the other acknowledging the negative role of human activity on the climate (Schuman et al. 2018). This may indicate that farmers who once thought of climate change as a natural phenomenon may seek other explanations through their affiliations with spiritual/religious beliefs that lead them to consider impacts as God's punishment (Havea et al. 2018).

The factors influencing farmers' perceptions and decision-making are largely driven by traditional leadership and are often more practically-focused, place-grounded and culturally-rooted rather than scientifically-based (Sachitra and Siong-Choy 2019; Hage and Posner 2015). This could be attributed to the nature of community leadership being rooted in shared visions, social networks, and pre-existing knowledge and skills associated with members of the communities (Martiskainen 2017). Religious belief is a more dominant view compared to scientific explanation of climate change. This may have resulted from challenges to translate the term 'climate change' into local languages, causing farmers to attribute the negative impacts to other causes more connected to their religious belief and natural environment. As a consequence, the negative impacts may be held liable as the work of God (Havea et al. 2018) with little mention of climate change or climate-related extreme events.

Connections Between Religion and Agriculture

Agriculture is identified as one of the growing competitive sectors around the world due to the rising demand for agricultural products (Ridha and Wahyu 2017; Sachitra and Siong-Choy 2019). Agricultural scientists have failed to exploit the potential of religious and spiritual traditions for advancement in agriculture. A study in the Bamenda Grassfields of Cameroon concluded that the delay in agricultural modernization is caused by failing to see the potential of how spiritual beliefs and religion can contribute towards agricultural advancement because it was never a key focus in any scientific study (Lang 2018).

In the PICs, agriculture, livestock husbandry, and fisheries are some of the most common practices providing rural households with food security and supplementary income (Iese et al. 2020). People's involvement in, and behaviours towards, these activities can also be explained by experience, belief, and knowledge of traditional or modern science, which may likewise influence their attitude towards the factors affecting their agriculture, livestock, and fishing activities. Havea et al. (2018) found that the spiritual wellbeing of the people in Tonga became negatively affected by climate change. While the study found that people viewed climate change impacts as a form of God's punishment, their faith also served as a means of adaptation, especially using the Bible scriptures as information or instruction. It is therefore common to hear people claim that climate-related extreme events are explainable by God's action, which they believe is the cause of all disasters and climate variations (Janif et al. 2016; Smiley 2019). Some scientists contend that limited understanding and knowledge of scientific information is one of the main barriers to effective climate-change adaptation in rural communities because of the incapability of the rural population to interpret the information disseminated (Mackay et al. 2019; Phuong et al. 2018). In another view, it was believed that only God could reverse the negative impacts thanks to maintained faith and prayer (Havea et al. 2018).

This paper reports on a study in rural communities on Malaita, Solomon Islands. It aims to investigate how religion influences farmers' perceptions of climate change in the To'abaita region of northern Malaita. The study evaluates two main perceptions among farmers—one where the negative impacts on crop production are attributed to nature and the will of God, and the other, where impacts are viewed as induced solely by human actions including cultural behaviour and sorcery.

Methodology

Study Site

This study was conducted in the two communities of Malu'u and Sikwafata in the To'abaita region of Malaita Island (Fig. 3.1). Malu'u has an estimated population of 4,410 comprising around 900 households with an average household size of 4.9.

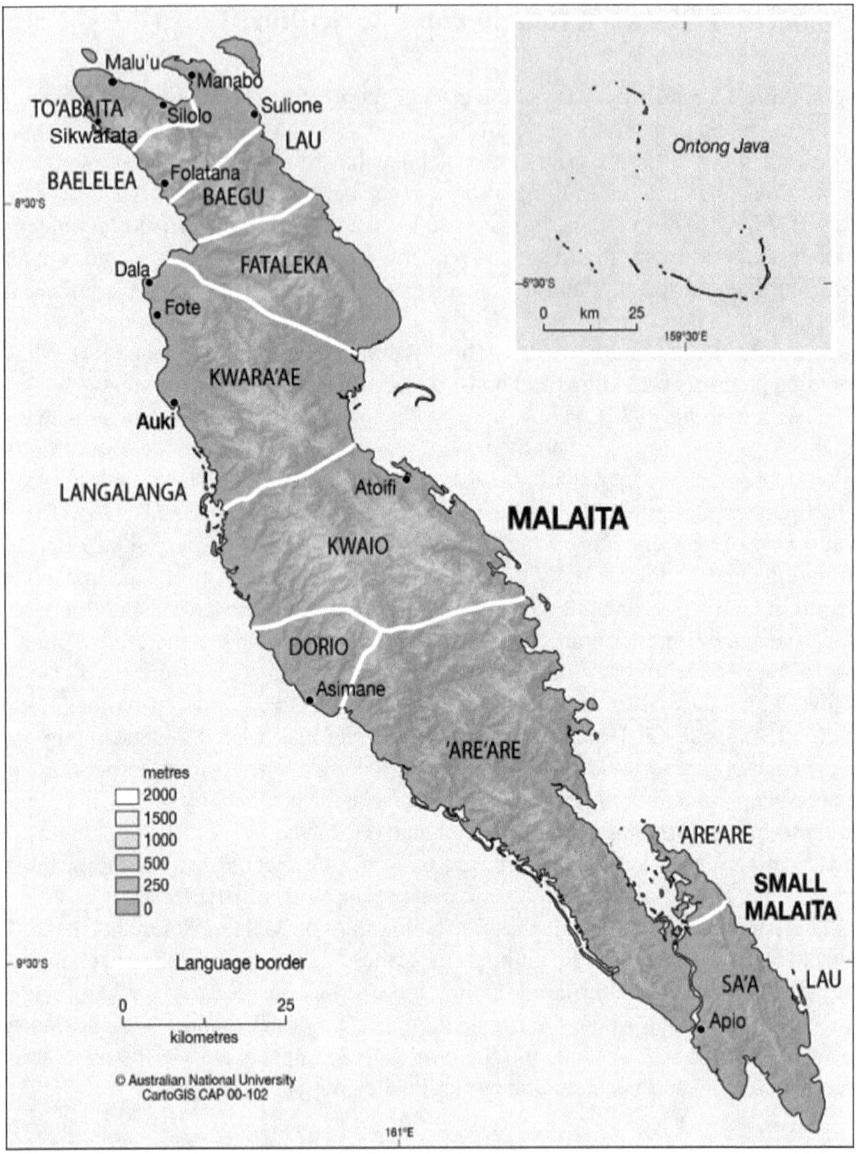

Fig. 3.1 Map of Malaita showing the location of the two study sites—Sikwafata and Malu'u in the To'abaita region. The outer islands of Malaita are shown in the small map (insert) (*Source* CartoGIS Services, College of Asia and the Pacific, The Australian National University)

Sikwafata has a population of 1,020 with 200 households and an average household size of 5.1. Population estimates are based on information collected during the study.

There are eight religious denominations in Malu'u and Sikwafata - Anglican; South Seas Evangelical Church (SSEC); Seventh-Day Adventist (SDA); Methodist; Jehovah's Witness; Baptist; Anglican Revival; and SSEC Revival. Malaita follows Solomon Islands' national six-level education system—Early childhood, Primary school (grades 1–6), Junior high (grades 7–9), Secondary high (grades 10–12), Training centres (rural, vocational and technical centres) and Tertiary (SINSO, SIMoHMS, and PC 2017). To'abaita region has 11 schools, two of which are located in Malu'u and Sikwafata.

Sampling, Data Collection

The sites were selected because of the principal researcher's familiarity with the location and its people, through tribal and family ties. A network system was used to identify farmers in each location, after which the snowballing method was applied. This is where the original farmer identifies the next potential farmer to participate in interviews. The interviews took place either at the farmers' homes or at the local market. Inclusive sampling ensured appropriate representation of genders, age groups, and coastal and inland farmers.

Primary data collection took place in April and May of 2019. A social-demographic survey was conducted to capture the household profile of the two sites and a perception survey was administered as a one-on-one interview using a semi-structured questionnaire (n = 98) and personal observation on farmers' behaviour towards the negative impacts on their food productions.

The farmers were asked four main questions.

- *What do you think are the main causes/problems affecting your food production?*
- *What are the crop types you noticed that were affected or have changed over time?*
- *What are the main constraints/difficulties you faced that are affecting your food production system?*
- *In your opinion can you describe the factors you faced in the last 30 years and the possible outcome in the next 30 years that may affect your food productions?*

The period of 1988–2050 was selected as the reference period for this study. According to IPCC (2018), there is high confidence that the average temperature is likely to exceed or reach 1.5 °C between 2030 and 2052. Representative Concentration Pathway (RCP) 8.5 scenario or the pathway with the highest greenhouse gas emissions (GHGs) was adopted in this study assuming that business as usual continue as normal leading to high concentration of carbon in the atmosphere (Riahi et al. 2011). This indicates rapid global warming. Based on the climate models, the climate change projection and the potential impacts and associated risk include "increased mean temperature in most land and ocean regions (high confidence), hot extremes in most inhabited regions (high confidence), heavy precipitation in several regions (medium confidence), and the probability of drought and precipitation deficits in some regions (medium confidence)" (IPCC 2018, p. 9). Thus, the time period of 1988–2050 was chosen to capture farmer's perception over the period of 62 years to assess their observation (1988–2019) and their perception or projection of climate change impacts on their food production by 2020–2050.

Data Analysis

Primary data for the field survey was transcribed and organised into Microsoft Excel, and then uploaded into the NVivo 10 software package. NVivo10 is a non-numerical software that handles most audio files (including WAV, MP3, and WMA), image files (including BMP, GIF, JPEG, and TIFF), and video files (including MPEG, AVI, WMV, MP4, and QT) (Hoover and Koerber 2009). The transcriptions were translated from Solomon Islands' Pidgin into English before they were entered into NVivo 10. Once entered, a thematic analysis approach was used to identify the common ideas, concepts, words, topics, or patterns, which were then organised into specific themes. These themes are presented in the results and discussion sections.

Results

Social-Demographic Profile

Of the 98 participants, there were 22 female and 29 male from Malu'u and 21 female and 26 male from Sikwafata. About 83.0% were primarily involved in farming activities, specifically crop farming while 17.4% were engaged in both farming and other activities like small businesses, trades like carpentry, and formal employment as teacher and other forms of employment. The analysis showed that 56.1% of the farmers were male whilst 43.9% were female. The participants fell within the following age groups: youth (18–35 years), middle-age (36–55 years) and elderly (above 56 years). The majority of the farmers were part of the middle-age group (35.7%) for both genders. Most had completed only primary education (43.8%) followed by early childhood (11.3%), junior high (11.3%), secondary high (7.1%) and tertiary (5.1%) while 18.4% had no formal education, which was slightly higher for the elderlies (11.2%) compared to youths (3.1%) and the middle-age group (4.1%). Three percent refused to mention their education status. For the elderly group, none of the farmers reached tertiary education. Around 48.0% of participants came from households with 5–7 family members. The mean household size of communities surveyed was 5.1.

Most of the farmers were low-income earners (65.3%) earning only USD2-35 on a weekly basis. Weekly household income ranged from USD2-212, depending on the sales of their products and household members' employment. The highest proportion of household expenditure was on store-bought foods (27.2% of youths, 26.1% of middle-age, and 35.5% of elderlies) followed by food bought from the market (19.6% of youths, 11.7% of middle-age, and 19.7% of elderlies). Other expenditure items include, school fees and stationeries (25.0% of youths, 30.0% of middle-age, and 21.1% of elderlies), clothing (17.4% of youths, 14.4% of middle-age, and 10.5% of elderlies), home repair (9.0% of youths and 5.3% of elderlies), and contributions to social events (5.4% of youths, 3.6% of middle-age, and 1.3% of elderlies). The remaining 2.0% of youths, 5.4% of middle-age, and 6.6% of the elderlies refused to respond. All age groups spent more on store-bought foods however; the proportion was slightly higher for the elderly age group.

Perceptions on the Causes of Negative Impacts on Food Production

Table 3.1 presents farmers' responses to the factors negatively affecting food production. The responses include climatic and non-climatic stressors, which farmers observed as a hindrance to optimal farming practices. The responses involving climatic and natural factors include disaster, cyclone, rainfall, drought or dry spell, temperature, humidity, and unspecified climate changes; the causes attributed by farmers to these factors included climate changes caused by God's will, by humans' impacts on the environment, and sometimes by traditional or culturally-grounded beliefs. Twenty-nine percent of youths, 25.8% of middle-aged, and 37.5% of the elderlies considered that God caused the negative impacts on their food production through changes to normal patterns or variations of climate/weather and disasters. Note that all farmers considered natural hazards causing disasters as normal rather than aberrant, including tropical cyclones, torrential rainfall, drought/dry spell, and fluctuations in temperature and humidity (54.1% of youths, 57.1% of middle-age, and 53.2% of elderlies).

Figure 3.2 presents the crop type ranked from the least to the most damaged due to the factors listed in Table 3.1. Figure 3.3 presents the constraints and difficulties limiting the ability of farmers to address the impacts on crops caused by the factors listed in Table 3.1. All crops and vegetables were affected but the two most affected were taro (78.6%) and sweet potato (77.6%) (Fig. 3.2). The analysis also found that the main constraints (21.7%) were the lack of equipment (tilling tools and planting materials) and labour (fewer than two people and old age) (20.4%) followed by other constraints as presented in Fig. 3.3.

Table 3.1 Farmers perception on the causes of negative impact on food production

Question	Age	Perception based on individual responses	Percentage
What do you think are the main causes/problems affecting your food production?	21–35	Natural factors	54.1
		Climate change and climate-related extreme events caused by God	29.0
		Folklore (cultural behaviours/sorcery)	12.9
		Human factors (impacts on the environment)	6.4
	36–55	Natural factors	57.1
		Climate change and climate-related extreme events caused by God	25.8
		Folklore (cultural behaviours/sorcery)	17.2
		Human factors (impacts on the environment)	14.4
	Above 56	Natural factors	53.2
		Climate change and climate-related extreme events caused by God	37.5
		Folklore (cultural behaviours/sorcery)	9.4
		Human factors (impacts on the environment)	6.3

Note Natural factors according to farmers refers to climate-related extreme events considered to be of natural cause or process

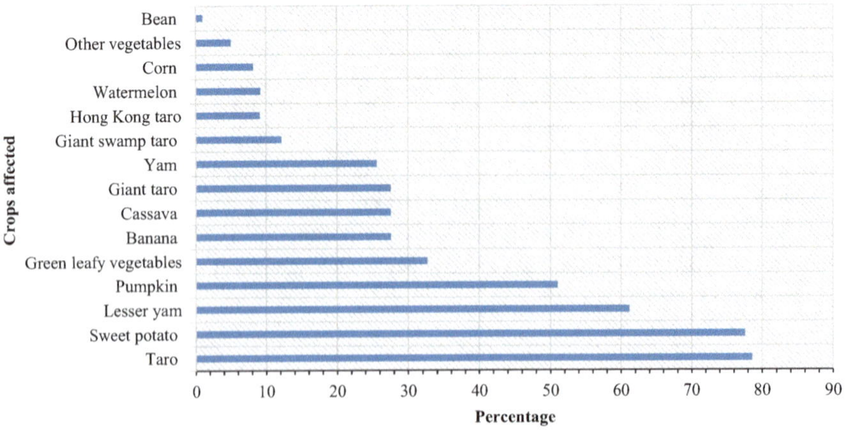

Fig. 3.2 Summary of farmers' response on crop type affected by climate and non-climate stressors

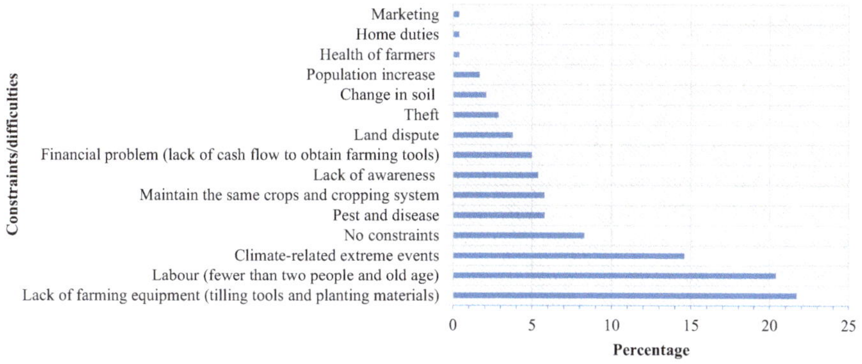

Fig. 3.3 Summary of farmers' responses on the constraints and difficulties limiting their farming activities

Farmers' Observation in 1988 and Perceptions for 2050 on Food Production

Fifty-six percent of farmers stated that by 2050, the population will continue to increase; demand on space for development will limit the availability of land for gardening, social conflict, theft, and land disputes will increase, and youth involvement in agriculture will decline due to lack of interest in farming. Compared to 1988 when climate change was not a major concern, there is more concern now and for the future, particularly with changing rainfall patterns and the increasing proportions of hot days. Although farmers show great concern, they are uncertain about climate change and its impacts on crop production in the long term. The farmers' concerns for the future include unsustainable farming practices such as removal of useful vegetation to improve soil fertility on garden sites, repeated (unbroken) cultivation on the same piece of land leading to loss of soil fertility, and loss of certain crop varieties due to a shift to fast-yielding varieties. Farmers also identified the lack of appropriate gardening tools, loss of planting materials, growing reliance on processed foods, and concern that future generation will lose interest in farming due to modern influences.

Discussion

Local farmers play a significant role in food production (home gardening and cash cropping) that contributes to household food security as well as the economy at local and national levels. In the Solomon Islands, farming of crops, livestock husbandry, fishing, and handicraft making are major sources of income (FAO and SPC 2019). Home gardening is a common practice in most developing countries, including the PICs and provides a source of food security and livelihood (Galhena et al.

2013). Farming is the main occupation of communities in North Malaita. *Social-demographic* for instance age, gender, education status, or farming experience does contribute towards farmers' perception on the factors affecting their food production leading to food insecurity (David and Abbyssinia 2017; Zhou et al. 2019). These factors, under the influence of culture, folklore, and Christian beliefs, will affect farmers' judgments or attitudes towards the causes of impacts on their food productions or subsistence farming.

Farmers' Social-Demographic Profiles

Farmers are also involved in various income-earning activities, both in the formal and informal sectors. The majority (82.6%) is fully engaged in farming (41.8% male and 40.8% female) while others (17.3%) are involved in both farming and other forms of cash-earning activities. The incomes of farming households vary depending on the sales of their products and can be as low as USD2 a week. About 48.0% of households consist of 5–7 members, with a mean household size of 5.1. According to SINSO, SIMoHMS, and PC (2017) national survey in 2015, the population of Malaita showed a balance between average income-earning (USD66) and expenditure (USD61). This is comparable to the mean income (USD49) in this study despite 65.3% of the farmers' income being well below the SINSO, SIMoHMS, and PC (2017) figures. A study in Fiji also showed a similar trend for a rural population engaging in farming (O'Meara et al. 2019). All age groups interviewed in this study frequently bought food from the store and markets, corresponding to findings in Bottcher et al. (2019). The middle-age category spent more on education because they were more likely to have children in school, compared to the youth and elderly age groups. The youth category did not mention spending on house repair as some live with their families or have newly built thatched houses, which can last for two to three years.

The farmers interviewed have an average of 29.7 years of farming experience. Older farmers are more likely to have experienced changes in the weather patterns and to have more practical knowledge about crop management. Farmers with 41–50 years of farming experience have significant farming knowledge, especially of the traditional methods of nurturing the land and optimising production, followed by those with 21–30 years of experience and 11–20 years of experience.

According to Bottcher et al. (2019), Horsey et al. (2019) and SINSO, SIMoHMS, and PC (2017), about 20.0% of the Malaita population has never received any form of formal education. This study found a corresponding figure, with 18.0% of the farmers (14.0% female and 4.0% males) having received no formal education. The majority of the farmers (24.4% male and 19.3% female) completed primary education. Thirty-four women and 33 men never completed primary school. All three age groups have a similar proportion of those who completed primary school whereas more youths received tertiary level education.

Demographic characteristics of farmers play an important role in shaping the perception that will determine whether to adopt or reject new ideas or practices

(Ghauri and Qureshi 1999). According to Phondani et al. (2020), the willingness to adopt new ideas is influenced by characteristics such as the education, age, and income. The analysis of farmers' perceptions in this study reflects a similar pattern.

Impacts Observed by the Farmers on Their Food Crop Production

Farmers recalled that crops affected by climatic and non-climatic stressors (Table 3.1) over the last thirty years include taro (*Colocasia esculenta*), sweet potato (*Ipomoea batatas*), yam (*Dioscorea alata*) and lesser yam (*Dioscorea esculenta*), cassava (*Monihot esculenta*), giant taro (*Alocasia macrorrhiza),* giant swamp taro (*Cyrtosperma chamissonis*), Hong Kong taro (*Xanthosoma sagittifolium*), and vegetables (Fig. 3.2). The impacts observed include tuber damage or rot, pest infestation and disease, wilting, more vine growth instead of tuber growth, loss of seedlings, small tuber sizes at harvest, and loss of taste. Sweet potato, one of the most important crops grown by the farmers, grows more vigorously during the rainy season but produces fewer tubers. Farmers' observations were similar to the results of previous research in the region by Iese et al. (2015, 2020) and Maeke (2013). Loss of traditional crop varieties was also highlighted as a problem. Crop varieties that were planted in the past are now difficult to find because farmers prefer early yielding varieties. These problems affected crop production and reduced incomes, as exemplified by the following statement:

> It affects my livelihood. I don't have enough food to eat and to sell in the market. (Malu'u, age 61)

Elderly farmers tend to have more historical knowledge of weather patterns giving them a better perspective on climate-related extreme events despite their lack of scientific understanding of climate change. For example,

> I think the negative impacts on my garden are natural. Because of no proper knowledge except our traditional knowledge passed on from our forefathers: like the planting times and which crops are suitable for different weather conditions like rainy and dry seasons so that it won't be affected. However, this is now challenging because the weather is changing. This month it should rain, it still is dry. It is very hot and our plants are dying so we are not sure what to plant because the weather is now confusing and changing. (Sikwafata, age 76)

Similar conclusions have been reached in comparable contexts in the PICs and beyond (Garay-Barayazarra and Puri 2011; Korovulavula et al. 2019; Lefale 2010). Although farmers in this study could not quantify crops harvested and sold, climate change may have affected crop production and contributed to the reduction of the income to USD2-35 weekly.

Analysis revealed that there was a major shift in farmers' perception of past (1988) climate change impacts compared to the future (2050). Economic, internal, external, and social factors have likely influenced this change in perception as concluded by Hallam et al. (2012). It was observed that, while there was confusion between climate change and climate-related extreme events, most farmers accepted that climate change would be of great concern in the future, based on experiences of change in the frequency and intensity of rainfall, dry spells, and cyclone over the past three decades. By 2050, population growth is also considered a concern because of the growing population density in the community and competition for land for cultivation to produce food. McCubbin et al. (2015) made a similar conclusion.

> Population will increase… the land we cultivated on the hills will soon have houses and we won't have a place to plant. (Malu'u, age 62)

Cause of Impacts: Climate-Related Events, Human Factor and Religious Beliefs.

Half of the farmers suggested that the negative impacts had natural causes and were not related to climate change. This indicates that climate awareness in the two communities is lacking. Despite that, farmers have witnessed changes in the weather patterns, they still considered this variability normal. However, more than half of the farmers interviewed felt that crop yield was climate-related. Farmers often perceived weather patterns, natural factors, and disasters as one entity, with religious and biblical scripture having a strong influence on people's lives. Previous studies by Haluza-DeLay (2014), Le Dang et al. (2014) and da Silva Oliveira et al. (2017) found that religious beliefs influence people's perceptions of, and responses to, climate threats. This explains why 92.3% of the farmers interviewed tend to have strong spiritual beliefs.

> I know that this is God's will and is natural. It was said in the Bible that we would face problems; the crops will die and fail. The other reason is that I do not give my tithes to the church during harvest. So when I do not give my tithes, then God punishes me by causing my crops to die and making me hungry because my gardens are affected. (Sikwafata, age 74)
>
> I know that it is God's doing if I forget or do not give my tithe to God or my first harvest, my garden becomes affected because it says in Proverb 3: 9–10 that I should give back my first harvest to God and he will bless my garden. But because I did not, my crops died or did not grow well and that is why God punishes my food crops to die. (Malu'u, age 26)

Similar sentiments are found in rural Fiji (Janif et al. 2016) and elsewhere, especially in faith-engaged culturally-grounded communities (Rowland 2019; Schuman et al. 2018; Skrimshire 2014). Havea et al. (2018) and Nunn et al. (2016) also demonstrated how religion influences people's attitudes towards unprecedented climate-driven changes, which can be interpreted as interventions by God in human affairs. Lack of information, communication, income and education influences behaviours and may have contributed to low or slow adoption of modern agriculture technologies. Thus, farmers may continue to bear the negative impacts without attempting to fully address the effects or to enhance their capacity to cope with the effects they face because of ignorance or faith in God (Paltasingh and Goyari 2018). Salite (2019) also described a similar situation in which the losses of crops were explained by the biblical principle that the prosperity of crops comes from God. This has shaped farmers' behaviour to accept the negative impacts on the crops as the work of God. In other studies, Glassman (2018), Konisky (2018) and Schuman et al. (2018) illustrate strong relationships between Christian beliefs and farmers' behaviours towards changes in the environment while choosing to remain faithful to their beliefs and retain trust in God, which was also observed in this study:

> We only follow what the leaders tells us to do... if they tell us climate change is the problem and advise us what we should do to protect our crops then we follow. So we only follow what was said and preached. (Malu'u, age 70)

Farmers feel that impacts from human activities in their communities affect nature. These activities include: increased disputes over land for cultivation and construction of new homes; removal of vegetation for farming resulting in soil erosion; increased runoff due to ineffective drainage systems; and cultivating the same area repeatedly causing soil infertility. They also highlight folk practices (cultural behaviours/sorcery) as a negative cause. Some farmers (39.5%) believed that sorcery, certain plants, foods, and activities (taboos) also negatively affect their crop production.

> There are certain activities which have to be followed to avoid damage to the gardens or losing the food crops. First, consuming certain foods before going to the garden can kill crops. They include turtle and Koa (mangrove fruits). It was believed that turtle has a strong odour and is stuck on your skin when you sweat so if you walk past or go to your garden, the plants will die the next day. This is true for taro and is still believed. I still believe in this especially eating turtle because I cooked it and gave it to my grandchildren, they went to their mother's garden, and the entire taro died, both the young and the matured plants died. So if you eat turtle, you have to wait 4 months before going back to the garden, it gives time for your body to flush out the turtle from your system as it was believed. If you eat Koa, you have to stay 2-3 months away from your garden. The second belief was a vine. If this vine is found growing in your garden, it will kill all the grown crops" I am not certain how but whenever this vine exists, none of the crops will survive. Third, the intimacy of couples was believed to be a taboo because it is "Nanaiabu" (producing children) and it is against the cultural belief to go to the gardens straight afterward because it will affect the planting process in the garden and no crops will survive. However, this belief is dying. Fourth, witchcraft or magic has long been associated with the supernatural belief which communities because of jealousy and land disputes. (Sikwafata, age 86)

The communities believe that consuming turtle and the fruit of mangrove *(Bruguiera gymnorrhiza)* makes them highly vulnerable to losing their crops, especially one of the main cultural and economic crops (taro). This traditional belief is prevalent since farmers have experienced such impacts by consuming these foods. Certain shrubs and vines are only found in the bushes and not on the coastal areas. They are poisonous and associated with bad luck if found in the gardens. Sexual intimacy was also forbidden before being involving in any garden activities since it was believed that it would affect the growth of crops. Additionally, sorcery was also considered a bad practice in the Melanesian countries (Foana'ota 2015; Humble 2013). In Malaita, sorcery *was seen as a problem* and it was said that farmers themselves suspected other people of killing their crops through various actions largely arising from jealousy.

A Way Forward for Farmers

Strongem Waka lo Community fo Kaikai (SWoCK), SPC and USAID project: 'Vegetation and land cover mapping and improving food security for building resilience to a changing climate in Pacific island communities, and 'Choiseul Province climate change vulnerability and adaptation assessment' are few of a many projects and assessments conducted to assess agricultural vulnerability encountered through climate-related extreme events to improve agricultural production, awareness and, promulgate positive attitudes towards food and agriculture resilience in the Solomon Islands (Mataki et al. 2013; SPC and USAID 2016; United Nations Development Programme 2016). About 5.4% of the farmers highlighted the lack of information on climate change and appropriate advice from the Government as a major gap. About 17.1% of the farmers mentioned that there is a need for respective government ministries to provide information and advice, and to conduct relevant training in their communities.

The Solomon Islands Ministry of Agriculture and Livestock (MAL) Sector Policy 2015–2019 highlights extension services as a key policy execution tool for the sector and its development programs. However, support for extension services is constrained by low budget allocations (Solomon Islands Government 2015). In 2018, MAL deployed 140 extension officers throughout Solomon Islands, but none had reached out to the farmers of the Malu'u and Sikwafata communities. Farmers are willing to learn and to acquire new knowledge and skills in understanding and responding to climate change impacts on crop production, yet access to information is one of the main barriers in rural communities. The integration of traditional and scientific knowledge is important for the improvement of food production under the impacts of climate change and it is good to engage church organisations as partners to explain climate change impacts to rural farming communities. There is also a need to assist farmers by providing farming equipment and planting materials.

> We need awareness … educate communities of climate change impact by the government/NGOs. (Malu'u, age 60)
>
> Government should address/provide clear awareness on climate change to local communities. (Sikwafata, age 39)

Conclusion

Climate change is one of the main factors affecting farmers' food production in Solomon Islands thus affecting food and income security. Yet the majority of the farmers tend to look at climate change and climate-related extreme events as the work of God, perhaps as punishment or the fulfilment of biblical prophecies. This is because information and awareness barriers disadvantage the farmers. Farmers' religious beliefs and formal education level play a significant role in influencing their perceptions. This may also reflect a lack of appropriate advice and information about climate change and its impact on food production. The majority of the farmers have limited formal education and although they mentioned climatic issues, they linked them to their Christian faith or natural factors. Farmers remain concerned that in the next 30 years, climate change will be their biggest challenge, followed by rapid population growth. They have raised their concerns about the difficulties they are facing and have indicated that appropriate advice, training, and awareness are necessary. Further study is required to add to the current knowledge. Farmers will also benefit from further support from government and other stakeholders to improve agriculture, health, and livelihood of rural farming communities.

References

Arbuckle JG, Hobbs J, Loy A, Morton LW, Prokopy L, Tyndall J (2014) Understanding farmer perspectives on climate change: toward effective communication strategies for adaptation and mitigation in the Corn Belt. JSWC Nov/Dec 69(6):505–516

Bottcher C, Underhill SJ, Aliakbari J, Burkhart SJ (2019) Food purchasing behaviors of a remote and rural adult Solomon islander population. Foods 8(10):464

da Silva Oliveira RC, Albuquerque UP, da Silva TLL, Júnior WSF, da Silva Chaves L, de Lima Araújo E (2017) Religiousness/spirituality do not necessarily matter: effect on risk perception and adaptive strategies in the semi-arid region of NE Brazil. Glob Ecol Conserv 11:125–133

David II, Abbyssinia M (2017) Factors affecting smallholder farmers' perception regarding their use of soil conservation practices: evidence from farming at Qamata irrigation scheme, South Africa. J Hum Ecol 59(2–3):82–91

FAO, SPC (2019) Country gender assessment of agriculture and the rural sector in Solomon Islands. Honiara

Foana'ota L (2015) 4. Sorcery and witchcraft as a negative force on economic and social development in Solomon Islands. Talking it through, 75

Galhena DH, Freed R, Maredia KM (2013) Home gardens: a promising approach to enhance household food security and wellbeing. Agric & Food Secur 2(1):8

Garay-Barayazarra G, Puri RK (2011) Smelling the monsoon: senses and traditional weather forecasting knowledge among the Kenyah Badeng farmers of Sarawak, Malaysia. IJTK 10(1):21–30

Ghauri MJCB, Qureshi AA (1999) Statistical hand book of Punjab forestry. Wildlife and fishers' department Punjab Forest Dept.-Lahore

Glassman LW (2018) "You help them out and god gets the glory:" social class and inequality in a fundamentalist Christian church. Soc Incl 6(2):127–139

Goklany IM (2015) Carbon dioxide: the good news. Available at SSRN 2674685

Hage J, Posner BZ (2015) Religion, religiosity, and leadership practices. LeadShip & Organ Dev J 36(4):396–412

Hallam A, Bowden A, Kasprzyk K (2012) Agriculture and climate change: evidence on influencing farmer behaviours. Social Research, Scottish Government

Haluza-DeLay R (2014) Religion and climate change: varieties in viewpoints and practices. Wiley Interdiscip Rev: Clim Chang 5(2):261–279

Havea PH, Hemstock SL, Jacot Des Combes H, Luetz J (2018) "God and Tonga Are My Inheritance!"—Climate Change Impact on Perceived Spritiuality, Adaptation and Lessons Learnt from Kanokupolu, 'Ahau, Tukutonga, Popua and Manuka in Tongatapu, Tonga. In: Climate change impacts and adaptation strategies for coastal communities. Springer, Cham, pp 167–186

Horsey B, Swanepoel L, Underhill S, Aliakbari J, Burkhart S (2019) Dietary diversity of an adult Solomon Islands population. Nutrients 11(7):1622

Hoover RS, Koerber AL (2009) Using NVivo to answer the challenges of qualitative research in professional communication: benefits and best practices tutorial. IEEE Trans Prof Commun 54(1):68–82

Humble GJ (2013) Sorcery and animism in a South Pacific Melanesian context. J Advent Mission Stud 9(2):002–020

Iese V, Maeke J, Holland E, Wairiu M, Naidu S (2015) Farming adaptations to the impacts of climate change and extreme events in Pacific Island countries: case study of Bellona Atoll, Solomon Islands. In: Impacts of climate change on food security in small Island developing states. IGI Global, pp 166–194

Iese V, Halavatau S, De Ramon N'Yeurt A, Wairiu M, Holland E, Dean A, Veisa F, Patolo S, Havea R, Bosenaqali S, Navunicagi O (2020) Agriculture under a changing Climate. In: Climate change and impacts in the pacific. Springer, Cham, pp 323–357

IPCC. (2018). Summary for policymakers. In: Masson-Delmotte V, Zhai P, Pörtner HO, Roberts D, Skea J, Shukla PR, Pirani A, Moufouma-Okia W, Péan C, Pidcock R, Connors S, Matthews JBR, Chen Y, Zhou X, Gomis MI, Lonnoy E, Maycock T, Tignor M, Waterfield T (eds) Global Warming of 1.5°C. An IPCC Special Report on the impacts of global warming of 1.5°C above pre-industrial levels and related global greenhouse gas emission pathways, in the context of strengthening the global response to the threat of climate change, sustainable development, and efforts to eradicate poverty. In Press

Janif SZ, Nunn PD, Geraghty P, Aalbersberg W, Thomas FR, Camailakeba M (2016) Value of traditional oral narratives in building climate-change resilience: insights from rural communities in Fiji. Ecol Soc 21(2):7–16

Karmakar R, Das I, Dutta D, Rakshit A (2016) Potential effects of climate change on soil properties: a review. Sci Int 4(2):51–73

Konisky DM (2018) The greening of Christianity? A study of environmental attitudes over time. Environ Polit 27(2):267–291

Korovulavula I, Nunn PD, Kumar R, Fong T (2019) Peripherality as key to understanding opportunities and needs for effective and sustainable climate-change adaptation: a case study from Viti Levu Island, Fiji. Climate and Development, 1–11

Kuleshov Y, Inape K, Watkins AB, Bear-Crozier A, Chua ZW, Xie P, Xie P, Kubota T, Tashima T, Stefanski R, Kurino T (2020) Climate Risk and Early Warning Systems (CREWS) for Papua new Guinea. In Drought—Detection and Solutions; Intech Open, 147–168.

Lang MK (2018) The role of religion in agriculture: reflections from the Bamenda Grassfields of Cameroon since pre-colonial times. Mgbakoigba: J Afr Stud 7(2):54–73

Le Dang H, Li E, Bruwer J, Nuberg I (2014) Farmers' perceptions of climate variability and barriers to adaptation: lessons learned from an exploratory study in Vietnam. Mitig Adapt Strat Glob Change 19(5):531–548

Lefale PF (2010) Ua 'afa le Aso Stormy weather today: traditional ecological knowledge of weather and climate. The Samoa experience. Clim Change 100(2):317–335

Mackay S, Brown R, Gonelevu M, Pelesikoti N, Kocovanua T, Iaken R, Mackey B et al (2019) Overcoming barriers to climate change information management in small island developing states: lessons from pacific SIDS. Clim Policy 19(1):125–138

Maeke J (2013) Vulnerability and impacts of climate change on food crops in raised atoll communities: a case study of Bellona community in Solomon Islands. Published Master of Science in Climate Change thesis. Pacific Centre for Environment and Sustainable Development (PaCE-SD). Faculty of Science, Technology and Environment. The University of the South Pacific

Mataki M, Solo G, Donohoe P, Alele D, Sikajajaka L (2013) Choiseul province climate change vulnerability and adaptation assessment report.

Martiskainen M (2017) The role of community leadership in the development of grassroots innovations. Environ. Innov. Soc. Transit 22:78–89

McCubbin S, Smit B, Pearce T (2015) Where does climate fit? Vulnerability to climate change in the context of multiple stressors in Funafuti, Tuvalu. Glob Environ Chang 30:43–55

Morton LW, Roesch-McNally G, Wilke AK (2017) Upper Midwest farmer perceptions: too much uncertainty about impacts of climate change to justify changing current agricultural practices. J Soil Water Conserv 72(3):215–225
Nunn PD, Mulgrew K, Scott-Parker B, Hine DW, Marks AD, Mahar D, Maebuta J (2016) Spirituality and attitudes towards Nature in the Pacific Islands: insights for enabling climate-change adaptation. Clim Change 136(3–4):477–493
O'Meara L, Williams SL, Hickes D, Brown P (2019) Predictors of dietary diversity of Indigenous food-producing households in rural Fiji. Nutrients 11(7):1629
Paltasingh KR, Goyari P (2018) Impact of farmer education on farm productivity under varying technologies: case of paddy growers in India. Agric Food Econ 6(1):7–25
Phondani PC, Maikhuri RK, Rawat LS, Negi VS (2020) Assessing farmers' perception on criteria and indicators for sustainable management of indigenous agroforestry systems in Uttarakhand, India. Environ Sustain Indic 5. https://doi.org/10.1016/j.indic.2019.100018
Phuong LTH, Biesbroek GR, Wals AE (2018) Barriers and enablers to climate change adaptation in hierarchical governance systems: the case of Vietnam. J Environ Planning Policy Manage 20(4):518–532
Rauff KO, Bello R (2015) A review of crop growth simulation models as tools for agricultural meteorology. Agric Sci 6(09):1098
Riahi K, Rao S, Krey V, Cho C, Chirkov V, Fischer G, Kindermann G, Nakicenovic N, Rafaj P (2011) RCP 8.5—A scenario of comparatively high greenhouse gas emissions. Clim Change 109(1–2):33
Ridha RN, Wahyu BP (2017) Entrepreneurship intention in agricultural sector of young generation in Indonesia. Asia Pac J Innov Entrep 11(1):76–98
Rowland P (2019) Ecosystem services and faith communities in Oceania. Ecosyst Serv 39:100994
Sachitra V, Siong-Choy C (2019) The moderating effect of religiosity on resource-capability-competitive advantage interaction. Int J Soc Econ 46(5):722–740
Salite D (2019) Explaining the uncertainty: understanding small-scale farmers' cultural beliefs and reasoning of drought causes in Gaza Province, Southern Mozambique. Agric Hum Values 36(3):427–441
Schuman S, Dokken JV, van Niekerk D, Loubser RA (2018) Religious beliefs and climate change adaptation: a study of three rural South African communities. Jàmbá: J Disaster Risk Stud 10(1):1–12
Skrimshire S (2014) Climate change and apocalyptic faith. Wiley Interdiscip Rev: Clim Chang 5(2):233–246
Smiley KT (2019) A polluting creed: Religion and environmental inequality in the United States. Sociol Perspect 62(6):980–1000
Solomon Islands Government (2015) National agriculture and livestock policy: 2015–2019. Honiara
Solomon Islands National Statistics Office (SINSO), Solomon Islands Ministry of Health and Medical Services (SIMoHMS) and Pacific Community (PC) (2017) Solomon Islands demographic and health survey 2015. Pacific Community, Noumea, New Caledonia
SPC and USAID (2016) Climate change vulnerability assessments for communities in six Pacific Island countries: Fiji, Kiribati, Samoa, Solomon Islands. Tonga and Vanuatu, Suva, Fiji
United Nations Development Programme (2016) Enhancing resilience of communities in Solomon Islands to the adverse effects of climate change in agriculture and food security: experience and knowledge from Strongem Waka lo Communiti for Kaikai (SWoCK) project. Honiara, Solomon Islands
Wairiu M, Lal M, Iese V (2011) Climate change implications for crop production in Pacific Islands region. Food production–approaches, challenges and tasks 67:67–86. InTech, London. https://doi.org/10.5772/18
Walsh KJ, McBride JL, Klotzbach PJ, Balachandran S, Camargo SJ, Holland G, Knutson TR, Kossin J, Lee T, Sobel A, Sugi M (2016) Tropical cyclones and climate change. Wiley Interdiscip Rev: Clim Change 7(1):65–89

Wang W, Zhao X, Cao J, Li H, Zhang Q (2020) Barriers and requirements to climate change adaptation of mountainous rural communities in developing countries: the case of the eastern Qinghai-Tibetan Plateau of China. Land Use Policy 95:104354

Zhou D, Shah T, Ali S, Ahmad W, Din IU, Ilyas A (2019) Factors affecting household food security in rural northern hinterland of Pakistan. J Saudi Soc Agric Sci 18(2):201–210

Part II
The Roots of Spirituality and Religion in the Pacific Islands Region

Chapter 4
Leveraging *Vanua*: Metaphysics, Nature, and Climate Change Adaptation in Fiji

Francisco Gelves-Gómez and Shannon Brincat

Abstract Conceptions of Nature are infused with metaphysical ideas across all cultural systems. This is no less so in Fiji, where ideas of nature are infused with meanings across a rich tapestry of traditional *iTaukei* beliefs/practices and faith systems. Yet the question of whether these metaphysical ideas offer a means by which communities can harmonise the human and nature relationship, especially in terms of climate change adaptation, remains unexplored. This chapter examines this question by, firstly, outlining the underlying metaphysical systems within Fijian communities, explaining how these metaphysical systems view human/nature relationships. Secondly, we look to how these metaphysical systems are currently acknowledged within Fiji's climate adaptation policies, focusing on the Fiji National Climate Change Policy (2018–2030) and the Fiji's Climate Change Bill proposed in 2019. We argue that a more consistent appreciation of the metaphysical content of *vanua* may help develop more socially relevant adaptation strategies for *iTaukei* communities.

Introduction

Conceptions of Nature are infused with metaphysical ideas across all cultural systems. That is, the abstractions of metaphysical thought serve as the footing for everyday practices, providing the foundation for worldviews within which communities mostly operate. Indeed, Alfred North Whitehead (1925) argued that every science belies a metaphysics, or something we could call more broadly a cosmology.

F. Gelves-Gómez · S. Brincat (✉)
School of Law and Society, University of the Sunshine Coast (USC), Maroochydore, QLD, Australia
e-mail: sbrincat@usc.edu.au

F. Gelves-Gómez
e-mail: fgelvesg@usc.edu.au

F. Gelves-Gómez
Environmental Social Sciences Research Group, School of Ecosystem and Forest Sciences, University of Melbourne, Melbourne, VIC, Australia

J. M. Luetz and P. D. Nunn (eds.), *Beyond Belief*, Climate Change Management,
https://doi.org/10.1007/978-3-030-67602-5_4

As the underlying basis for these shared worldviews, metaphysics can therefore play a crucial role in promoting ecological thought (Tucker 1994) and, by extension, climate change adaptation (Pearson et al. 2020). In Fiji, *iTaukei* conceptions of Nature and local ecological knowledge are infused with meanings, values, perceptions and cosmologies derived from a rich tapestry of traditional belief systems, myths, and various forms of Christian theology (especially Methodist), and other faith-systems such as Hinduism and Islam. This is in conversation with modernist notions of science, philosophy, and policy—each of which carry with their own metaphysical assumptions—also introduced to the islands as part of the steady ingress of globalisation For instance, as evidenced by France (1966), part of this tapestry includes the colonial mindset that manipulated and altered key aspects of *iTaukei* metaphysics related to land and community for its own purposes of ownership and alienation. In everyday life, these ideas about Nature and the human-Nature relationships developed from them, become cumulative practices, beliefs, knowledges and relations with the environment (Berkes 2018) in which conceptions of Nature form a worldview together with an ethic that are to be understood holistically (McGregor 2004).

The question of whether the complex metaphysical system/s within Fiji offers a means by which its communities can harmonise human-Nature relations, especially in terms of future climate change and adaptation, remain largely unexplored. Ravuvu's (1987) comprehensive interrogation of the Fijian ethos thoroughly describes the traditional *iTaukei* worldview, including its metaphysical basis. Similarly, Tuwere (2002) reveals the metaphysical basis of Fijian society in a theology that links spirit, place, land, custom, and community in complex and overlapping layers of lived practice. Others like Kaplama (2016) or Miyazaki (2004) have compared aspects of Fijian metaphysics with or against dominant Western forms. Others still have looked to the metaphysics behind specific historical practices in Fiji (see e.g. Obeyesekere 2005). Most recently, Fache and Pauwels (2020) have highlighted the relational ontology and knowledge system in *iTaukei* thought, looking to combine such traditional knowledge systems and western ecological science. These explorations serve as the basis for our examination of the metaphysical content of the belief-system and cultural practices in traditional Fijian (*iTaukei*) communities[1] as a potential way to increase effectiveness of climate adaptation policy and communication.

Fiji, along with many other Pacific Islands, is highly vulnerable to climate change and requires environmental policies and adaptive strategies amenable to its context, inclusive of its philosophical and cultural systems. Environmental policy aspires to be neutral and objective, framing solutions to environmental problems and decision-making processes in secular terms and language. In reality, neither environmental policy nor the science underpinning and driving such policymaking are objective or

[1] We are aware of the religious plurality in Fiji. However, to limit the scope of our chapter we focus on *iTaukei* population (around 57% of the population) that are predominantly Christian (around 90%) and nearly 2/3 of which are Methodist. See https://www.statsfiji.gov.fj/index.php/statistics/social-statistics/religion.

neutral (Latour 2004; Leipod et al. 2019; Pascoe et al. 2019). National approaches to climate policy are captured within a few dominant framing discourses that largely exclude the worldviews of traditional societies and indigenous communities. In the last 25 years, for instance, environmental policies related to climate change have been typically articulated under the flag of 'development' and economic rationalist discourses (Pascoe et al. 2019). Such discourses on climate change policy have a tendency to either ignore or supress local realities (Brightman and Lewis 2017). Climate change policy in Fiji is no exception (see for example Republic of Fiji 2018). In the Pacific Islands, including Fiji, Luetz and Nunn (2020) and Nunn et al. (2016) have argued that climate change initiatives have failed due, in large part, to the lack of consideration of characteristics of contemporary Fijian society including cultural attributes related to spirituality and faith. Aligned to this denial of spirituality, climate change adaptation initiatives and strategies are largely technocratic, science-driven, and command-and-control approaches that efface local knowledges and worldviews (Luetz and Nunn 2020; Oakes 2019).

In our chapter, we are particularly concerned with exploring indigenous Fijian metaphysics that are deeply implicated within spirituality, faith and cosmologies of Nature that may have been downplayed in climate adaptation policy formation. Without sufficient acknowledgement of the complexity of *iTaukei* metaphysical ideas that permeate the archipelago's culture and thinking, any adaptive strategy may be ineffective—to use a *iTaukei* metaphysical expression, it would lack '*mana*', that is lacking proper ordering, relationship, relatedness, connectedness and thereby remain inconsistent with the understandings of the community (see Tomlinson 2009; Tomlinson and Bigitibau 2016; Tuwere 2002).[2] Without being embedded and derived from this metaphysical system and its interconnectedness with Nature (and spirit and community) as encapsulated in the notion of *vanua*, we contend that such policies will remain external impositions with little traction or relevance. As stated by Leveridge (2009, p. 25), following Nabobo-Baba's anthropological studies in Fiji:

> The philosophies and beliefs of a community define the nature of their knowledge and their way of knowing. They also shape environmental perception, [a]ffecting how people interact and learn to survive within their environment.

Taking this claim at face value, what we suggest is an alternative way of communicating adaptive climate change strategies commensurable with local metaphysical systems across Fiji. This could make climate policy and science understandable and actionable by the community in which it is to apply. Yet, given the size of scope of metaphysics and cosmologies, our chapter can only outline these issues, identifying areas where more research should be undertaken—including the need for a much wider study that includes the cosmologies and metaphysics of the other important faith-systems in Fiji, including Hinduism and Islam.

This chapter is divided into three parts. Firstly, we outline contours in the *iTaukei* metaphysical systems within Fiji and their conceptions of Nature as well

---

[2]There are many various definitions of mana that we have adapted from. See Na iVolavosa Vakaviti (Fijian Dictionary). Suva: Institute of Fijian Language and Culture. Also see Tomlinson (2009).

as human-Nature relationships. Secondly, we analyse how such metaphysical ideas are expressed within the most recent Fijian climate policy—both the *Fiji National Climate Change Policy* (2018–2030) (hereafter The Policy) and the proposed *Fiji's Climate Change Bill*—showing how these are largely de-linked from the *iTaukei* cosmological worldview. In our conclusion, we summarise key points in our discussion and speculate on how a more consistent appreciation of *iTaukei* metaphysics articulated through the idea of *vanua* might be used to co-create adaption strategies from within these belief-systems and therefore reach *mana*, 'achieving its intended purpose' (Tomlinson 2009).

Metaphysics and the Cosmology of Nature in **ITaukei** *Fiji*

Metaphysics is the branch of philosophical enquiry into the fundamental nature of reality—it grapples with ideas above/over (meta) the physical realm. All cosmologies (or worldviews) have an ontology within them, an assumption of 'what there is', containing metaphysical ideas as well. Whilst at a philosophical level, metaphysics is focused on first-order principles (basic proposition or assumption that cannot be deduced from any other proposition or assumption—see Aristotle 2018), at the social level of the every-day, it is the common assumptions of one's world—its origins, types of being, and purposes—that give shape to forms of acting, being, meaning, and thinking. What stems from this broad conception of metaphysics is that traditional societies have a deep metaphysical basis for their worldview that is no less complex merely because it may be expressed in myth and custom, cultural practices, and ritual, rather than weighty tomes. Yet Western science, philosophy, and policy are only beginning—and slowly—to engage with these metaphysical traditions, let alone understand them. For instance, environmental policy is willing to take from traditional ecological knowledge but largely unwilling to (re-)frame itself from within this discourse and practice (Berkes 2018).

Metaphysical traditions must be understood in their own terms, inclusive of culture and historical context. Yet, given the shared concern of metaphysics in determining the nature of existence, different systems can dialogue with each other despite the fact they also compete. Our chapter is not intended as deploying non-Western metaphysics to challenge Western metaphysics or vice versa, but to see how these relate to the formation of climate adaptation policy within the particular cultural context of the *iTaukei* of the present that, as we will see, is a hybrid system (as the vast majority of metaphysical ideas are today). One of the key areas in which metaphysics meets climate policy relates to the cosmological understandings of Nature itself: what there is in the universe, what are its origins, and its development (and related questions as to 'why' the universe is the way it is). Facets of these concerns are reflected in Fijian origin myths, like that of Degei.[3] Cosmology, then, is not just the scientific study of the origins of the universe but also includes belief systems which straddle Nature, the spirit and social worlds. The outer boundary of any cosmology is ultimately metaphysical because it hypothesises beyond the scope of scientific knowledge at any given time rather than knowing through demonstration. At this edge cosmology is speculative. Yet there is a stark contrast between the types of explanation available to a cosmology in physics (Unger and Smolin 2015), between the evidentiary burden of 'testable hypothesis' and the epistemological limits of our empirical means for doing so. Its search for 'physical laws' (i.e. universal truths) is akin to a first order principle of metaphysics and it is here that the need to explain meets assumptions that are not yet explainable—the problem of infinite regress of causes or non-dependent First Cause of the universe being the exemplar. Yet focusing on explanations of cosmology is only one aspect of metaphysics. Of equal importance are the values and interests that a community places around such understandings. Just as the physical conditions necessary for our existence impose a selection effect on what we observe as cosmology, so too does the philosophical and cultural system influence *how* this is interpreted and for *what* purposes. As stated by Nabobo-Baba (2006, p. 24) "knowing and knowledge are not accultural, but are products of, and thus influenced by, particular cultures, and should be understood through research techniques which reflect that culture."

Our argument is focused on this cultural aspect of cosmology, the way in which "metaphysics in practice" (Law and Lien 2018, p. 152) shapes climate adaptation policy and the ways it *could* be aligned and enacted more effectively. The historical context in which climate adaptation practice takes place is entangled with metaphysical understandings that reproduce a particular kind of message. Immanent to this message is the positioning of humans, Nature and spirit (including religious belief). To describe the genealogy of such entanglements is beyond the scope of this chapter. The way that certain assumptions are instrumentalised in climate policy, however, can be presented within examples of legislation (that we trace in section two). When analysing the relation between metaphysics and Nature—and the community values/interests of these in the context of *iTaukei* peoples today—the scope of research is obviously vast. Nevertheless, some relations between metaphysics and *iTaukei* conceptions of Nature do stand out as paramount:

1. The interconnectivity and deep intersubjective relationalism between ancestors, community, land, and spirit in the concept of *vanua*, and;
2. the relation between this traditional belief-system and the colonisation of this metaphysics under a hegemonic Christian theology, and, colonial land-owning system.

Studies have shown there is "considerable unanimity" on the understandings of the postulates of pre-Colonial Fijian cosmology (Parke 2014, pp. 38–39), some

[3] According to the myth The Great Serpent Degei was the the creator of the islands and people, and, controlled aspects of natural process, for instance, day and night with the opening and closing of his eyes or earthquakes (see Reed and Hames 1997).

of which continues to exist in socio-cultural and politico-legal forms, albeit now cross-fertilised with other more recently introduced metaphysical ideas. Pre-colonial *iTaukei* practices gave the highest respect to Nature that was linked across its cosmology: in material terms of flora/fauna and the physicality of land and sea; in social terms, such as ancestors, customs, morality, and the community itself; and to spirits, theological principles, and creation stories. The primary examples are the Fijian myths, especially of Dakuwaqa and Degei, that reveal the prominence of Nature across this cosmology, including origin stories, ancestral migrations, and ongoing protections of the community (Reed and Hames 1997). These are not just mythical stories of the past but remain lived metaphysical traditions within community norms, customs, and morality—despite colonisation, Christianisation, and modernisation. For the *iTaukei*, as across most of Melanesia generally, there is no separation between Nature and society, nor are either of these realms separate from spirit (Foale 2006). The social realm does not have control or authority over the natural world either in terms of its power or in normative or spiritual senses. Instead, the *iTaukei* are nested in relations across and between humans *and* spirits *and* Nature. This relational ontology as Poirier (2008) describes it, is fundamental for any effective climate change policy and climate change action.

This relational ontology is exemplified in the lived concept of *vanua* that is clearly visible in all daily practices of *iTaukei* from the structure of the community, to agricultural cultivation, to storytelling. Whilst sometimes reduced in translation to 'land', *vanua* is at its core a metaphysical postulate. For Ravuvu (1987), it includes common values and beliefs about life in the natural and spiritual worlds. For Kaplama (2016, p. 161), it is at once cultural and philosophical, physical and metaphysical, an "idea-object" that "constitutes not only the physical land on which Fijians live, eat, dwell and regenerate but also their particular worldview" (*Weltanschauung*). He summarises *vanua* as "the representation of the unity of earthly-physical and spiritual-metaphysical life as well as the unity of all members of the *vanua*" (2016, pp. 160–162). This encompasses socio-ecological environments (e.g. lands, oceans, rivers) and the realms of spirits and life forces (e.g. spirits of the ancestors, deities) "who are believed to have the power to affect, direct and sustain life in earthly realm" (Kaplama 2016 pp. 160–162).

One of the keys to unveiling—and leveraging—the complexity of relations between metaphysics, *vanua* and climate change policy and action, lies in acknowledging the interconnectedness of this cosmology. As Parke (2014, p. 35) explains, there is no disjunction "between the realm of people and the realm of spirits. Spirits and people shared a common world, of which the geographical component included places with which both people and spirits were closely and directly associated." This is affirmed also by Nabobo-Baba (2006, p. 88) who observes that "the spiritual and the material worlds are interconnected; respect for people, resources, the ancestors, and God, govern all important behaviours and values." There is neither a duality between realms, nor is either realm autonomous. At the metaphysical level these form an interconnected whole. This cosmology includes relations of dependency and mutual recognition with Nature. That is, within the relational cosmology of *vanua* is intersubjectivity and not just between human-human within bonds of community but including all ancestors and non-human spirits as well. Nature is very much subjective: places are subjects with agency, as are living creatures. The roles that spirits play are environmental just as much as they are social. For example, Parke (2014, p. 32) describes so-called *digiwai* spirits related to natural processes:

> ...one [is] responsible for successful food crops, another for success in hunting, and a third for fishing. These spirits could in turn expect to be respected and looked after by the living, by suitable presentations of food and *yaqona* [kava]. Failure on the part of the living to fulfil their reciprocal obligations could result in drought or in a person being bitten by a wild pig or by a shark.

Similarly, Ravuvu (1987, pp. 14–15. Emphasis added) states that:

> ... the cultural *vanua* permits insights into people's actions and how they bear on their efforts to live and resolve the exigencies of existence both in this earthly world and in the world of the spirit beings who can protect and destroy, or direct and make possible what is required for life. For to live well in this world *and* in the other world, one has to live according to *vanua* beliefs and values. Thus the concept of *vanua* is an encompassing one; *it is the totality of a Fijian community*. Used in various contexts, it can refer alternatively to the social and physical environments, or to the supernatural world, or to all the elements which make life occur.

These passages show how the levels of Nature, spirit, and human are spatially and temporally linked as the totality of community.

This co-existence requires the human community to pay customary respects to the other realms, including respect for Nature and to the ancestral spirits (*kalou-vu*) who fulfil a similar role—that is, in this relational cosmology, Nature and spirit are inseparable from each other and from community. The teaching and learning of important traditional knowledges, for example, are deemed to be "guided by familiar *vanua* spirits or God" (Nabobo-Baba 2006, pp. 117, 127). These ancestral spirits are concerned for the wellbeing of the *yavusa* group descendants and, importantly, these relations between the living and the ancestral spirits are "reciprocal"—with the community to frequently present *isevu* (first-fruits) and various other customs to both placate them but also renew connections with these worlds (Parke 2014, p. 29). Both ancestral and natural spirits watch for breaches of customary codes of conduct or respect that mirror social norms of the community regarding reciprocity. Consequently, these "cords" between people and *vanua* are constantly replayed in Fijian ceremonies—not only solidifying existing social bonds of community but also those of ancestral spirits and of the natural world around the locality. A s stated by Kaplama (2016, p. 158), "the people that constitute it [the community] must continually strengthen the *vanua* in order to preserve their values not as individuals but as the members of the *vanua* which gives them their identity." He explains that the "Fijian ethos" is defined by: "reciprocation, recognition, respect, appreciation, strengthening of the bond, incorporation, unity, continuity" and the lack of any of these is seen negatively as leading to weakness and eventual discontinuity of the *vanua* bond (see Kaplama 2016, pp. 160–161).

Tomlinson and Bigitibau (2016) analyse a range of performative practices that link spirits, kava, and various speech practices with traditional, pre-Christian society.

There is even ongoing communication between the community and ancestral spirits through the *Bete* (the traditional priestly group). The *yaqona* ceremony for new visitors is not just to seek formal inclusion in the community from the head of the *yavusa* and under the relevant *mataqali* (the village-based, agnate grouping) but also for the supplication of the spirits—both ancestral spirits and those of the local natural environment—to admit the visitors as well. The offering, acceptance and mutual consumption of kava is the gateway or portal between realms, though these realms are at all times co-present. Such actions of reciprocity fulfil a range of social functions, from promoting group cohesion, performing ancestral ties, maintaining learned traditions/stories, upholding norms of conduct, and designating land entitlements. Through these deliberate performances—whether stories, rituals, or moral practices/customs—comes the constant recognition of ancestors, Nature and spirit. It means there is "continuity" in *vanua* as a lived tradition that, despite its metaphysical basis, remains the core socio-political basis of *iTaukei* community within their daily lived practices as well (Farrelly 2010; Ravuvu 1987). There is no separation, then, between the realm of people and the realm of spirits, whether natural or ancestral. Spirits and people share a common world of which the geographical component—Nature itself—includes those places within which both are "closely and directly associated" (Parke 2014, p. 35).

The emphasis of interconnectivity and reciprocity within this cosmology would suggest a philosophical and cultural system more conducive to ecological views of stewardship of Nature, given that spirits and ancestors are attached to the physicality of places—a powerful impetus for the protection of natural attributes associated with them. Many studies have found that indigenous traditions often foster pro-environmental practices and ways of being. For example, after a survey of the literature on religious belief and environmental attitude, Taylor et al. (2016, pp. 1000–1001) concluded that "indigenous traditions may be more likely to be pro-environmental than other religious systems and that some nature-based cosmologies and value systems function similarly." The *iTaukei* cultural practices surrounding *vanua* seem consistent with this finding. In particular, the historical and mythologised relations of the *vanua* are revitalized to enable "current problems to be faced jointly" (Ravuvu 1987, pp. 247–248) by the *vanua* community itself. There is intentionality and agency within this force. This raises the fundamental question that, if the *vanua* is threatened by climate change, how the community could adapt in accordance not only to preserve *vanua* but to ensure *vanua* traditions are upheld and respected within adaptation approaches. Linking adaptation as part of reciprocal duties is crucial then to effective policy formation, which we will explore shortly.

But before that, one crucial question is how does the *iTaukei vanua* tradition co-exist today alongside other introduced metaphysical systems, particularly of colonialism and the now dominant Christian theology? Is *vanua* subservient, subsumed, or co-productive with these systems? Turning first to Christianity, like other Pacific Islands, historical processes led to the Christianisation of Fiji (Luetz and Nunn 2020). What emerged was a hybrid of traditional beliefs that interconnected Christian faith with historical, cultural, and political processes and metaphysical understandings of belonging to the land in space and time (see Ryle 2005, p. 77). For Ryle, *vanua* blended with Christianisation, as she writes:

> Christianisation entailed the redefinition of *vanua*, adding the Christian God as supreme power among the Fijian cosmology of deities, yet retaining most other elements and understandings of *vanua* as an ancestral shadow-land of place, kinship relations and spiritual power that exists alongside Christian belief and practice (*lotu*).

This influence is clearly reflected in the three main ideological pillars in Fijian society today: '*Vanua, lotu* and *matanitu*', broadly translated as "land and tradition, Church and governance" (Ryle 2005, p. 58; 2010). This metaphysical hybridisation bears the complexities of melding ancient local beliefs with Christian religious precepts (and of course political interests at the time of colonisation and independence). Christianity proved effective at its missionary work because of its capacity to absorb traditional practices and acknowledge these within its structure, both theological and liturgical. Here, the 'message of God' resonated strongly with Fijian people in terms of their existing metaphysics of social hierarchy and the derived notion of the *mataqali* for land ownership purposes (see France 1966). The latter was an expression of the military might, underpinning missionary activities that was used politically by various interests across the archipelago. The predominantly Fijian/Methodist religio-cultural structure that came into being during the colonial period (1874–1970) (Ryle 2005) bears testimony to the construction of national or group identity, along the lines of Benedict Anderson's (2006) 'imagined communities' and, as Ryle suggests, "in people's minds belongs to a not very distant past and is an essential part of their cultural and historical sense of identity" (2005, p. 61).

The arrival of Christianity did not just involve a spiritual change but entailed the contestation of existing metaphysical ideas and rewriting of worldviews. In terms of spirit, the most significant was the subsumption of animism and polytheism, the many localised gods and myth-stories, under an overarching monotheism. While these stories remain, they are being less and less told, and are not seen as efficacious (*mana*) though still constitute a strong belief-system (Tomlinson 2009). The older generations impart these stories rarely to the younger generations, deem them inconsistent with Christian beliefs, and yet they still co-exist—especially the spirits of Nature in particular places that remain *tabu*. As Nabobo-Baba so aptly describes it, Christian perspectives are often associated with Indigenous elements of spirituality though the latter are less "publicly or loudly declared" (2006, p. 87). Despite their subsumption these stories hold potential for a better understanding the metaphysical ideas regarding specific phenomena including the origins and authority of social customs and explaining aspects of Nature itself (Nunn 2001). Indeed, these forms of spirit seem to adapt to the changing make-up of the community at any one time, perhaps losing favour and/or mana relative to other temporal actors yet nevertheless always secure within this cosmology and, in this way, remain the core traditions of belief that the community may retreat to during crises.

Given the fundamental relation between all three levels (human, spirit and Nature) in the Fijian metaphysics, the contestation of the understanding of the spirit world by Christian theology meant a shift in the conceptions of Nature too. For example, when he conceded in 1854, Cakobau was compelled to '*lotu*' (enforced religious conversion) alongside the destruction of the traditional temples in Bau and cutting down the sacred *nokonoko* trees. This was a visceral example of how social take-over and colonial order went hand-in-hand with spiritual *and* natural destruction. Another story from Kadavu in 1863 recounts of when a chief ordered a sacred *vesi* to be cut down declaring: "If the God is in the tree, the axe will be spoilt, but if the axe cuts it, that will be proof the tree is only wood. The trees were felled, and thus the 'religious test went in favour of Christianity" (Thornley 2002, p. 23; Tomlinson 2009). This was a shift between human powers tested through the divine. Such political actions challenged the very foundation of the traditional metaphysical assumptions, significantly undermining the 'old gods' and Nature with it. Jehovah "towered" over the local gods: such symbolism was particularly resonant in Fijian metaphysics and characteristics of hierarchy (Toren cited in Tomlinson 2009, p. 86). But the old gods were sidelined, not abolished or disbelieved. This was part of a sustained campaign by missionaries to shame aspects of traditional Fijian culture by over-emphasising "warfare, cannibalism, torture, infanticide, widow strangling, and treacherous murder" (Brantinger in Kaplama 2016, p. 157). Aspects of this shaming continue today in some sermons. However, this was a two-edged sword for whilst the Missionaries recast local spirits as 'demons' and denounced them, it nevertheless affirmed their very presence and power in their respective "bailiwick" (Tomlinson 2009, p. 67). The missionary trod a contradictory path between justifying gods on one hand and denying them on the other, leaving ample space for the old metaphysics and its spirits to exist alongside Christianity, though subordinate to it. This hybrid spiritual world is captured poetically by Tippett (Cited in Tomlinson 2009, p. 67) when he describes in Kadavu:

> One can walk round the coastline for a fortnight and never run out of stories and legendary epics and origin myths.... Then suddenly out of this tradition one is immediately confronted by some pastoral issue that shines through in a currently felt spiritual need... One is immersed in a world view not his own; and then strangely one is led back into some strange place in the Old Testament... It is a conflict between black and white magic... The trees one knows, the cleared space, the little formation of stones. One knows that someone has returned to his 'Canaanite gods' and that the pastor or the catechist in this next village has to deal with it... You stop saying 'This is just superstition'.

One can feel in this passage the ongoing tension between two metaphysics, theologies, and competing worldviews. Their very contestation confirms their co-presence and hybridisation. With the almost total Christianisation of the *iTaukei* population the question is how these views toward Nature related to the *iTaukei* view that preceded it? How do they blend in this hybrid today? And how do the answers to these questions affect the 'response-ability'(Haraway 2008)[4] of Fiji's people in view of climate change ecological and social impacts?

[4]Haraway (2008) describes 'response-ability' from an ethical and political standpoint that is deeply implicated and entangled with the ways we engage, know and be in the word. Haraway suggests the reshaping of relations in terms of actors' (human and nonhuman) capacity to respond and attune to each other. In this sense, being 'response-able' to climate change also takes on a metaphysical dimension.

The dominant conceptions of nature in the Judeo-Christian tradition reveals a single real nature (Williams 1980; Cronnon 1995). Amongst other attributes, this conception of nature asserts the human dominion over the rest of the material world as a moral right (Strang 2017). Similar values are found to underly some of the common approaches to climate change mitigation and even adaptation strategies such as 'payment for ecosystem services' or carbon trading. This ontological base of man's assumed dominion over nature (Brincat and Gerber 2015; Castree 2005, 2012; Escobar 1999; Gerber 2019) is the root of many of the problems contributing to climate change (Moore 2016, and importantly for an argument on religion see Berry 1988). There are a number of studies that have compared religions to ascertain the link between environmental concerns and belief-systems. In terms of Christian beliefs, White's (1967) famous claim that the Judeo-Christian tradition has promoted anthropocentric attitudes and environmentally destructive behaviours has been confirmed by a number of contemporary studies (Taylor et al. 2016). There are range of reasons for this: Christian anthropocentrism and the desacralization of nature observed in Genesis; scepticism of science (including climate science) as denying Biblical truth; humanity's dominion over Nature bequeathed by God. Of course, there are counter-hegemonic discourses within Christianity also, such as man's stewardship of Nature, "creation care" and Liberation Theology to name but few. These, however, are subordinate discourses at best. Hand and Van Liere (1984) showed how religiously conservative Christian traditions (e.g., Baptists and Mormons) were more likely to endorse the idea of mastery over nature than liberal denominations (e.g., Episcopal and Methodist) that have "a value orientation compatible with the demands of a limited world" (Hand and Van Liere 1984, p. 568). Greeley (1993) has shown Catholics were just as likely as non-Christians to support environmental spending. However, other studies show religious views are a weak predictor of environmental attitudes (Boyd 1999), and others show it is the more secular a person the "more pro-environment" they tend to be (Guth et al. 1995, p. 377; Roser-Renouf et al. 2016). As Taylor (2015) surmises, the hope that religions may come to 'the environmental rescue' requires careful scrutiny.

If these findings are indicative for Fiji, it throws additional complexities when it comes to metaphysics in practice, specifically human-Nature relationships. The predominance of Christianity may imply a dissonance between it and the capacity of the traditional *iTaukei* notion of spirit for dealing with environmental problems. That is, given that the vast majority of ethnic Fijians identity as Christian may impede the pro-environmental *iTaukei* traditions. As an exemplar, Currenti et al. (2019) have documented how after Cyclone Winston residents of an interior village worked towards making buildings more resilient with government and NGO support. Importantly, they showed how the "belief in the Bible and Divine providence" was "a factor influencing attitudes towards dealing with changes in climate and socioeconomic conditions" (Currenti et al. 2019, p. 74). They identified the problem of those that relied "exclusively on divine intervention to save them from the impacts of climate change" and the potential to "undermine cooperative inclinations" between church denominations (Currenti et al. 2019, p. 74) Yet, at the same time, with Methodists predominating across the archipelago—a denomination that studies elsewhere have shown has a more pro-environmental stance in comparison to other Christian derivations such as Baptist or Mormons (Hand and Van Liere 1984)—may provide a counterweight to this. At the most basic level, the importance of Christian metaphysics to existing *iTaukei* society would suggest the need to have churches included in the formation of climate policy and its communication/dissemination (Luetz and Nunn 2020; Nunn 2013). For example, having the Assembly of Christian Churches in Fiji (ACCF) (which covers about around 50% of the population) support could be a significant driver for adaptation messaging. Regardless of such opportunities, the issue highlights that a survey of religious attitudes to the environment and climate change in Fiji needs to be undertaken to understand this linkage more precisely. Nunn and colleagues' (2016) study offers a primer for this, showing that the majority of church attendees across the Pacific have spiritual values that explain their feelings of connectedness to Nature. How these views differ between each denomination and *iTaukei* community, how these relate to other faith-based systems across Fiji, and how they differ across environmental issues, all need to be determined.

Vanua must also be read in the context of land ownership (which is its literal translation), with all the historical antecedents, both traditional and colonial, and the metaphysical contestations of property, territory, community (etc.) as well. This is even more relevant considering that almost 80% of land in Fiji is currently owned under communal titles (Iati 2009) and managed under customary principles of governance (Scheyvens and Russell 2012) that, in effect, directly impact peoples' capacities to respond to climate changes (see for example Gharbaoui and Blocher 2016). As we have seen, ancestral spirits are concerned primarily with the prosperity and continuity of the *yavusa* which includes agricultural land, forest land, and fishing areas owned communally by those people descended from a recognised common ancestor (Parke 2014). But this was not necessarily the traditional view. This system, like the hybridisation of the metaphysics related to spirit, was a result of contestation between Fijian traditions, the chiefly elite, and the new colonial power emerging in the 19th Century. It is important to note that along with the spiritual and religious shifts in Fiji at this time came also the entanglements of economic and political ambitions not just of the European missionaries but the commercial interests of the planters, and, the concerns of the Crown's administration (Sutton 2005). With them came the notions of property, land tenure, and the market that have their own metaphysical basis that were essentially Lockean and stadial at the time (and remain so). The Lockean view of property held that land had to worked to be alienable was clearly a key assumption behind The Native Lands Commission and lay behind the interests of the planters too. This conception of land was far removed from notions of ancestry and spirit that had no bearing on ownership in terms of this metaphysic of land and its alienability. Indeed, traditions had to be created and reintegrated back into this system by Governor Gordon and the chiefs to ensure some land was left for the Fijians (France 1966). Gordon's stadial view of history held that Fijians remained at a pre-feudal level of development and had reached the stage when the inalienability of land was deemed an essential feature of its progress (if not for their very survival). Gordon therefore made "traditional inalienability" central to his theory of Fijian land tenure (France 1966, p. 236). This was based, not only on the insistence of the chiefs who would profit from it, but also a particular reading of Fijian history by a range of anthropologists at the time. In particular, Fison held that the voice of the people (commoners) represented ancient law "made by the ancestors of the people, who were their gods," under which all lands were traditionally owned by *mataqali* and were, according to this "immemorial" custom, "inalienable" (France 1966, p. 236). This 'version' of history continued despite the fact that the Commission soon found alienation of land had been a common practice among Fijians before colonisation. In the late 1880s the Council of Chiefs revealed "complete confusion among the chiefs as to the names of social divisions and units of ownership." Nevertheless, they claimed the *mataqali* (the clan-level) was a "firm, and ancient, and understood by everyone" and to prevent disputes within the *mataqali* these lands were to be "subdivided into family land, to be held according to hereditary succession, and that registration of these lands should convey the legal ownership" (France 1966, p. 209).

Metaphysical systems related to land and property were manipulated on all sides by elite interests: chiefs, planters, and colonial administrators, leading to the hybridised tradition that underpins the land system that prevails today. In what France calls (1966, p. 300) "the classical exposition of Fijian social structure," it was claimed that the *yavusa* consisted of the direct agnate descendants of a single ancestor god, whose sons became the founders of component *mataqali*. The Fijian 'paramountcy of interests' doctrine that prohibited further sales of land (although it could be leased) was developed on this basis. Gordan adapted the *yasana* as an administrative unit to correspond with traditional *matanitū*, the major socio-political confederations (see Routledge 1985). So, the land remained with the clan, under hierarchical chieftainship, though increasingly turned toward commercial usage, shifting from subsistence agriculture and household production to those more amenable to foreign commercial and imperial interests. Demarcating ownership through the *mataqali* meant larger tracts of land could be leased, something of keen importance to both the planters and commercial interests of the Crown. Regardless of whether this was in accordance with traditional Fijian life or not, this view became the bedrock of hybridisation of land, tradition, and ongoing native title. By 1914, the Governor could threaten that if *mataqali* or *tokatoka* (family unit that are part of a *mataqali*) did not appear before the Native Lands Commission it would be considered landless, and land without such an owner would become government owned (France 1966). In the end, rather than attempting to discover the 'traditional' system of land ownership, the commission simply constructed one by cherry-picking that most suited for its purposes, and then sought going about recording the *mataqali* boundaries. This system has since become "hallowed and sacrosanct," "a protective device" of the "Fijian ethos" though it "depends less on its historical accuracy than on its social significance" (France 1966, p. 312). For Walter (1978, p. 351), this is a dual system, imposed by a "simplified., administratively modelled organisation… at considerable variance locally with the pre-existing version". The fact that the constitutional crises of 2009 reaffirmed this system, only seeking to clarify land boundaries is telling (see Fijian Government 2010).

To this date, the customary systems of land tenure are perceived as a hindrance to 'development' (economic) championed by neoliberal and Eurocentric logics that see in individual land holdings the best way to develop a territory (Fingleton 2005; Scheyven and Russell 2012). They have also played a peculiar function in protecting the metaphysics of spirit in *iTaukei* lands also. That is, despite the rupture involved in the change from traditional to colonised peoples, and the contrived retrofit of land law to *iTaukei* practices, the land policy and the *yasana* administrative unit ensured the 'old gods'—located in the ancestral lands and natural places of each group—were never lost. Arguably, more so than Christianity's uneasy 'confirmation by subordination' of local gods and ancestral spirits, it was these colonial land reforms that kept aspects of the traditional metaphysics alive though in hybridised form (and arguably in ways unrecognisable to *iTaukei* culture now largely lost to history). The spirit world was always there and remained so, subordinate but *not* subsumed to Christianity or Crown lands. In the next section, we will see how this hybrid metaphysical system has influenced the more recent policies concerning climate change adaptation in Fiji.

Disconnects Between Vanua *and Climate Change Policies in Fiji*

In Fiji, top-down climate change action is driven by political measures and strategies that are often aligned with the global discourses of development, sustainability, and resilience – key components of what has elsewhere been described as the 'economic rationalist' discourse on climate change (Pascoe et al. 2019). Such elements come in a package and cannot be disassociated from supposedly global conceptions of Nature, what society is, or how religion and spirituality are to be located (Latour 2017). These functional assumptions of economic rationalism have their roots in the human/Nature binary inherent to Western and Christian metaphysics, already discussed, and the hierarchical relationship between humans and Nature that they assume. As such, they seem widely at odds with the deep relationalism of *vanua* especially in terms of ongoing processes of recognition/respect owed to ancestors and spirit in *iTaukei* communities.

One of the most obvious examples of this tension between *vanua* and climate policy is found in the Republic of Fiji National Climate Change Policy (2018–2030).[5] This legislation follows directly from Western (and Judeo-Christian traditions) ideals expressed in the Sustainable Development Goals (United Nations 2015a) and the Paris Agreement (United Nations 2015b). It focuses on discourses of development, sustainability, and resilience that largely ignore alternative worldviews/metaphysics that frame human-Nature relation as reciprocal, thereby restricting the capacity of *iTaukei* to respond to climate change. At the same time, it locks in adaptation policy to external systems of validation. For example, Fiji's current National Climate Change Policy is supported via investments and financial resources that together complement

[5]The Republic of Fiji National Climate Change Policy 2018–2030 replaces the previous policy launched in 2012.

Fiji's aspirations to develop and build resilience. These rely on international aid and a recently established (2017) National Green Bond to finance its endeavours to face climate change and which have been commended for their efforts to achieve a sustainable financial base (Anantharajah 2019). The links to the overriding economic rationalist discourse are here obvious.

A more complex example of the disconnection to *vanua* within The Policy is the dominance of the themes of 'resilient development' as the pathway to follow in planning and responding to climate change. The Policy furthers Fiji's commitments in relation to achieving inclusive socio-economic development, on the premise that such development cannot happen without addressing climate change concerns. Outlining a "woven approach" to climate change and resilient development, the policy recognises the importance of the connections between the many factors that contribute to climate change, adaptation and resilience. The Policy therefore acknowledges Fiji's need to strengthen its actions to achieve climate change goals, but also the importance of promoting ways geared towards transformation of multiple sectors (e.g. transitioning to 100% renewable electricity by 2030). Yet, The Policy defines resilience as "the ability to cope with shocks and disturbance with minimal disruption and successfully return to a stable state" (Republic of Fiji 2018, p. 17). In other words, resilience is the ability of Fiji, its people and ecosystems to 'bounce-back' *after* climate shocks and disturbances. This definition arguably adheres to the most common vocabulary used in policy instruments on resilience and climate (Davidson et al. 2016). It is grounded on the premises of development and economic growth emanating from dominant ideas in the technocratic worldview perspective that antagonises humans and Nature. This definition, however, has been widely criticised for its instrumentalism regarding how to act and respond in order to create resilience (Cretney 2014). By articulating policy in terms of the capacity of socio-ecological systems to 'bounce-back', policy and actions are focused on the social factors that might contribute to returning functionality to particular (and probably damaged) institutions and structures (Cote and Nightingale 2012). This does not take into account how responding to climate change is enmeshed within broader categories such as the environment, economics, spirituality and religion at the local level. Nor does it adequately account for how some climatic changes, such as rising sea levels, may impact on ancestral lands and spirits worlds in ways which prevent these communities to return to a 'stable state'. Moreover, such instrumentalist approach to policy may be trading political expediency for effectiveness by challenging long-term values related to *vanua* and by restructuring the environment and society to accommodate new, external definitions of what is desirable or appropriate adaptation (Cote and Nightingale 2012; Cretney 2014). To the degree these adaptation mechanisms do not mesh with local interests, belief-systems, or community needs they will remain ineffective.

Fiji's Climate Change Bill, proposed in 2019 (hereafter The Bill), is the Fiji Government's most recent legislative proposal articulating its intentions related to climate change. Though in draft form, it shows a greater—though still limited—acknowledgement of themes more consistent with *vanua*. The Bill provides a regulatory framework to support the planning, development and implementation of long-term climate polices and actions to protect Fiji's people and their territories in view of the climate change emergency. The Bill is embedded in the context of international obligations set by the Paris Agreement. Divided into seventeen (17) parts, The Bill is wide ranging in its implications but also in its aspirations. From establishing legislative guides for research, defining commitments and obligations for climate change mitigation, to formulating mechanisms for engagement and financing that will support climate change governance mechanisms and arrangements, The Bill impacts all sectors of Fiji. The Bill is comprehensive, and its overall intent is to establish ways in which Fiji can mitigate its own climate change impacts, while promoting responses, adaptation, and resilience to climate change. The Bill lays the foundations for a future act that will shape institutional and governance arrangements which will permit Fiji's institutional response to climate change through national legislation.

Perhaps most importantly, the 'Principles' make direct reference for the recognition of the *iTaukei* (and the Rotumans) including "their respective ownership" of *iTaukei* lands and Rotuman "lands, their unique culture, customs, traditions and language" (Article 5(i)). It refers to traditional knowledge, and *iTaukei* needs in terms of mitigation and adaptation. Moreover, Article 81 recognises traditional knowledge of the *iTaukei* and Rotumans and the multi-cultural and multi-faith composition of the Fijian population. This recognition is more hortatory than real, however, as it makes no mention of the traditional land ownership forms (outlined above) that would seem fundamental for implementing this type of policy at the local level. The *iTaukei* Land Trust Board and Act (1940) and Register of leases is referred to and must give consent in terms of strategies related to REDD+ programmes and sequestration rights but this is not outlined in greater detail. The institutional and governance structures that the act aims to create (identified in Article 4(d)) should be consistent with *vanua* and land systems. Part 4, on 'Governance', is a real missed opportunity because rather than affirming local traditions, it adopts a hierarchical approach of Ministerial powers, with a Head of the Climate Change and International Cooperation Division, and Inspectors rather than an emphasis on local, bottom-up mechanisms of consultation. The National Climate Change Coordination Committee makes no mention of interactions with local communities other than a vague power to make "consultative groups as required" (13(j)), which seem again the prerogative of the head rather than emanating from the community. This leads to many questions around the objective of climate resilience and enhancing adaptive capacity in respect to Fiji's environment and ecosystems (Article 4(m)) if they are not tethered to the conditions or needs of local communities. For example, The Bill aims to provide "for the relocation and rights of at-risk communities" but these rights are not determined clearly and the surrounding text of the act seems more concerned with relocation than wider notions of human rights (especially of the collective community). The language of its principles (Article 5) are all made in the terminology of international instruments on climate change and not in terms of *vanua* that could be a complementary, shared narrative. For example, principles of intergenerational equity (Article 5(c)) could be made in reference to ancestral spirits, local Natural and sites of cultural significance, and community norms regarding past and future generations, that would then resonate across both discourses.

As legal instruments, both the Climate Change Bill 2019 and the National Climate Change Policy embody an alien conception of Nature to that of *iTaukei* metaphysics. The referent of security is the nation-state (Article 4(p)) that eschews any wider notion of bringing in climate change as part of an ecological security framework, in which *vanua* could be more centrally placed. Ontologically speaking, this resonates with the ideas of *a* Nature that can be managed and controlled, serving a purpose to people. For instance, in the context of The Bill, adaptation is defined as the "adjustment in natural or human systems in response to actual or expected climatic stimuli or their effects which moderates harm or exploits *beneficial opportunities*" (p. 6, emphasis added). In this technocratic approach, although implicit, the subjects of such benefits are solely human. Oceans are the only thing that are offered as a separate area worthy as being an objective of "consideration" (Article 4(o) but are then subsequently reduced to an anthropocentric lens in the principle of (Article 5(j) where the ocean is defined merely as "critical to identity and livelihood of Fijians." Whilst an important concession seemingly recognising the relationship between people and the environment in ways more than just economic, it does not reflect Nature in its own terms. Similarly, The Policy through its 'woven approach', explicitly accepts the tight "relationship between social and ecological systems and that human prosperity is dependent on environmental integrity and health" (Republic of Fiji 2018, p. 40). By centring on the economic, it lacks acknowledgement of the intertwined "spiritual and emotional understandings of belonging to the land in space and time" (Ryle 2005, p. 73) of *iTaukei* communities. Whilst The Policy further states the need to "ensure that all development actions adequately address the impacts, linkages and dependencies between human wellbeing, ecosystem services, biodiversity, oceanic systems, and the global climate" (p. 41), it gives the false impression of a socio-ecological arrangement in which spiritual and metaphysical constructs are absent. In this technocratic approach, the subjects of any benefits are solely human, making invisible the relations between *vanua,* territory, Nature and spirituality.

Conclusions

In this chapter we have argued for a better understanding and recognition of metaphysics, as they relate to culture, everyday life, and importantly in the context of this book, to religious belief, faith and spirituality. We have shown how existing and proposed policy frameworks related to climate change adaptation largely operate outside many of these factors. Exploring the metaphysical grounds and discourses associated with climate change policy in Fiji, we emphasise the way in which policy instruments are reliant on a discourse lacking content that speaks to the belief systems that are fundamental in *iTaukei* society (Luetz and Nunn 2020). These policy measures are bound to a set of principles that are focused on hard measures (such as development of infrastructure and new energy networks), notions of sustainable development, and resilience that, while important, overlook critical variables within the philosophical and cultural systems of the *iTaukei* (and other Fijians) that could otherwise contribute to effective and sustainable climate change adaptation (McNamara et al. 2020). These policies are also framed within a particular metaphysical conception in which humans are disconnected from Nature, a Western view that contrasts markedly with that of most Fijians. The consequence is that such climate change policies cannot leverage social institutions and their metaphysical basis that could potentially allow for more effective climate management and governance interventions (Chandler 2014). *Vanua* must be leveraged: its interconnection of Nature, and ancestor spirits, are powerful tools for group action and adaption to climate change and a more consistent appreciation of this metaphysical system within policy may help develop more socially relevant adaptation strategies. The metaphysical dimensions in *Vanua,* together with the associated everyday practices it entails, have the potential to redefine climate change responses and adaptation. These could trigger ethical, political and knowledge practices bearing the onus of a convivial and relational practice cognisant of the entanglement between the social and ecological worlds.

Our argument about the importance of metaphysical basis in the creation and delivery of policies, is relevant elsewhere. Indeed, Fiji's colonial history, processes of hybridisation, and metaphysical systems are unique but find parallels in other communities that could make for future comparative research. Climate change adaptation is spatially mediated, but its consequences will be global, crossing political and cultural boundaries, straddling belief systems and worldviews. A heightened attention to metaphysical bases of those outside Western-centric worldviews, should be hermeneutically understood if successful climate change adaptation is sought for those who need it most. We believe that these are sufficient grounds to develop a research agenda that works not only toward the understanding of metaphysical basis as they relate to spirituality, everyday action and climate change, but also works towards deciphering how to better apply this knowledge into policy making and practice.

References

Anantharajah K (2019) Governing climate finance in Fiji: Barriers, complexity and interconnectedness. Sustain 11(3414)

Anderson B (2006) Imagined communities: Reflections on the origin and spread of nationalism. Verso, London

Aristotle (2018) Physics (trans: Reeve C). Hackett Publishing Company, Inc., Indianapolis and Cambridge

Berkes F (2018) Sacred ecologies, 4th edn. Routledge, New York

Berry T (1988). The dream of the Earth. Sierra Club Books, San Francisco

Boyd H (1999) Christianity and the environment in the American public. J Sci Study Relig 38:36–44

Brightman M, Lewis J (2017) Introduction: the anthropology of sustainability: beyond development and progress. In: Brightman M, Lewis L (eds) The anthropology of sustainability: beyond development and progress. Palgrave Macmillan, New York, pp 1–34

Brincat S, Gerber D (2015) The necessity of dialectical naturalism: Marcuse, Bookchin, and dialectics in the midst of ecological crises. Antipode 47(4):871–893

Castree N (2005) Nature. Routledge, London and New York

Castree N (2012). Making sense of nature. Taylor and Francis, London

Chandler D (2014) Beyond neoliberalism: resilience, the new art of governing complexity. Resil 2(1):47–63

Cote M, Nightingale A (2012) Resilience thinking meets social theory: situating social change in socio-ecological systems (SES) research. Prog Hum Geogr 36(4):475–489

Cretney R (2014) Resilience for whom? Emerging critical geographies of socio-ecological resilience. Geogr Compass 8(9):627–640

Cronon W (1995) The trouble with Wilderness; or, getting back to the wrong nature. In: Cronon W (ed) Uncommon ground: rethinking the human place in nature. W. W. Norton & Co, New York, pp 69–90

Currenti R, Pearce T, Salabogi T, Vuli L, Salabogi K, Doran B, Kitson, R, Ford J (2019) Adaptation to climate change in an interior Pacific Island Village: a case study of Nawairuku, Ra, Fiji. Hum Ecol 47:65–80

Davidson J, Jacobson C, Lyth A, Dedekorkut-Howes A, Baldwin C, Ellison J, Holbrook N, Howes M, Serrao-Neumann S, Singh-Peterson L, Smith T (2016) Interrogating resilience: toward a typology to improve its operationalization. Ecol Soc 21(2):27

Escobar A (1999) After nature: steps to an antiessentialist political ecology. Curr Anthropol 40(1):1–30

Fache E, Pauwels S (2020) Tackling coastal "overfishing" in Fiji: advocating for indigenous worldview, knowledge, and values to be the backbone of fisheries management strategies. MaritE Stud 19:41–52

Farrelly T (2010) Indigenous and democratic decision-making: issues from community-based ecotourism in the Boumā National Heritage Park, Fiji. J Sustain Tour 19(7):817–835

Fingleton J (2005) Introduction: the debate over land tenure. In: Fingleton J (ed) Privatising land in the Pacific: a defence of customary tenures. The Australian Institute, Canberra, pp 6–16

Foale S (2006) The intersection of scientific and indigenous ecological knowledge in coastal Melanesia: implications for contemporary marine resource management. Int Soc Sci J 58(187):129–137

France P (1966) The charter of the land: a study in the cross-fertilisation of Fijian tradition and British colonial policy. Australian National University, Canberra

Gerber D (2019) The distortion of nature's image: reification and the ecological crisis. State University of New York, United States of America

Gharbaoui D, Blocher J (2016) The reason land matters: relocation as adaptation to climate change in Fiji Islands. In: Milan A, Schraven B, Warner K, Cascone N (eds) Migration, risk management and climate change: evidence and policy responses. Springer, Switzerland, pp 149–173

Greeley A (1993) Religion and attitudes toward the environment. J Sci Study Relig 32:19–28

Guth J, Green J, Kellstedt L, Smidt C (1995) Faith and the environment: religious beliefs and attitudes on environmental policy. Am J Polit Sci 39:364–382

Hand C, Van Liere K (1984) Religion, mastery-over-nature, and environmental concern. Soc Forces 63:555–570

Haraway D (2008) When species meet. University of Minnesota Press, Minneapolis, MN

Iati I (2009) Reconsidering land reform in the Pacific. Research paper. Council for International Development, Wellington

Kaplama E (2016) Heraclitean critique of Kantian and Enlightenment ethics through the Fijian Ethos. Cosm Hist: J Nat Soc Philos 12(1):143–165

Latour B (2004) Politics of nature: how to bring the sciences into democracy. Harvard University Press, Cambridge, MA

Latour B (2017) Anthropology at the time of the anthropocene: a personal view of what is to be studied. In: Brightman M, Lewis J (eds) The anthropology of sustainability: beyond development and progress. Palgrave Macmillan, New York, USA, pp 35–50
Law J, Lien M (2018) Denaturalizing nature. In: de la Cadena M, Blaser M (eds) A world of many worlds. Duke University Press, Durham and London, pp 131–171
Leipod S, Feindt P, Winkel G, Keller R (2019) Discourse analysis of environmental policy revisited: traditions, trends, perspectives. J Environ Planning Policy Manage 21(5):445–463
Leveridge T (2009) Divergent paths: The ontology, epistemology, and pragmatics of community-based conservation in locally-managed marine reserves, Fiji (MSc). Oxford University
Luetz J, Nunn P (2020) Climate change adaptation in the Pacific Islands: a review of faith-engaged approaches and opportunities. In: Filho W (ed) Managing climate change adaptation in the Pacific Region. Springer, Hamburg, pp 293–312
McGregor D (2004) Traditional ecological knowledge and sustainable development: towards coexistence. In: Blaser M, Feit H, McRae G (eds) In the way of development: indigenous peoples, life projects and globalization. Zed Books, London
McNamara K, Clissold R, Westoby R, Piggott-McKellar A, Kumar R, Carke T, Namoumou F, Areki F, Joseph E, Warrick O, Nunn P (2020) An assessment of community-based adaptation initiatives in the Pacific Islands. *Nature Climate Change*
Miyazaki H (2004) The method of hope: anthropology, philosophy and Fijian knowledge. Stanford University Press, Stanford, California
Moore J (2016) Introduction: anthropocene or capitalocene? Nature, history, and the crisis of capitalism. In: Moore J (ed) Anthropocene or capitalocene? Nature, history, and the crisis of capitalism. PM Press, Oakland, CA, pp 1–13
Nabobo-Baba U (2006) Knowing and learning: an Indigenous Fijian approach. Institute of Pacific Studies, University of the South Pacific, Fiji
Nunn P (2001) On the convergence of myth and reality: examples from the Pacific Islands. Geogr J 167(2):125–138
Nunn P (2013) The end of the Pacific? Effects of sea level rise on Pacific Island livelihoods. Singap J Trop Geogr 34(2):143–171
Nunn P, Mulgrew K, Scott-Parker B, Hine D, Marks A, Mahar D, Maebuta J (2016) Spirituality and attitudes towards nature in the Pacific Islands: insights for enabling climate-change adaptation. Clim Change 136:477–493
Oakes R (2019) Culture, climate change and mobility decisions in Pacific Small Island Developing States. Popul Environ 40(4):480–503
Obeyesekere G (2005) Cannibal talk: the man-eating myth and human sacrifice in South Seas. University of California Press, Berkeley
Parke A (2014) Degei's descendants Spirits: place and people in pre-cession Fiji. ANU Press, Canberra, Australia
Pascoe S, Brincat S, Croucher A (2019) The discourse of climate change science: scientific reporting, climate negotiations and the case of Papua New Guinea. Glob Environ Change 54:78–87
Pearson J, McNamara K, Nunn P (2020) iTaukei ways of knowing and managing mangroves for ecosystem-based adaptation. In: Filho W (ed) Managing climate change adaptation in the Pacific Region. Springer Nature, Switzerland, pp 105–127
Poirier S (2008) Reflections on indigenous cosmopolitics-poetics. Anthropoligica 50(1):75–85
Ravuvu A (1987) The Fijian Ethos. Institute of Pacific Studies, University of the South Pacific, Fiji
Reed A, Hames I (1997) Myths and legends of Fiji and Rotuma. Literary Productions Ltd., Fiji
Republic of Fiji (2018) Republic of Fiji National Climate Change Policy 2018–2030
Roser-Renouf C, Maiback E, Leiserowitz A, Feinberg G, Rosenthal S (2016) Faith, morality and the environment: portraits of global warming's six Americas. Yale University Program on Climate Change Communication and George Mason University, New Haven, USA
Routledge D (1985) Matanitū: the struggle for power in early Fiji. Institute of Pacific Studies in association with the Fiji Centre Extension Services, University of the South Pacific, Fiji

Ryle J (2005) Roots of land and church: The Christian state debate in Fiji. Int J Study Christ Church 5(1):58–78
Ryle J (2010) My God, my land: Interwoven paths of Christianity and tradition in Fiji. Ashgate, Aldershot
Scheyvens R, Russell M (2012) Tourism, land tenure and poverty alleviation in Fiji. Tour Geogr 14(1):1–25
Strang V (2017) The Gaia complex: ethical challenges to an anthropocentric 'common future'. In: Brightman M, Lewis J (eds) The anthropology of sustainability: beyond development and progress. Palgrave Macmillan, New York, USA, pp 207–228
Sutton V (2005) Custom, tradition and science in the South Pacific: Fiji's new Environmental Management Act and Vanua. J S Pac Law 9(2)
Taylor B (2015) Religion to the rescue (?) in an age of climate disruption. J Study Relig, Nat Cult, 7–18
Taylor B, Van Wieren G, Zaleha B (2016) Lynn White Jr. and the greening-of-religion hypothesis. Conserv Biol 30(5):1000–1009
The Fijian Government (2010) Clarification on native lands commission. Retrieved from https://www.fiji.gov.fj/Media-Centre/News/Clarification-on-Native-Lands-Commission
Thornley A (2002) Exodus of the ITaukei: The Weslayan Church in Fiji: 1848–74. Institute of Pacific Studies, University of the South Pacific, Suva
Tomlinson M (2009) Efficacy, truth, and silence: language ideologies in Fijian Christian conversions. Comp Stud Soc Hist 51(1):64–90
Tomlinson M, Bigitibau S (2016) Theology of Mana and Sau in Fiji. In: Tomlinson M, Kāwika T (eds) New Mana: transformations of a classic concept in Pacific languages and cultures. ANU Press, Acton, Australia, pp 237–256
Tucker M (1994) An ecological cosmology. In: Chapple C (ed) Ecological prospects: scientific, religious, and aesthetic perspectives. State University of New York Press, Albany, USA, pp 105–126
Tuwere I (2002) Vanua: towards a Fijian theology of place. Institute of Pacific Studies, University of the South Pacific and College of St John the Evangelist, Suva
Unger R, Smolin, L (2015) The singular universe and the reality of time: a proposal in natural philosophy. Cambridge University Press, Cambridge
United Nations (2015a) Transforming our world: the 2030 agenda for sustainable development
United Nations. (2015b). The Paris agreement
Walter M (1978) Analysis of Fijian traditional social organization: the confusion of local and descent groupings. Ethnol 17(3):351–366
White L (1967) The historical roots of our ecologic crisis. Science 155:1203–1207
Whitehead AN (1925) Science and the modern world. Cambridge University Press, Cambridge
Williams R (1980) Problems in materialism and culture. Verso, London

Francisco Gelves-Gómez is a PhD (Geography) student and sessional academic at the School of Social Sciences, University of the Sunshine Coast. He is also a researcher with the Environmental Social Science Research Group from The University of Melbourne. Francisco's main academic work is at the interface between the natural and social sciences, particularly focused on critical aspects of biodiversity conservation and human and more-than-human dimensions of conservation sciences and climate change. He also studies complex socio-ecological systems and the application of these forms of thinking for management and governance of land and sea.

Shannon Brincat is a Senior Lecturer in Politics and International Relations at the University of the Sunshine Coast, Australia. His most recent project, *The Spiral World*, traces dialectical thinking in the Axial Age. He has been the editor of a number of collections, most recently *From International Relations to World Civilizations: The contributions of Robert W. Cox* and *Dialectics and World Politics*. He is also a co-editor of *Global Discourse*.

Chapter 5
Gods, Spirits and Natural Hazards: Ontologies and Epistemologies of Natural Hazards and Climate Change in Kiribati and Papua New Guinea

Guy Jackson and Annah Piggott-McKellar

Abstract Pacific Island countries are already experiencing severe impacts from climate change and these will likely increase in the future. In response, a plethora of climate change adaptation projects have been implemented across the region with the aim of assisting those most affected. Yet these projects have largely failed to reduce vulnerability to climate change in a meaningful way. This chapter addresses a commonly overlooked, but argued here to be crucial, reason for failure: how religion shapes local perspectives of climate change, and the lack of acknowledgement with these by external actors. We address this gap through exploring the changing ontological and epistemological understandings of climate change and hazards among two distinct cultural groups from the Pacific Islands—the Bedamuni of Western Province, Papua New Guinea and the I-Kiribati. To structure our diachronic investigation, we explore the drivers of socio-cultural change across three periods: pre-contact, colonial and missionary, and contemporary. A complex story of interaction with exogenous ideas and syncretism is portrayed within both societies. Yet today we find the Bedamuni continuing to draw heavily from traditional belief systems, especially the presence and influence of spirits, to explain hazards while simultaneously being devout Christians. Despite noting changes in weather and seasonality, Bedamuni people have had little exposure to global climate change discourses. Conversely, the I-Kiribati, as the occupants of one of the countries most exposed to climate change, have had relatively high exposure to scientific climate change discourses. Notwithstanding this, I-Kiribati mostly draw from Christian beliefs to explain natural phenomena while maintaining traditional beliefs in gods, spirits and magic. Like others in the Pacific, these cultures have unique histories, ontologies and epistemologies so, for climate change adaptation to be successful in the region, governments, agencies, and NGOs must acknowledge that no two places are alike

G. Jackson (✉)
School of Earth and Environmental Sciences, The University of Queensland, St Lucia, QLD, Australia
e-mail: guy.jackson@LUCSUS.lu.se

Lund University Center for Sustainability Studies, Lund University, Lund, Sweden

A. Piggott-McKellar
School of Geography, University of Melbourne, Melbourne, VIC, Australia

J. M. Luetz and P. D. Nunn (eds.), *Beyond Belief*, Climate Change Management,
https://doi.org/10.1007/978-3-030-67602-5_5

and that understanding recipients' ontological worldviews and accommodating these into adaptation strategies is an essential starting point for the development of effective future strategies.

Introduction

The need for effective, proactive and sustainable climate change adaptation is increasingly urgent in Pacific Island Countries (Nunn and Kumar 2019a). Current, ongoing, and escalating impacts of climate change including sea-level rise, increasing intensity of storm and cyclone activity, prolonged droughts, and rising atmospheric and ocean-surface temperatures are interacting with increasingly vulnerable socioecological systems (Barnett 2010).[1] In recent years, billions in international funding has poured into national governments, INGOs, regional organisations, and aid agencies across the Pacific with the stated goal of reducing vulnerability to climate change and disasters (Donner et al. 2016). Despite this, limited progress has been observed. The failure in producing sustained outcomes is generally because projects are often driven by externally defined goals, they naively consider communities as homogeneous entities, and are too narrowly focused on biophysical climate impacts rather than the wider human-societal factors that generate and drive vulnerability (McNamara et al. 2020; Nunn and Kumar 2019a, Piggott-McKellar et al. 2020). To move toward a more comprehensive understanding of reasons for failure, a deeper analysis of the role of religious and cultural perceptions of hazards and climate change is helpful.

Research into the role of religion in the Pacific as either conducive to or hindering climate change adaptation or disaster risk reduction is limited (see e.g., Cox et al. 2018; Cox et al. 2019; Haluza-DeLay 2014; Luetz and Nunn 2020 for exceptions). When it is discussed, religion has often been simplistically framed as a barrier to meaningful risk reduction (Luetz and Nunn 2020). While this may be the case in some circumstances (Ayeb-Karlsson et al. 2019), there is a need to critically engage with local worldviews and religious institutions or else we will likely continue to see the failure of external interventions. Acknowledgement of the importance of religion is becoming more common, for instance, with initiatives that aim to engage with religious dogma to spread messages of climate resilience and disaster preparedness in the Pacific (see Banfield 2019).

With this context in mind, the present chapter investigates the changing ontological and epistemological understanding of environmental hazards and climate change in two Indigenous Pacific cultures. Through this we show the importance

[1] Pacific Island Countries have always been highly exposed to natural hazards owing to their unique geography, extensive coastlines, and self-provisioning livelihoods. Yet increasing vulnerability has been driven by centuries of colonisation and more recently neoliberalism. Related processes such as capitalism, missionisation, and the provision of aid have contributed to the replacement of traditional livelihoods, environmental knowledge, and social protection systems (Barnett 2010; Campbell 2014; Jackson 2020).

of tailoring information and adaptation imperatives around climate change adaptation and disaster risk reduction to target cultural contexts which typically interpret natural and social phenomena through faith based and spiritual lenses. It is argued that a lack of nuance in policy hitherto has hindered potential transformational change throughout Pacific Island Countries and beyond.

The rest of the chapter is as follows. First, we review existing research from the Pacific on traditional and introduced religious beliefs and perceptions of climate and hazard causality. Second, a brief section on the methodology and case study backgrounds is presented. Third, the ontological and epistemological understanding of hazards and climate change across the two case study sites are detailed, leading into the identification of themes underlying the commonalities and differences of the study sites. Lastly, we discuss the importance of these findings for the future of climate change adaptation and disaster risk reduction in the Pacific.

Changing Religious Perceptions in the Pacific

Throughout history, natural hazards have been perceived as supernatural events caused by gods or spirits, often in response to human actions that were mistaken or intentionally perverse (Ahmad 2019; Furendi 2007; Hardwick and Stephens 2020). This understanding first began to shift in Europe during the enlightenment when rational 'naturalistic' explanations superseded religious ones (Hardwick and Stephens 2020). Today the dominant scientific epistemology is logical positivism, which requires that only what can be proven through logic and replicated through empirical observation can be 'true'. Despite the ubiquity of logical positivism and its dominance in the development of global ideas and initiatives, many people around the world favour faith-based understandings of natural hazards and climate change (Ahmad 2019; Haluza-DeLay 2014; Luetz and Nunn 2020). These understandings vary immensely and are hard to generalise about, in part due to the subjectiveness of people's interpretations of their religion (Gaillard and Texier 2010).

There is a scientific consensus that climate change is occurring and that this is likely attributable to an increase of greenhouse gas emissions (IPCC 2018). Within Pacific Island Countries, observable changes in the climate are widely recognised by local communities, yet often causality is attributed to God (Donner 2007; Luetz and Nunn 2020). Indeed, Indigenous and Christian rationalisations exist even at the governmental level (Kempf 2017; Luetz and Nunn 2020; Tomlinson and McDougall 2012). Logical positivism means little to most Indigenous people in the region because they are enculturated with their own unique Indigenous epistemologies and strong Christian faith. Unfortunately, and perhaps unnecessarily, this has created a conflict based on values and beliefs between Pacific Islanders' worldviews and those of scientists and other secular climate change adaptation actors (Luetz and Nunn 2020). We suggest this conflict is partly responsible for the failure of externally driven climate change adaptation to date.

Ethnographic investigations have highlighted the central role spirits play within Pacific Island societies (Bird-David 1999; Lindstrom 1990). While no two cultures are the same, there are commonly three forms of spirits: creator spirits (or gods in some Polynesian societies) which are linked to genesis stories; nature spirits which can influence the weather and other physical processes; and ancestral spirits with whom the living can communicate and who often intervene in human affairs (e.g., sorcery, crop growth, protection) (Lindstrom 1990; Sørum 1982). Recognition of the centrality of spirits is crucial to understanding Indigenous ontological worldviews.

It has been reported that spirits and sometimes gods play a role in causing natural hazards while also intervening to protect societies if called upon (Falck 2018; Laughlin 2018; Sillitoe 1993). The belief in spirits being responsible for phenomena is a widespread human trait which features in pagan and monotheist religions throughout history in all geographic locations (Donner 2007). Narratives that seek to explain natural hazards by attributing them to spirits have two key purposes. First, they provide solace through rationalising the seemingly chaotic and random nature of disruptive phenomena (e.g., Barrios 2015; Eisenbruch 2020; Taylor 1995). Second, they have a more practical purpose in providing warnings and lessons about the physical risks to which societies are exposed (Janif et al. 2016), which then informs strategies such as early warning and food system redundancy (e.g., food storage, extensive kinship networks, drought tolerant cultivars). Many Pacific Island cultures developed both spiritual (e.g., moral codes, ritual) and practical adaptations to periodic environmental perturbations (Campbell 1990; Jackson 2020).

With the introduction and dissemination of Christian teachings throughout the Pacific Islands, the old spirits were not always fully replaced but sometimes reinterpreted within a Christian framework (Falck 2018; Macdonald 2018; Tomlinson and McDougall 2012). Many Pacific Island cultures had a creator being in their genesis stories and this sometimes became God, while spirits sometimes became angels and demons. Many Melanesian cultures have strong eschatological beliefs or millennialist tendencies (Frankel 1986; Robbins 2001; Sillitoe 1993). Stories of the Second Coming and Last Judgement in the Bible were sometimes used by missionaries to show that Christianity was a more developed and universal expression of their own eschatological beliefs (Jebens 2000). Some cultures comprehensively did away with many of their Indigenous practices and beliefs (Robbins 1995; Tomlinson and McDougall 2012), while others maintained both traditional beliefs alongside strong religious faith despite contradictions in stories and meanings (Rudiak-Gould 2010).

There is no denying the profound impact of Christianity on social formations, political power, land use, and broader socioecological relationships in the Pacific Island region. Perhaps most profoundly was the observed transformation of Nature-as-subject to Nature-as-object in some socities, which, for example, justified doing away with previous ecological taboos (e.g., sacred areas, trees, animals) and thus led to a greater human impact on ecological systems (Robbins 1995). The changing contours of socioecological relations are critically important to understanding changing exposure and, at times, vulnerability to hazards and climate change (Jackson et al. 2020). Additionally, whereas ritual was formerly the primary means of appeasing spirits to ensure climatic stability and food security, Christianity has led to prayer being the favoured method of achieving these in some societies (Jackson 2020; Jackson et al. 2020). While '*Kastom*' and other traditional systems of religion and political organisation are omnipresent in the Pacific Islands region, churches are equally important institutions today (Tomlinson and McDougall 2012). For example, although chiefs are still supreme leaders in chiefly societies, the pastor can be equally, if not more, influential today in Pacific Island Countries or indeed be both simultaneously (White 2012). Yet elsewhere in the region there is a clear demarcation between church and state as with the Urapmin in Papua New Guinea (Robbins 2012).

The influence of Indigenous religion and Christianity, complicated and intertwined as they are in the Pacific region, means for any external climate change adaptation strategy to be successful it should understand and acknowledge unique ontologies and traditional (non-western) power structures. Writing about Christianity in the Pacific, Tomlinson and McDougall (2012: 12) provide the following perceptive warning for external interventions throughout the Pacific:

> While projects of political and social improvement that are not motivated by religion are hardly doomed to failure, projects that do not recognize Pacific Islanders' religious aspirations probably are.

Beyond tokenistic involvement of the church, we suggest understanding what Indigenous people believe (i.e., ontology) and how they know it to be true (i.e., epistemology) regarding both natural hazards and climate change is a crucial first step in improving the effectiveness of external interventions in Pacific Island Countries. This chapter provides examples from two very different cultures, one from Papua New Guinea and the other from Kiribati, and will demonstrate that one uniform secular approach to climate change adaptation will never work in the deeply religious Pacific.

Methods

The data used in this study derives from extensive ethnographic fieldwork in the authors respective study sites. In PNG, research was conducted over four months in July through October 2018. In Kiribati, research took place over two months in March through May 2017. Both authors had different original research aims, namely Indigenous food system disaster vulnerability in PNG (see Jackson 2020; Jackson et al. 2020) and the effectiveness of external climate change adaptation strategies in Kiribati (see Piggott-McKellar et al. 2020). Nonetheless, significant data on people's ontological and epistemological understanding of hazards and climate change were obtained during fieldwork. Indigenous epistemologies were foregrounded in these studies using a variety of group and individual interviews and empathetic *talanoa* discussions premised on local norms regarding knowledge and information sharing (see Vaioleti 2006). All oral data was transcribed verbatim and along with extensive notes was analysed using NVivo.

Bedamuni, Western Province, Papua New Guinea

The Bedamuni occupy dissected foothills in the very remote and economically disconnected Strickland-Bosavi region of Western Province, PNG (Fig. 5.1). Bedamuni who live within their customary territory now number ~11,000, a fourfold increase since sustained contact in the early 1960s (Hoad 1963). While Australian government patrols interacted with the Bedamuni in the 1940s it was not until 1962 that sustained contact occurred after a widespread campaign of pacification. Almost all Bedamuni are now Christian, with Evangelical, Seven-Day Adventists (SDA), and Catholic denominations—in declining order of prevalence—established throughout the territory. Today, livelihoods are primarily self-provisioning with shifting cultivation of bananas, taro, yams, and sweet potato, hunting, fishing, and the wild foraging of wild and planted sago the most important components of their food system (Jackson et al. 2020). The cash economy remains poorly developed due to lack of roads, high cost of air freight, and limited paid work opportunities. A recent increase in the underlying drivers of vulnerability has led to Bedamuni being more susceptible to the impacts of recurring hazards such as El Niño associated droughts (and subsequent wildfires), earthquakes, flooding, and outbreaks of pests and diseases (Jackson et al. 2020).

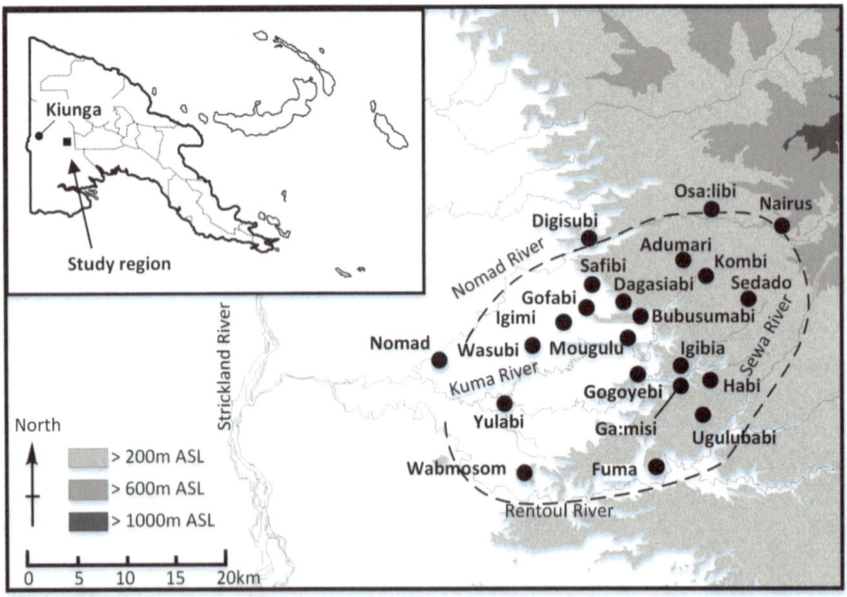

Fig. 5.1 Bedamuni territory location and boundary with major settlements highlighted (prepared by Bradd Witt)

Kiribati

Kiribati is an atoll island nation situated in the Pacific Ocean spanning the Equator and the International Date Line (see Fig. 5.2). There are 33 islands which comprise Kiribati within three island groups (Gilbert, Phoenix, and Line), having a combined land size of 810 km^2 spanning over 3 million km^2. Of the 33 islands only 20 are inhabited and all save one are less than 3 m above sea level. These factors (small land size, isolation, and lowness) have resulted in the country being recognised as one of the most physically vulnerable nations to climate change, receiving significant international attention in recent years (Ives 2016; Johnson 2019; Thomson 2013). Livelihoods in Kiribati differ significantly between the main island, South Tarawa, which is home to over half the country's ~110,000 total population and outer islands. Livelihoods on South Tarawa are driven by the cash economy while people on outer islands, such as Abaiang, where most fieldwork was undertaken, live largely subsistence-based, relying on fishing and harvesting marine resources, and cultivations of crops including coconut, banana, and taro. Such people often generate a small income from the sale of handicrafts, surplus foods, and copra (dried coconut kernel).

Religion, which first came to the region in 1850s through Protestant missionaries, plays a significant role in the daily lives of the I-Kiribati with over 95% of the country affiliating with a Christian denomination (57% Roman Catholic Church, 31% Kiribati Protestant Church with other denominations making up the remaining) (Kiribati National Statistics Office 2016).

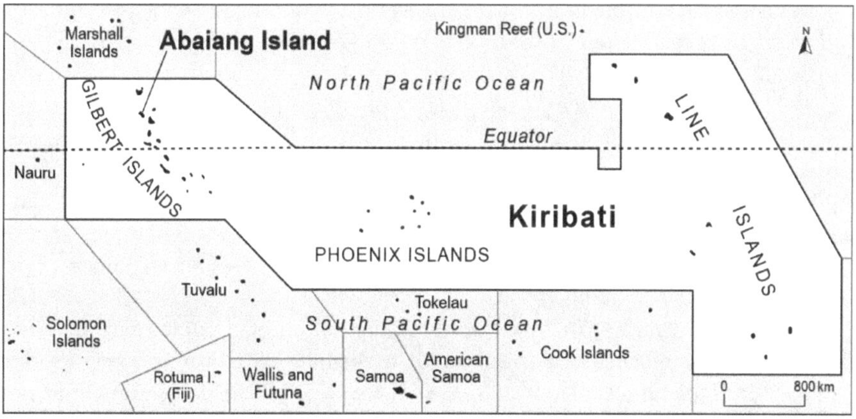

Fig. 5.2 Map of Kiribati, showing Abaiang island (prepared by Chandra Jayasuriya)

The Changing Perceptions of Hazards and Climate Change in Three Stages

To understand how modern Bedamuni and I-Kiribati interpret the causes of natural hazards and climate change, it is important to explore the drivers and consequences of socio-cultural change over time. To achieve this, we discuss below the following three stages, (1) pre-contact, (2) colonial and missionary, and (3) contemporary. Each period is not perfectly demarcated with considerable overlap and strong multi-linearity evident. For instance, we recognise that Indigenous societies were never static, as culture is constantly constructed, contested, and adapted to new social and biogeophysical influences. As such, these 'stages' serve as an analytical tool to create 'meaningful sequence' (Ballard 2003) of change. The similarities and differences between the two case studies will be explored using common themes.

Pre-contact Period

The first period that we discuss is pre-contact—Indigenous perspectives and understandings. Unique yet comparable origin stories involving creator spirits exist in both study sites. Whereas I-Kiribati are more likely to attribute the title of gods as well as spirits, in the Bedamuni case, they are considered a type of spirit.[2] We first turn to the Bedamuni.

The Bedamuni believe in creator spirits that no longer play a role in the human world. The Bedamuni creation myth suggests there was once no life on the earth "*only mud, mud everywhere*" (Sørum 1991: 95). As an elderly informant stated, out of this mud came "*one man called Awamuni, he made all the plants, earth and all this, and they* [ancestors] *were living* [there]" (Iguabi village 11-08-2018). Awamuni was a giant and created the first woman Dunumuni by copulating with a tree. Dunumuni and Awamuni had four children, two males and two females who married. Their children became the ancestors of all people (Sørum 1991: 95). The lineage of clans (e.g., dog, pig, crocodile) can be traced to these ancestors. In Bedamuni genesis stories their ancestors were considered the first people that then became many of the surrounding cultures (Sørum 1991).

Apart from creator spirits there are three tiers of spirits: nature, ancestral, and *sagisu* (offspring of ancestral spirits or of a shaman and spirit woman) (Sørum 1982:44). In Bedamuni society, nature spirits are widely considered dangerous and responsible for many biophysical processes including droughts, floods, and earthquakes. Spirit mediums called upon ancestral spirits to seek guidance and foreknowledge of hazards and the future more generally. Through correct rituals, ancestral spirits could intervene and bring rain to break a drought or ensure crops grew well. For example, an elder male in Igimi village (19-08-2018) suggested,

[2]We acknowledged the challenges involved in achieving this type of comparison considering the translation from local languages into English in this study.

When they [ancestors] *see that the bush is very dry, or burnt by the fire, nothing to eat, then they would go to these people* [spirit mediums] *and ask them to make rain to fall. And then these people make bule bule* [magic] *and then they will hear the sound in the sky. And then they will say: "Ok, it's going to rain tonight".*

Rich in traditional ecological knowledge, spirit mediums could predict hazards through observation of indicators and communication with spirits. Participants suggested there was at least one spirit medium in each village and they served important religious and political roles in pre-contact Bedamuni society. During fieldwork in 2018 Bedamuni informants claimed there are no practicing spirit mediums left due to Christian conversion and missionary influence.

The overarching (and somewhat unsatisfactory) concept of animism appears to have defined pre-contact Bedamuni society. Everything had an essence, from mountains, trees, animals, and people, and nothing happened randomly, without cause. Ancestral spirits lived in their sections of the forest and continued to live much as the Bedamuni, for example, gardening and marrying. This suggests there was no separation of physical and metaphysical worlds as seen in Judeo-Christian religions. When asked if there was a heaven or hell in traditional religion, an elder from Igimi village (13-08-2018) responded, "[we] *were not told about these two places. But we were told not to create problems. When you create a lot of problems you will die*". Similar to many cultures, true knowledge of the world was not widely disseminated, but passed down by spirit mediums when they find a "*quiet man, who will not spread this knowledge*" (Journal 06-09-2018).

Turning to the I-Kiribati, as with the Bedamuni and most Indigenous cultures, there exist origin stories. These stories differ slightly depending on which islands and families they come from. The principal creation story is that of *Nareau the Creator*. *Nareau*, meaning 'the spider', was a god who had no known origin and who could do human as well as godly things. It was said that he once floated in space all alone but then eventually came across *Te Bomatemaki*, which is the earth and the sky sealed together. After trying and failing to open *Te Bomatemaki, Nareau the Creator* created *Nareau the Wise* who was then tasked with creation of the earth and sky, sun, moon and stars (Beiabure et al. 1979). Spirits were then formed ranging from *Uka*, the essence of moving air, to *Nei Kika*, the octopus. Spirits were created which eventually evolved into human beings. The I-Kiribati believed that all their ancestors were spirits at some point (Beiabure et al. 1979).

Rituals also played a role in pre-contact life in Kiribati as documented in the seminal book *Tungaru Traditions* by Arthur Grimble (Grimble 1989) which provides a detailed ethnographic exploration of life during the early 1900s. For example, people in certain clans would perform various agricultural rituals that would promote productive yields, or protect harvests and crops from theft with spells and magic often included throughout these rituals. Through these stories told and rituals performed, insights can be gleaned into how I-Kiribati in pre-contact time related and interacted with the world around them when misfortune and even death were commonly attributed to inappropriate or ill behaviour that angered the gods (Beiabure et al. 1979).

Colonial and Missionary Period

The second period we will discuss is missionisation, the introduction of Christian religion. Missionisation within the Pacific region was ubiquitous and had a significant, profound, and sustained impact on Indigenous peoples' ontological and epistemological understandings (Garrett 1982). While ubiquitous, the impact of missionaries differs significantly across the region, often resulting in a unique syncretisation of Indigenous and religious worldviews. This is evident for our two case study sites.

In Kiribati, the first missionaries were Protestant and arrived in the northern islands of Kiribati in 1852. Upon arrival, these early missionaries faced a significant challenge converting I-Kiribati owing to their underlying, strong, and persisting Indigenous beliefs. This was especially challenging by the time Roman Catholic missionaries arrived two decades later not least as many I-Kiribati had converted to Protestant by this time (Gheuens 2017). As time progressed, Christian messages and their attendant interpretations of the world began increasingly to permeate the worldviews of local people. Not only did the church spread new teaching, they also urged and persuaded people to change aspects of culture including practices of magic and polyamory (Beiabure et al 1979; Gheuens 2017).

While Christian beliefs and affiliation with churches became ubiquitous across the region the importance of ancestral spirits and magic rituals persisted, albeit to a lesser extent. Today, the impact of early missionaries has left a permanent and persistent mark on the I-Kiribati, shaping the lives of people across the country. For example, church donations are one of the biggest expenses incurred in many villages, demonstrating how large of an influence the church has, especially considering the limited cash available. Today Christian beliefs dominate with over 95% of people associated with a Christian denomination.

This missionisation of the I-Kiribati has impacted their ontological worldviews. While attribution was previously given to Indigenous gods for ills, and misfortunes, this has now been transferred to God in a biblical context. This extends to climate change and natural hazards, as people now generally hold deterministic Christian views surrounding climate change and hazards. The wave of influence of missionisation experienced in Kiribati leading to slow decline and/or evolution of Indigenous ontological and epistemological beliefs was similarly experienced by the Bedamuni. While there are significant differences, in that while in Kiribati religion has primarily (but not completely) taken the place of Indigenous ontologies, with the Bedamuni it has resulted in a more syncretised ontology between Indigenous and religious views.

Within the Bedamuni, Australian Evangelical missionaries followed government patrols in the then Australian Territory of Papua and New Guinea and established a permanent mission at Mougulu in 1968. The Australian government first pacified the Bedamuni who were perceived as warlike because of their practice of undertaking cannibal raids and the resultant blood feuds that these, among other conflicts, led to (Hoad 1963). The Australian colonial government imbued many Judeo-Christian values before the arrival of missionaries such as respect for authority, law and order, the cash economy, sedentary settlements, and the proscription of witchcraft and black magic (Jackson, forthcoming). When the missionary, Tom Hoey from Brisbane, arrived in the Bedamuni territory, his main intention was to learn the language, translate the bible, and convert the Bedamuni, who he regarded as fallen people needing to be saved from the clutches of the devil.

After learning many customary beliefs, Hoey identified a similar Indigenous story of a great historical flood and initially drew from this and the example of Noah's Ark to show similarities within the Bedamuni and western belief systems (Mackay and Hoey 1984). Furthermore, the Bedamuni had many moral stories which detailed how to live and behave; they had a single creator spirit Awamuni who created a woman and populated the earth, a story that aligns with the Christian belief in Adam and Eve. Over time most Bedamuni, like almost all Papuans, became practicing Christians. When GJ spoke to the now elderly Gidama, the first Bedamuni pastor, he suggested that Christianity did not conflict with his understanding of traditional religion. Yet speaking with elders who were closest to the missionary, they admitted an element of opportunism (material benefit) to be gained by this arrangement. Many Bedamuni elders' admitted to still holding Indigenous beliefs concurrently but saw no significant moral contradiction with this.

Between the establishment of a mission in 1968 and 2018, when fieldwork was undertaken, significant changes have occurred to Bedamuni religion: there was limited transmission of knowledge regarding spirit communication, initiations are no longer practiced, knowledge of early warning indicators of hazards were lost, among many other changes. For example, while still believing in spirits, most people now pray to God to help with any hardship, whether from hazards, growing food insecurity, or illness (Jackson 2020). Gidama expressed it this way, through a translator, "*he believes that when he prays, they pray, help will come. Now they pray, when the problem came in, they prayed, and* [missionary's daughter] *came into assist* [during the 2015/16 El Nino]" (13-08-2018). While unique, the three Christian denominations—Evangelical, Seventh Day Adventist, and Catholic—, are now very important institutions within Bedamuni society, apparently taking over some of the role of traditional social protection, which was once solely the responsibility of clans and kinship networks.

The Contemporary Period

The third stage is characterised by the introduction of the scientific perspective of hazards into case study sites and the influence this has (or has not) had on people's ontological and epistemological understanding of hazards. As Kiribati has received global attention and funding for its universally acknowledged exposure to climate change, this case study provides a contrast with that of the Bedamuni who, while experiencing the impacts of climate change (especially in regard to precipitation patterns, increasing temperatures, and pest increases) have had less exposure to these discourses.

In Kiribati, climate change discussions and messaging are widespread. People are aware of the term 'climate change' and can point to particular changes in the environment and associate them with climate change impacts. For example, I-Kiribati often link climate change to their experiences of hotter temperatures, less predictable rainfall, changing coastal dynamics, and saltwater intrusion. This was illustrated during fieldwork where people detailed how they previously grew *bwai bwai* (a local root crop) for decades in traditional *bwai bwai* pits in certain parts of Abaiang Island. Owing to saltwater inundation, they could no longer grow them in these areas (see Fig. 5.3). People also related significant impacts from coastal erosion, which has destroyed many trees, directly relating these changes and experiences to climate change (Fig. 5.3).

Despite this practical and experiential understanding of climate change impacting the local environment and people's livelihoods, as well as scientific information disseminated to people regarding the causes of climate change, attribution of these changes is often filtered through people's religious beliefs. In discussions with local government workers and NGO representatives who have regular contact with communities working on climate change adaptation projects, the following example responses were given about how people perceive climate change information:

> The Christians, we believe that God looks after everyone and decides what happens so they just put their faith in God and say there won't be any climate change because God will be kind to us.

> We don't believe climate change will happen, we don't believe there is a second flood coming because in the Bible it says there was only one and that was in the time of Noah.

This shows that beliefs about the religious determinism of climate change are prevalent in Kiribati today, as indeed they are in neighbouring island countries (Kempf 2017). These underlying worldviews have been identified by external actors working with local communities as one of the biggest challenges to overcome in successfully implementing climate change adaptation projects in the region. This situation contrasts with that of the Bedamuni who, while also not subscribing to

Fig. 5.3 Left: Erosion evident along the coastline of Abaiang Island; Right: abandoned *bwai bwai* pits that became filled with saltwater

scientific ontologies, have stronger ties than I-Kiribati to spiritual explanations of natural hazards and climate change.

While the Bedamuni have been influenced by secular knowledge through mission-facilitated education and healthcare, as well as colonial and Papua New Guinea government contacts, community informants intimated that the first sustained non-religious advice they ever received came during and after the 2015/2016 El Niño drought response. Even then, the advice was given mostly by the missionary's daughter, a Christian. The Highlands earthquake in February 2018 again saw significant numbers of INGOs—and religious groups such as Seventh Day Adventist—enter Bedamuni territory, many explaining the hazard in scientific terms which meant little to the inhabitants. The Bedamuni were already very aware of the risks posed by both extended droughts (or '*dry seasons*' as they are known) and earthquakes because of their long history of exposure to them.

Despite the spread of Christianity, many declared Bedamuni Christians drew on their customary beliefs to explain the 2018 Highlands earthquake rather than invoking the Christian God or science. The main explanatory story relates to the mountain spirit's wife leaving and travelling down the Nomad River (Fig. 5.1). GJ first heard this story in Fumanibi village (09-08-2018) when a woman named Nanebe claimed she dreamt the spirit's wife was travelling down the river saying:

> *Namone* ["I'm sorry"], *as the fish spoke our language,* [it continued to say] *it's me the mother. I'm going to die today. If it was the father, there would have been big changes but because it's me, I'm not going to do anything.*

Many other participants, from villagers, young and old, to the vice president of the Local Level Government believed this to be the cause of the earthquake. Yet a few pastors and other Christians suggested the drought of 2015/2016, the earthquake, and the overall decline in food security (see Jackson et al. 2020) were punishment for past practices (e.g., cannibalism, warfare). Conversely, in Adumari and Safibi villages, informants suggested that sometimes earthquakes are caused by people who had lost loved ones, with one mother suggesting "*they start playing around to make earthquakes, some of them know how to make earthquakes*".

Regarding climate change, almost every Bedamuni informant recognised that the dry season (traditionally August and September used as a time for clearing gardens and planting) is starting later or not at all, and that the temperature, particularly between October and February, is becoming much warmer than customary. There has been a significant increase in garden pests and diseases, although whether this is driven by climate change is uncertain (Jackson et al. 2020). When asked if they knew about climate change, only the few who studied at high schools outside the Bedamuni territory suggested they had heard of that concept. As part of the earthquake response, CARE International ran several DRR sessions but only teachers and other educated people attended. Speaking with some teachers afterwards, they claimed to understand some of the natural causes of hazards. GJ also discussed the causes of drought with many participants and, although many suggested that God and/or spirits controlled all phenomena, they seemed very interested in the discussion of scientific knowledge regarding climatic cause and effect. It appears Bedamuni may hold many competing understandings concurrently (see Atkinson-Nolte and others, *this volume*), and, as such, may be receptive to external information when it is communicated in a culturally informed and situated manner.

The Missing Link? Acknowledging Unique Ontologies and Epistemologies and Other Lessons for the Future

This chapter has explored some changes to Bedamuni and I-Kiribati ontologies and epistemologies through time and considered how this has influenced their perceptions of hazards and climate change. Through exploring three ontological stages of pre-contact, missionary, and contemporary, we argue that broadly understanding what people believe is a necessary first step in any climate change adaptation or disaster risk reduction project or any external intervention for that matter. While secular (science-based) discourses dominate within aid agencies, they often do not at the local level (Cox et al. 2018, 2019; Luetz and Nunn 2020).

In both case studies, traditional and Christian teachings and beliefs are important factors shaping the worldviews of people, both in the past and today. Yet how this impacts people's understanding of climate change and hazards is strikingly different. I-Kiribati attribute all environmental change, recent and future, to God, drawing on Biblical teachings and promises to inform understandings of climate change. In contrast, the Bedamuni have deep faith in God on the one hand—they attend church and perform Christian burials—yet on the other hand they continue to draw from traditional religion to explain environmental perturbations.

While some have implicated Christianity as imbuing passivity and reducing the capacity for autonomous risk reduction (Ayeb-Karlsson et al. 2019), cultures throughout the Pacific have, throughout history, developed many risk reduction strategies (e.g., surplus creation, elaborate kinship and trading networks) (Campbell 1990, 2014) while simultaneously holding strong traditional faith-based beliefs. Indeed, in both case studies explored in this chapter, people are implementing risk reduction strategies informed by faith and by precedent to both experienced and anticipated climatic and environmental changes. The Bedamuni continue to practice a wide range of absorptive and adaptive strategies in response to El Niño precipitated droughts such as transplanting rhizomes from gardens to swamps, creeks, or rivers to preserve their planting material for when rains return, experimenting with new cultivars and crops, among many other examples (Jackson 2020). People throughout Kiribati are adapting to changing coastal conditions by planting mangroves to stabilise coasts, retreating homes, and planting crops locations of lower exposure. Through utilising this understanding and practice, aid agencies and other development actors can leverage religious beliefs and institutions to identify culturally acceptable climate change adaptation, an approach which has germinated in other Pacific Island contexts (Granderson 2017; Nalau et al. 2018; Nunn et al. 2017; McMillen et al. 2017).

This chapter argues that as a first step, the context-specific worldviews of local people must be broadly understood if climate change adaptation has a chance to have a long-lasting impact in communities. Appropriate pathways and solutions can be decided upon from that point as it is clear a one-size-fits-all approach has largely failed (Korovulavula et al. 2019; Nunn and Kumar 2019b). In the same way, a one-size-fits-all religious approach might be just as likely to fail owing to the variations in culturally grounded ontological perspectives across the region, as shown in this chapter. Rather a deeper understanding of the localised context and worldviews of the people in which climate change adaptation is being implemented is essential as a first step to move toward success and optimal outcomes for those most vulnerable in the Pacific and around the world. Yet we recognised the already constrained budgets of aid agencies and governments which may into the future be further reduced (Nunn and Kumar 2019a).

To conclude this chapter, we suggest opportunities and questions for an evolving research agenda. First, as there is significant diversity in faith-based beliefs within and between Pacific Island cultures, as shown in this chapter, we suggest that projects to build resilience must be specifically tailored to these. To achieve this, a guide detailing the specifics and commonalities in faith-based understandings and approaches, would be a useful first step, especially as more research is conducted throughout the Pacific on faith-based approaches. Additionally, a wealth of ethnographic data already exists in the literature which could be drawn upon for comparative purposes. Future research should seek to devise such an approach in consultation with aid agencies and governments and collate findings from research in different cultures in a comparative fashion. Second, Indigenous cultures throughout the Pacific are already adapting to a changing climate through their unique faith-based frameworks. As such, there presents an opportunity to leverage and learn from these existing strategies. A preoccupation with physicalist adaptations is hindering more nuanced and effective society wide adaptations that have been recognised for many years as more effective. Further, comparative research and a synthesisation of findings into academic and plain language policy documents will be essential. Finally, we pose two questions which could be addressed in future research: how can faith-based understandings and science be reconciled to lead to scientifically informed yet religiously appropriate strategies? And, what do these strategies look like? We suggest that community-led adaptation and experimentation will provide the answers to these questions where external actors can play a facilitation role, rather than driving and informing adaptation. However, the findings cannot be sealed behind paywalls and in obscure journals, agencies, governments, religious institutions, and communities must be made aware of any solutions.

References

Ahmad AN (2019) Disaster cosmologies in comparative perspective: Islam, climate change and the 2010 floods in Pakistan's Southern Punjab. J Hist Sociol 32:311–330

Ayeb-Karlsson S, Kniveton D, Cannon T et al (2019) I will not go, I cannot go: cultural and social limitations of disaster preparedness in Asia, Africa, and Oceania. Disasters 43(4):752–770
Ballard C (2003) Writing (pre)history: narrative and archaeological explanation in the New Guinea Highlands. Archaeol Ocean 38:135–148
Banfield A (2019) Theology of disaster risk management: baseline study: Fiji. Report sponsored by UnitingWorld, Church Agencies Network, and Australia Aid
Barrios RE (2015) Environment as trickster: epistemology and materiality in disaster mitigation. Aust J Disaster Trauma Stud 19(SI):47–51
Barnett J (2010) Dangerous climate change in the Pacific Islands: food production and food security. Reg Environ Change 11:229–237
Beiabure M, Teraku T, Uriam K (1979) Creation: the work of the Gods. Kiribati: Aspects of History. Ministry of Education, Training and Culture, Tarawa, Kiribati
Bird-David N (1999) Animism revisited: personhood, environment, and relational epistemology. Curr Anthropol 40(s1): S67–S91
Campbell JR (1990) Disaster and development in historical context: tropical cyclone response in the Banks Islands, Northern Vanuatu. Int J Mass Emergencies Disaster 8(3):401–424
Campbell JR (2014) Development, global change and traditional food security in Pacific Island countries. Reg Environ Change 15:1313–1324
Cox J, Glen F, Kant R et al (2018) Disaster, divine judgment, and original sin: Christian interpretations of tropical cyclone Winston and climate change in Fiji. Contemp Pac 30:380–411
Cox J, Varea R, Finau G et al (2019) Disaster preparedness and the abeyance of agency: Christian responses to tropical cyclone Winston in Fiji. Anthropol Forum. https://doi.org/10.1080/00664677.2019.1647833
Donner SD (2007) Domain of the Gods: an editorial essay. Clim Change 85:231–236
Donner SD, Kandlikar M, Webber S (2016) Measuring and tracking the flow of climate change adaptation aid to the developing world. Environ Res Lett 11(5):054006/1–054006/9
Eisenbruch M (2020) Why disasters happen: cultural framings from the Diamond Island stampede in Cambodia. Int J Disaster Risk Reduct 44:101423
Falck C (2018) The (re-)appropriation of spirit beings- Spirits of the dead and spirits of God in a Sepik community. Oceania 88(1):107–126
Frankel S (1986) The Huli response to illness. Cambridge University Press, Cambridge
Furedi F (2007) The changing meaning of disaster. Area 39(4):482–489
Gaillard JC, Texier P (2010) Religions, natural hazards, and disasters: an introduction. Religion 40:81–84
Garrett J (1982) To live among the stars: Christian origins in Oceania. World Council of Churches in association with the Institute of Pacific Studies, University of the South Pacific, Fiji
Gheuens J (2017) Landscape and Identity in Kiribati. Dissertation Uppsala University
Granderson AA (2017) The role of traditional knowledge in building adaptive capacity for climate change: perspectives from Vanuatu. Weather Clim Soc 9(3):545–561
Grimble A (1989) Tungaru traditions: writings on the Atoll Culture of the Gilbert Islands, University of Hawaii Press, Hawaii
Haluza-DeLay R (2014) Religion and climate change: varieties in viewpoints and practices. Wiley Interdiscip Rev: Clim Chang 5:261–279
Hardwick J, Stephens RJ (2020) Acts of god: continuities and change in Christian responses to extreme weather events from early modernity to the present. WIREs Clim Chang 11(2)
Hoad RA (1963) Patrol Reports. Western District, Nomad, 1963–1964. National Archives of Papua New Guinea, Accession 469
IPCC (2018) Global Warming of 1.5°C. An IPCC Special Report on the impacts of global warming of 1.5°C above pre-industrial levels and related global greenhouse gas emission pathways, in the context of strengthening the global response to the threat of climate change, sustainable development, and efforts to eradicate poverty. Available via https://www.ipcc.ch/sr15/download/. Accessed 11 May 2020

Ives M (2016) A remote Pacific nation, threatened by rising seas. Available via https://www.nytimes.com/2016/07/03/world/asia/climate-change-kiribati.html. Accessed 14 May 2020
Jackson G (2020) The influence of emergency food aid on the causal disaster vulnerability of Indigenous food systems. Agric Hum Values 37:761–777
Jackson G, McNamara KE, Witt B (2020) "System of hunger": understanding causal disaster vulnerability of indigenous food systems. J Rural Stud 73:163–175
Janif SZ, Nunn PD, Geraghty P et al (2016) Value of traditional oral narratives in building climate-change resilience: insights from rural communities in Fiji. Ecol Soc 21(2):7
Jebens H (2000) Signs of the second coming: on eschatological expectation and disappointment in Highland and seaboard Papua New Guinea. Ethnohist 47:171–204
Johnson K (2019) Climate Change could drown Kiribati, but the nation looks for Noahs Ark. https://www.abc.net.au/news/2019-07-21/kiribati-hopefull-in-face-of-climate-change-rising-sea-levels/11228658?nw=0. Accessed 14 May 2020
Kempf W (2017) Climate change, Christian religion and songs: Revisiting the Noah story in the Central Pacific. In: Dürr E, Pascht A (ed) Environmental transformations and cultural responses: ontologies, discourses, and practices in Oceania. Palgrave Macmillan, New York, pp 19–48
Kiribati National Statistics Office (2016). 2015 Population and housing census. Available via http://www.mfed.gov.ki/statistics/documents/2015_Population_Census_Report_Volume_1final_211016.pdf. Accessed 10 Jan 2019
Korovulavula I, Nunn, PD, Kumar, K et al (2019) Peripherality as key to understanding opportunities and needs for effective and sustainable climate-change adaptation: a case study from Viti Levu Island, FiJi. Clim Dev. https://doi.org/10.1080/17565529.2019.1701972
Laughlin CD (2018) Mana: psychic energy, spiritual power, and the experiencing brain. Time and Mind 11:409–431
Lindstrom L (1990) Big men as ancestors: inspiration and copyrights on Tanna (Vanuatu). Ethnol 29(4):313–326
Luetz J, Nunn PD (2020) Climate change adaptation in the Pacific Islands: A review of faith-engaged approaches and opportunities. In: Filho WL (ed) Managing climate change adaptation in the Pacific region, climate change management, 1st edn. Springer, Switzerland pp 313–320
Macdonald F (2018) 'God was here first': value, hierarchy, and conversion in a Melanesian Christianity. Ethnos 84:525–541
Mackay J, Hoey T (1984) The Biami legends of creation and Noah's flood. Available via https://answersingenesis.org/creationism/creation-myths/the-biami-legends-of-creation-and-noahs-flood/. Accessed 15 Jan 2018
McMillen HL, Ticktin T, Springer HK (2017) The future is behind us: traditional ecological knowledge and resilience over time on Hawai'i Island. Reg Environ Change 17(2):579–592
McNamara KE, Clissold R, Westoby R et al (2020) An assessment of community-based adaptation initiatives in the Pacific Islands. Nat Clim Chang
Nalau J, Becken S, Schliephack et al (2018) The role of indigenous and traditional knowledge in ecosystem-based adaptation: a review of the literature and case studies from the Pacific Islands. Weather Clim Soc 10(4):851–865
Nunn PD, Runman J, Falanruw M et al (2017) Culturally grounded responses to coastal change on islands in the Federated States of Micronesia, northwest Pacific Ocean. Reg Environ Change 17(4):959–971
Nunn PD, Kumar R (2019a) Cashless adaptation to climate change: unwelcome yet unavoidable? One Earth 1(1):31–34
Nunn PD, Kumar R (2019b) Measuring peripherality as a proxy for autonomous community coping capacity: a case study from Bua Province, FiJi Islands, for improving climate change adaptation. Soc Sci 8(8):225
Piggott-McKellar AE, McNamara KE, Nunn PD (2020) Who defines 'good' climate change adaptation and why it matters: a case study from Abaiang Island, Kiribati. RegNal Environ Chang. https://doi.org/10.1007/s10113-020-01614-9

Robbins J (1995) Dispossession the spirits: Christian transformations of desire and ecology among the Urapmin of Papua New Guinea. Ethnol 34(3):211

Robbins J (2001) Secrecy and the sense of an ending: narrative, time, and everyday millenarianism in Papua New Guinea and in Christian fundamentalism. Comp Stud Soc Hist 43(3):525–551

Robbins J (2012) Why is there no political theology among the Urapmin? On diarchy, sects as big as society, and the diversity of Pentecostal politics. In: Tomlinson M, McDougall D (ed) Christian politics in Oceania, Berghahn Books, New York, pp 198–210

Rudiak-Gould P (2010) Being Marshallese and Christian: a case of multiple identities and contradictory beliefs. Cult Relig 11(1):69–87

Sillitoe P (1993) A ritual response to climatic perturbations in the Highlands of Papua New Guinea. Ethnol 32(2):169–185

Sørum A (1982) The seeds of power: patterns in Bedamini Male Initiation. Soc Anal: Int J Anthropol 10:42–62

Sørum A (1991) The Forked Branch: A Study of Meaning in Bedamini Ceremonial. Dissertation, University of Trindheim

Taylor PW (1995) Myths, legends and volcanic activity: an example from Northern Tonga. J Polyn Soc 104:323–346

Thomson B (2013) Kiribati doomed to disappear. Available via https://www.sbs.com.au/news/kiribati-doomed-to-disappear. Accessed 14 May 2020

Tomlinson M, McDougall DL (2012) Introduction: Christian politics in Oceania. In: Tomlinson M, McDougall DL (eds) Christian politics in Oceania. Berghahn Books, New York, pp 1–21

Vaioleti TM (2006) Talanoa research methodology: a developing position on Pacific research. Waikato J Educ 12:21–34

White G (2012) Chiefs, church, and state in Snata Isabel, Solomon Islands. In: Tomlinson M, McDougall DL (eds) Christian politics in Oceania. Berghahn Books, New York, pp 171–197

Chapter 6
Climate Justice and God's Justice in the Pacific: Climate Change Adaptation and Martin Luther

Richard A. Davis

Abstract Climate change activism in the Pacific is often narrated through the concept of justice. It is unjust, it is claimed, that the small island developing states of the Pacific should face so much harm and even eradication from the effects of climate change when they have contributed so little to global carbon emissions. This is an issue of climate justice. Yet islanders, who are mostly Christian, often narrate the effects of the climate change through the concept of God's justice and judgement. In their disaster-prone region on the Pacific rim of fire, with tropical cyclones, tsunamis, and the threat of sea-level rise and coastal erosion, disasters invariably carry religious meaning. This alternative narrative can affect climate change adaptation for several reasons, including the idea that fighting climate change is fighting God's judgement, which must be humbly accepted. And of what use are fortifications against sea-level rise, if God is judging you? To shed light on these issues from a theological angle, the chapter engages with the medieval theologian Martin Luther and how he understood the relationship between God's justice and human action, with surprising resonances in the Pacific today.

Introduction

One of the most pressing issues in the Pacific region is the effects and expected effects of global climate change. The Pacific region is facing in climate change a major existential crisis, with sea-level rise expected to submerge Tuvalu, Kiribati, and many other islands and low-lying regions, and the destruction of other places due to increasingly powerful tropical cyclones (Walsh et al. 2012). The Pacific region is also a mostly Christian region and faith plays a significant part in how people comprehend and react to this existential crisis. Justice plays a major yet ambiguous part in these debates—Pacific peoples see themselves as victims of anthropogenic climate injustice, while also seeing themselves as being subject to divine justice through God's judgement and punishment (Holmgaard 2019).

R. A. Davis (✉)
Wesley House, Cambridge, UK

J. M. Luetz and P. D. Nunn (eds.), *Beyond Belief*, Climate Change Management,
https://doi.org/10.1007/978-3-030-67602-5_6

Climate injustice has been well documented by secular agencies and scholars (Barker 2008; Barnett and Campbell 2010; Fair 2019; Parks and Roberts 2006). This work shows that regions like the Pacific, and countries in those regions which have among the lowest greenhouse gas emissions in the world, but are facing serious consequences if their islands or some coastal areas become uninhabitable or even be completely submerged. Relating God's justice to climate change is less common in global discussions, but cannot be ignored in those regions and nations where Christian adherence is strong and continues to play a key role in how social, political, and in this case, existential issues are conceptualized (Nunn et al. 2016).

In discussing these issues in relation to climate change, I wish to demonstrate a method of bringing theology into this discussion, through focusing on the medieval reforming theologian Martin Luther. In 2017, the church celebrated or commemorated 500 years since the beginning of the Lutheran Reformation in Germany. When Martin Luther nailed his 95 theses to the church door in Wittenberg he sparked a revolution in the Church and changed the direction of history. The effects of his Reformation are still debated today, but Luther's place in history is assured with Lutheran churches having some 72 million adherents around the world. Believing that Luther's theological writings can still speak to contemporary issues, this chapter will ask what Luther can say to Pacific people facing the effects of climate change and climate injustice. In a wider perspective, this forms an example of what could form an approach for bringing theologians and philosophers into climate change discourse. When we look back through history, we can ask: What existential crises did our ancestors face and how did they think them through and adapt to them?

As a theologian located in the Christian Pacific from 2014 to 2020, I have been engaging with students, theologians, and church leaders exploring theological angles on climate change, examining both the Bible and theological traditions for resourcing the churches' teachings and actions on climate change. This research began in a paper presented to the Pacific Theological College's celebration of Luther's quincentenary in 2017. In seeking the relevance of Luther to today's Pacific, I chose to examine his relevance to climate change. While Lutheranism is not widespread in the Pacific, Lutherans are amongst the largest Christian grouping in the Pacific thanks to their numbers in Papua New Guinea.

Luther's Theology of Creation and Climate Change in the Pacific

My hope for this chapter is that Martin Luther's writings might speak to those in the South Pacific grappling with the effects and expected effects of a rapidly changing climate, and the need to adapt to this. First, it is important to have a grasp on how Luther understood the natural world as God's creation and humanity's place in it. Second, I will offer some theological insights based on Luther's writing on creation that might inform a contemporary theological engagement with the issue of climate change. These will be placed alongside more contemporary Christian reactions to climate change events.

Martin Luther was a thoroughly medieval man and a monk in the medieval church. This fact can sometimes be obscured because he initiated a Reformation that changed the world in the direction of the modern age. Yet, this is important to understand as we live in a completely different era. The medieval understanding of the world was that God created an ordered world that operated in a regular way, with a predictable rhythm to life. What scientific investigation there was in his time was conducted on the basis of understanding God's ordered creation, whose truths were accessible to reason. Luther's understanding of creation conformed to this understanding of God's created order. He wrote about the cold winter that would return to summer in an endless cycle. He never expected this to change.

Today climate change is changing this ordered cycle of nature on such a scale that humans are now affecting the geology of Earth. Our worldview is changing as a result, leading some to call our new age the Anthropocene (Deane-Drummond et al. 2017). Humans can affect creation to a degree unimagined in Luther's time. Yet we must concede that climate change does not completely invalidate the notion of an ordered universe. It may be that things will correct themselves over time, but perhaps at the cost of mass extinctions and the mass starvation of human beings.

For Luther, God's ordering of creation is not only limited to an unchanging creation, but also God's providence of creatures. In a lecture on Genesis 35, Luther demonstrated what the doctrine of creation implies for God's providence in times of trouble. In Genesis 35:5, Jacob is traveling to Bethel through the hostile land of Canaan. The verse reads: "As they journeyed, a terror from God fell upon the cities all around them, so that no one pursued them". This, Luther claimed, demonstrates the Creator's protection of Jacob (*LW* 6:241). Luther said that if we learn the doctrine of creation correctly, which is that "all things are in God's hands", then we should trust our Creator in all things. Luther presses the point home: "For if we firmly concluded that God is our Creator, we would certainly believe that He has heaven and earth in his hands and all things which are contained by these" (*LW* 6:242–243). Luther thought these words offer comfort—harm will occur only if God wills it. Here Luther cited Psalm 31:15 ("My times are in your hand; deliver me from the hand of my enemies and persecutors.") to emphasize that all things are up to God (*LW* 6:243).

Throughout his corpus, Luther often used images of winds and threatening seas to talk about dangerous situations faced by humanity. Here he likened the potential of a Canaanite attack against Jacob to rising and threatening seas. Luther also compares God's protection to his power to calm waters, citing Psalm 65:7:

> You silence the roaring of the seas,
> the roaring of their waves,
> the tumult of the peoples. (*LW* 6:243)

Luther drives this metaphor further: "We live in the midst of the sea itself, and as the waters of the Flood were 15 cubits higher than the loftiest mountains, so also now the ocean is three ells [cubits] higher than the earth" (*LW* 6:243). Why aren't we, asked Luther, overwhelmed by this trouble? "Because", Luther continued, "God has set a limit for the sea (cf. Prov. 8:29). Its waves, indeed, are bellowing and crashing against the shores just as if they were about to rush out onto the land, but their attack is checked and restrained by God" (*LW* 6:243). This was Luther's commonly used metaphor for God setting a limit to personal afflictions or social chaos.

Among those who may doubt that God has set a "limit for the sea", are those Pacific peoples, who, threatened with rising seas, have built seawalls or other fortifications around their village or island. On this topic, Luther said that we must have faith in God in the face of threats, writing that, "We should learn to place our trust in God our Creator and Protector, not in fortifications, guns, wisdom, and power" (*LW* 6:245). Luther said that fortifications are worthless if God wills our destruction. So instead of building seawalls, planting mangrove belts, or relocating villages, should Pacific Islanders instead leave it up to God to protect them? From Luther's perspective, such adaptations can be pursued, but they will only work as intended if God wills them to; without God's blessing, such adaptations to sea-level rise may prove to be a complete waste of money.

This Christian view of managing risk is a traditional one, but one invariably at odds with the secular technological solutions proposed by the international community and Pacific governments as they try to adapt to climate change. If climate change induced sea-level rise is seen as God's divine judgement in the Pacific, then, in the Christian mindset, no fortification will be able to defend against it, unless God is appeased. While this may appear to be a conflict of worldviews, it need not be so in practice. Some Pacific Christians believe that "spiritual preparedness" for climate changes is important (Cox et al. 2019, pp. 7–8), and there seems no reason why this cannot be linked to material preparedness. For example, it would be possible to build a sea-wall or plant mangroves as defences against sea-level rise and to have it prayed over by local ministers of religion, petitioning God that they live lives worthy of God's protection.

The Flood

Facing rising seas, some Christians in the South Pacific find comfort in the narrative of Noah (Genesis 6–9) and the idea that islands *cannot* be flooded because of the 'Rainbow Covenant' God made with Noah and his descendants (Mortreux and Barnett 2009; Davis 2015). Confidently held, this perspective could hamper further climate change adaptation efforts in the Pacific, as it could be claimed that the Pacific is already under divine protection. The Covenant reminds humanity of the promise to Noah that his children will populate the earth. Noah was sure about that. He was also reminded by the Rainbow, as humanity is, of the past destruction of the Earth. Christians are not only reminded of it, but also know that they will be spared that particular judgement of God in the future. Christians believe that they may, however, have to suffer other judgements of God on their sin.

Luther would agree that the whole Earth will not be flooded again. He commented that we are fortunate that we have escaped the flood, but we remain under God's judgement, and in future he thinks fire will consume the earth (*LW* 2:145, 149). To those Pacific Islanders confident that the Rainbow Covenant means that God will not allow rising seas to flood their land, Luther would caution that we should not be so confident in simplistic readings of God's promises. In his commentary on Genesis chapter seven, Luther entertained the idea that throughout scripture people have maintained false confidence in the true promises of God. This warning should make Pacific Christians wary of any hope placed in the Rainbow Covenant (*LW* 2:98–102).

To illustrate Luther's point, imagine that you are living between the times of Adam and Noah. At that time, could you believe that a flood would destroy the earth? In light of God's words given to Adam and Eve to rule the earth (Genesis 1:26–28), one would take for granted that the world must survive for it to be ruled by humans. For Adam's other descendants, Noah's actions in building the ark in preparation for the flood must have seemed strange if not heretical (*LW* 2:98). People would have been confident that the earth could never be completely flooded. And yet according to Genesis it was.

In the light of this, Luther asked how the idea of God's creation and man's dominion over the earth and commandment to fill and subdue the earth (Genesis 1:26–30) can be reconciled with the fact that God destroys humanity and its work except for the eight souls in the Ark (*LW* 2:98). Can we harmonize the flood with God's intention for humanity's domination over creation? (*LW* 2:99). Luther's answer was that God preserves things in ways "that the world neither sees nor understands" (*LW* 2:99). In the flood, God removes humanity's dominion, "not forever but for a time and then not even fully" (*LW* 2:100). According to Luther, therefore, God's promise was not a lie, but was kept in God's way. Similarly, those relying on the promise of the Rainbow Covenant could be in for a surprise when they realize that God may flood their part of the earth and still preserve peoples in other places as exiles or refugees.

Part of the punishment of the flood, according to Luther, was the human losing dominion over the creatures given in creation (*LW* 2:58–59). This has direct relevance to our environmental crisis and climate change. Scientists have described how climate change is increasing extinction rates and the loss of biodiversity (Taylor and Kumar 2016). Accordingly, with climate change the sphere and scope of humanity's dominion over creation (Genesis 1:28) is reducing. Ironically, as humanity gains further technological, military, and economic domination over the biosphere and atmosphere, and as we enter the Anthropocene, we are, from Luther's perspective, losing power, in so far humanity's God-given dominion is over an ever-shrinking pool of species (*LW* 2:100).

Together with contemporary Pacific theologians frustrated by unhelpful readings of the Noah narrative (Davis 2015; Rubow and Bird 2016, p. 159), Luther would have little time for Pacific Islanders who do not appreciate the need to adapt to climate change based on simplistic readings of the Rainbow Covenant. Even if Pacific misreadings of the Noah narrative are an obstacle to climate change adaptation, it is not possible to suppress the Noah narrative. Instead it might be an exegetical strategy for to try to shift the focus of the lesson here away from the protection that the

Rainbow Covenant gives to survivors of the flood, to the preparations Noah made in order to adapt to the impending disaster (Fair 2018).

Part of the reason why Noah's flood is so important in the Pacific is the strong cultural attachment Pacific peoples have to their land. This helps to explain why God's protection of that land is sought through the Rainbow covenant. Yet as some scholars have noted, this attachment to place can be an obstacle to village relocation to less vulnerable locations (Nunn and McNamara 2019, p. 32). Theology can help here, as Luther demonstrates in his promotion of appropriate Christian focus on being heavenly over being earthly. In his lectures on the letter to the Hebrews, as just one example, Luther writes

> to be heavenly is to despise the visible things, yes, even their images, and to cling to God alone, the divine Good, that is, the divine will, in prosperity as well as in adversity, in life and in death. To be earthly is to despise the invisible things, that is, the divine will, and to cling to the visible things, to set one's mind on the good fortune of the world. (*LW* 29:216)

Luther clearly condemns clinging to the things of the world, which would include one's land. This theme is seen elsewhere in Luther when he comments on the following relevant scriptures of 2 Corinthians 5:1 (see *LW* 9:157), Ephesians 2:19 (see *LW* 11:124; 22:203), Philippians 3:20 (see *LW* 12:81, 105; 22:203; 26:439–40; 39:219), and Hebrews 13:14 (see *LW* 11:124). Among the Christians of the Pacific, teaching and preaching on these passages of scripture could encourage a heavenly detachment from place which will allow for easier transformational adaptations, such as village relocation.

Creation's Suffering and Climate Change

I now wish to consider climate change related suffering experienced by human beings and other creatures. Climate change has already caused human suffering and projections are that sufferings will only increase in the future (Ripple et al. 2020). The South Pacific can expect sea-level rise causing land inundation and erosion, stronger cyclones, other impacts related to the spread of disease, and increased heat stress on humans, plants, and animals. Pacific peoples are already losing their homes and islands to sea-level rise. Some lost their lives in Tropical Cyclones Pam (2015), Winston (2016), and Harold (2020).

As climate change-related events, strong winds, floods, and rising seas were unknown to Luther. But he did experience storms; his decision to become a monk was made during a thunderstorm which threw him to the ground (Bainton 1950, p. 21). In Luther's writings, suffering due to floods and winds often became metaphors for understanding human suffering in the Bible and in his lifetime. In a sermon on Romans 8:18–22, Luther agreed with Saint Paul that we should put our current sufferings in perspective, as they are nothing compared with our future glory. Luther commented that:

> If one could conceive the nature of this reward he would be compelled to say: "Were it possible for me to suffer ten deaths by fire or flood, that would be nothing in comparison to the future life of glory. What is temporal suffering, however protracted, contrasted with eternal life? It is not worthy to be called suffering or to be esteemed meritorious." (Luther 1909, p. 99)

At this point, Luther may have anticipated the objection of Pacific leaders who claim that the people of the South Pacific nations are victims of a great climate injustice—they suffer the most from climate change, even though they contribute the least to it. This is based on the existential crises faced by many Pacific Islands nations and their relative climate innocence (based on their low per capita greenhouse gas emissions). On the face of it, this claim seems justified, but problems arise here when this narrative of innocence is articulated theologically; for if it is God who is punishing the Pacific with climate change, then this questions God's justice, for God appears to be punishing an innocent people. Furthermore, we might question, from a theological perspective the implicit assumption that a just God must match the punishment (climate change) with the sin (carbon emissions). This conundrum will be discussed further below; in the meantime we might note two ways around this. First, get rid of the idea of God's punishment altogether, or; second, note that throughout church history theologians and preachers have matched all manner of sins with all kinds of natural and anthropogenic disasters used by a wrathful God to punish humanity. For the believer in God's wrath, then, the soundness of the Pacific Islanders' complaint of injustice is found in the relationship between God's divine wrath and the sins which prompt it.

Luther wrote this of those suffering and crying out about injustice done to them:

> It ill becomes them to make extravagant complaint and outcry about injustice. "But," you say, "I have truly suffered injustice." Very well, so be it. But why do you make so much of your sufferings and never give a thought to what awaits you in heaven? Why not exalt the future glory also? If you desire to be a Christian, truly it will not do to conduct yourself in this impatient manner. If you must air your grievances, surely you may do it quietly and decorously. (Luther 1909, pp. 99–100)

What does this mean for the victims of climate injustice? First, according to Luther, that they should put their suffering in perspective of the coming redemption of creation and themselves. However, such an otherworldly perspective on current human suffering will not be terribly popular today. Even if contemporary Christians can agree with Luther, that in the end we will be better off in heaven, many will rightly ask whether God's justice extends to this life too. The second lesson is that they *should* air their grievances, even in subdued ways. Luther does not advocate quietism in the face of injustice.

Furthermore, according to Luther, it is not only humans who suffer the consequences of climate change, other creatures do as well. The land, sea, coral, plants, and animals are all under stress and suffering from loss of habitat or the risk of extinction. Creation suffers quietly and Luther pointed out that humans can learn from the quiet suffering of other creatures. In a sermon on Romans 8, he wrote:

> Paul holds up as an example to us the condition of the whole creation. He exhorts us to endure patiently, as the creature does, all the violence and injustice we suffer from the devil

> and the world, and to comfort ourselves with the hope of future redemption. Remarkable doctrine this, unlike anything elsewhere found in the Scriptures, that heaven and earth, sun, moon and stars, leaf and blade, every living thing, waits with sighing and groaning for the revelation of our glory. (Luther 1909, p. 103)

In another sermon on Romans 8, Luther named the suffering of creatures as that service they provide to the unjust and the wicked (Luther 1909, p. 109). Using the sun as an example, he suggests that this "noble and admirable work of creation" is "fit to serve only God, angels and pious Christians, who thank God for it." "But", he continued, "it must serve those who blaspheme and dishonor God and who are guilty of all wickedness and lawlessness" (Luther 1909, p. 109). So while creation serves God obediently, it suffers the indignity of serving the Devil and the wicked (a similar viewpoint is seen in Edwards 1741, p. 13). Enduring all sorts of injustices, Luther called creatures martyrs (Luther 1909, p. 110). Luther's lesson is that if the innocent sun can endure an injustice, then so can we. Yet merely accepting this injustice, and suffering patiently as creatures do, may lead to a fatalistic attitude in the face of climate change and hamper adaptation efforts, of which humans alone are capable of.

Similar to Luther, most Pacific Christians believe in the afterlife and the end to earthly suffering. This may discourage climate adaptation of a technological kind that preserves life on earth here. After all, the earthly attachments Christians have, and the desire to protect and secure them, distracts them from appropriate focus on future glory. As a spiritual adaptation to climate change, this is out of step with the care for creation focus of ecotheology, and might be called fatalistic fundamentalism (Rubow and Bird 2016, p. 159). In addition to Luther's point on what humans can learn from the patience of creatures, it might help adaptation efforts in the Pacific to emphasize that creatures suffer from our injustice toward them. If, therefore, Pacific Christians take seriously their cosmological and theological notions of their interconnectedness with all creation, and their responsibility to other creatures as brothers and sisters (cf. Kasulis 2002, p. 123), then adaptive behavioral change could be encouraged in order to reduce the suffering of other creatures, such as the air, land, sun, and sea due to human sin.

Fighting Climate Change, Fighting God

For the church in the Pacific and elsewhere, the worst case scenarios related to the effects of climate change make for apocalyptic reading. Sea-level rise and more frequent and stronger cyclones will be devastating to Pacific island nations. Elsewhere, heat waves and drought could have catastrophic effects on the global population and its food supply.

What might Luther have to say to this doomsday outlook? To answer this, we would ask whether Luther and his society faced their own existential crisis analogous to what Pacific nations face with climate change, and how he responded to that crisis. The answer is yes, he faced the existential threat of invasion by Ottoman Turks who loomed on Europe's horizon and threatened to overthrow Christendom. Europeans of the time feared the Turks unleashing upon them an unprecedented reign of terror, violence, and bloodshed. In light of this threat, Luther continually wrote about the Turks. Here I will consider whether these writings offer any help to Pacific peoples facing a very different threat to their existence.

In response to requests to write about the issue, Luther penned a small treatise "On War against the Turk" which mainly deals with the issue of Christians waging war on the Turk (*LW* 46:153–205). A part of this work deals with the question of what this threat is and where it comes from. Before examining these questions, it is worth exploring briefly Luther's notion of punishment by God. In his "Explanations of the Ninety-five Theses" (1518), Luther outlined several kinds of punishments which faithful Christians can suffer. They include "eternal punishment" in hell, purgatory, "spiritual penance" (in accordance with 1 Corinthians 11:31), "God's correction and scourging", and the "canonical punishment" of the church (*LW* 31:90–93). It is the fourth punishment, "God's correction and scourging", that interests us here. Based on various scriptures, Luther showed that God's punishment will be visited on the sinner (Psalm 89:30–33) and righteous alike (Revelation 3:19).

He went on to say that Christians need to recognise "plagues, wars, insurrections, earthquakes, fires, murders, thefts, as well as the Turks, Tartars, and other infidels" as "the lash and rod of God" (*LW* 31:91–92). Accepting this, then, church leaders who wished to fight against the Turks (or any other of these things) took on a serious fight, not only against the Turks, but also against God. Luther wrote: "They want to fight, not against iniquities, but against the lash of iniquity and thus they would oppose God who says that through that lash he himself punishes us for our iniquities because we do not punish ourselves for them" (*LW* 31:92).

In the face of the perceived threat of the Turk, Luther was determined that Christians should not surrender their faith. He urged prayer as a weapon against this threat. Surprisingly, Luther, in responding to the justified fears of ordinary people to this threat, did not draw attention to God's providential protection, instead he urged preaching on the wrath of God in punishing people (*LW* 46:171). To be effective, this preaching, Luther suggested, ought to focus on examples from the Bible that shows God's punishment toward his people, such as the flood (Genesis 7:1–24) and the punishment of Sodom and Gomorrah (Genesis 19:24–28). And since, Luther said, people now sin worse than people in biblical times, worse now are the punishments of God (*LW* 46:171).

Luther's counsel to those who faced this threat from the Turk was that the Turk must be disempowered by decoupling their military might from God's wrath. So long as the Turk was a tool of God's wrath, they could not be defeated. God's wrath can, however, be quelled through repentance and prayer. Luther wrote: "This fight must be begun with repentance, and we must reform our lives, or we shall fight in vain" (*LW* 46:171). The wise Christian will "recognize the Turk as God's rod and wrath which Christians must either suffer, if God visits their sins upon them, or fight against and drive away with repentance, tears, and prayer" (*LW* 46:184).

Applying this to climate change is not straightforward, but it is worth noting that the Turks were not the only instrument of God's wrath that Luther recognized during his lifetime. According to Luther, God uses a wide range of natural events and forces of nature to discipline sinful humanity. We have already seen that Luther wrote that God uses earthquakes and fires to discipline humanity. For example, Luther, commenting on Psalm 1:4 ("The wicked are not so, but are like chaff that the wind drives away"), described the wind as "the force of God's wrath and the breath of His anger" (*LW* 10:23). The same goes for hail as in Isaiah 28:17 ("And I will make justice the line, and righteousness the plummet; hail will sweep away the refuge of lies, and waters will overwhelm the shelter") (*LW* 16:231). Luther wrote that "big thunderstorms, lightning, floods, and fire in the skies and on earth are signs", but he warned against too much speculation into the meanings of these natural phenomena:

> But how this happens, or what natural powers are at work here, or what secret processes are active, these are questions that the magi investigate and juggle with; they are of no use to you and not necessary for you to know. It is enough that you should recognize God's wrath in these things and lead a better life. (*LW* 52:168)

So, in this life, according to Luther, God uses creation and natural disasters to urge humanity to repent and turn toward God. Furthermore, Luther was not interested in scientific analyses of the causes of weather events or *how* these things occur. Neither was he much interested in the question of *why* they occur. He already knew that God caused these events and that, given their destructive force, God caused them in order to discipline humanity.

For many people today, it would be mistaken to attribute the problems of climate change to God's wrath. For example, theologians and church leaders (such as Cliff Bird) attributed Cyclone Winston to nature's freedom or scientific explanations (Bhagwan 2016). The Pacific regional ecumenical body also rejected the notion that God lies behind climate change. In their Otin Taai Declaration of 2004, the Pacific Conference of Churches wrote:

> We would like to say a word about God's promise to Noah not to flood the earth again. Some Christians view this covenant as a guarantee that they are not at risk of flooding from climate change. But the sea level is rising and threatening Pacific Islands with flooding from high tides and storm surges. This is not an act of God. It is a result of human economic and consumer activities that pollute the atmosphere and lead to climate change. Most of these polluting emissions come from highly-industrialized countries. (Pacific Conference of Churches 2004, p. 3)

If this analysis is correct then these human activities can become the focus of political and social action. Indeed, one popular slogan among climate change activists in the Pacific is "We are not drowning. We are fighting" (https://350pacific.org), which challenges the narrative that Pacific people are victims who can do nothing but accept their fate as sea-level rise destroys their lands and cultures. After Cyclone Winston the slogan in Fiji was "Stronger than Winston", which aimed at boosting people's resilience in face of the destruction that the cyclone left in its wake (Finau et al. 2018, p. 130). Not everyone embraced these slogans. For those Christians who believed that Cyclone Winston was God's punishment, to be stronger than Winston was claim that it stronger than God's instruments of wrath and even God himself (Cox et al. 2018, pp. 389, 392). For those Christians, this attitude does not show the appropriate humility in the face of God's discipline, and threatens an even worse catastrophe.

If the Otin Taai Declaration was intended to change the thinking of Pacific Christians about the causes of climate change, it has not been successful. The notion that God did not directly cause the Cyclone, and that climate change was responsible, were dismissed by some Pacific Islanders who suggest that God works through climate change (Cox et al. 2019). An article in *The Contemporary Pacific* records the disdain one Fijian had for the *Fiji Sun* newspaper, which apparently expressed contempt for the divine wrath narrative:

> Some religious-minded people, like me, are beginning to or had already analysed the two natural disasters that have befallen Fiji as driven by the wrath of God. The unbelieving *Fiji Sun* and many others have called this thought idiotic and blamed it on climate change. Lest these unbelievers do not realise that even climate change is the wrath of God given that mankind has disturbed the balance of nature by his stupid manipulations of the environment. Surely the *Fiji Sun* is not as naive as it proclaimed. However, from a biblical context these calamities are similar to what befell the Israelites when they turned away from God and worshipped idols and sinned in front of God. (cited in Cox et al. 2018, p. 400)

This is not an uncommon view among Pacific Christians. *The Contemporary Pacific* article (Cox et al. 2018) records many other similar perspectives to this. Focusing on the aftermath of the Category 5 Tropical Cyclone Winston, which struck Fiji on 20 February 2016, the authors document some of the many explanations for the Cyclone. Even though the Pacific nations are apparent victims of climate injustice, Pacific Islanders often fully blame themselves for such natural disasters. To many Fijians, the causes for Cyclone Winston and other disasters were human sin. Within the context of Fiji, ordinary Fijians highlighted the following sins which they believed precipitated the Cyclone and its selective impacts:

- the current Bainimarama government's 2013 constitution, which makes Fiji a multi-racial and secular state (pp. 390, 397, 398)
- grog or kava drinking along with sabbath violations on Koro Island before it was heavily devastated by Cyclone Winston (p. 391)
- "breaking of a particular agreement or covenant between God and Fijians" (p. 393)
- general rebellion against God (p. 393)
- the non-removal of unfaithful leaders (p. 395)
- misuse of power for self-benefit instead of returning Fiji to the correct path (p. 398).

There is a clear political or moral interpretation given here. It is notable that there are no ecological sins mentioned, thereby disconnecting ecological sin from ecological punishment.

In another example that illustrates this, the authors mention the floods in the city of Nadi in 2012. Positioned near resorts frequented by vacationing Australians and New Zealanders, some Fijians attributed the flooding to God's punishment of "the permissive sexual behavior allowed by the tourist industry" and especially homosexuality (Cox et al. 2018, p. 390). At the grassroots level, then, much more work needs to be done if the church wishes to convince Christians that God is not responsible for the disastrous effects of climate change.

As the authors of the article note, there is a narrative of blame and divine punishment in explaining disasters, which may be articulated in the language of justice. Clearly some Fijians believe that God is just in punishing Fijians for their sins and using a cyclone to do so. Whereas some people might think that the Fijians are being harmed by a climate injustice—caused by the industrialized counties, Fijians might focus on themselves as being justly punished by God for their transgressions, or those of their leaders.

Another narrative in the Pacific, perhaps related, is that Pacific peoples are innocent and climate change results from the ecological sins of others. Why would God punish the innocent relatively harmless Pacific peoples, whose greenhouse gas emissions are very low? It makes no sense. For some Christians this view might appear impudent, as it challenges the wisdom and justice of God. Luther would remind us saying that all have sinned and deserve God's punishment (LW 31:347). Some Pacific leaders may need to consider, as some have, that the Pacific peoples, while they are victims of a global injustice, are not completely innocent either. One thing is sure, we can all do better to reduce our ecological footprint. Indeed, some Pacific peoples have taken this approach, accepting God's discipline, recognizing that they need to make ecological amends (Rudiak-Gould 2014, p. 368).

Across the Pacific, God's wrath has been a popular explanation for events such as Cyclone Winston and the Samoa Tsunami of 2009. Clearly many Christians in the Pacific would agree that Luther's approach is largely correct, both in diagnosis and in way to combat climate change. Christians who agree with this broad approach could take up arms against climate change, but they might ask, in agreement with Luther, of what use is political campaigning and climate change adaptation if we do not repent of our sins first? Based on Luther's reasoning, if this is not done before fighting climate change, then the fight is also against God. Once nations repent of their sins, they would argue, then God will stop using climate change events to punish them, and climate change can be battled without also, at the same time, battling God.

Critics of this popular theological view cannot explain why the Cyclone hit where it did and with such intensity. It is God's wrath? Clearly some Pacific Islanders, agreeing with Luther, would say that it is, meaning that the effects of climate change, such as cyclones, are tools of God's wrath on the sin-filled lifestyles of Pacific Islanders. If this is in fact so, then urging repentance to combat climate change disasters is a sound strategy. Pacific Christians could be encouraged to "reform" their lives. Many of them would agree that only then can the fight against climate change be effective and adaptation meaningful. To make a more direct impact on climate change people would have to convinced to change their environmentally damaging activities. However, in the hands of conservative church leaders, such an approach might well cause other social harm, such as against members of sexual minorities and against individual human rights, such as religious freedom.

All this seems at odds with the modern scientific view. Nowadays it is not magi who ponder the mysteries of the natural world, it is scientists who have largely discovered the processes at work in climate change, even if their work is incomplete in understanding our complex climate. To be sure, they have done so using a rational approach that excludes divine explanations and the factors of sin and repentance. For Christians in the Pacific who believe in God's power to act in history through climate change, scientific or meteorological explanations alone are not enough (Cox et al. 2019, p. 10). It is true that we have better scientific understanding of our climate than in Luther's day, but I'm not convinced that our theology has advanced at the same rate. Nevertheless, given the theological state of affairs in the Pacific, as discussed in this chapter, it would be unwise to discount the beliefs of Pacific Christians who continue to believe that God is involved in our weather patterns. Christianity remains a very strong influence in the community and political life of Pacific nations. To be effective in dealing with climate injustice and embracing climate change adaptation, one will have to take account of people where they are and will need to appreciate how theology influences the thoughts of Pacific Christians toward climate change and how these might influence adaptative actions.

Believing, as do many Christians in the Pacific, that God controls the weather and climate and can use climate change to discipline sinners, is not necessarily fatalistic or fundamentalist, it can be a demonstration of a "humble disposition in relation to God" (Cox et al. 2019, p. 13). There are practical actions that follow from this disposition, such as repentance from sin of all kinds (including ecological sins) and also actions that reinforce other adaptative actions, such as praying for guidance and waiting on God. "Embedded biblical theologies", as Rubow and Bird call them (2016, p. 164), therefore do not necessarily stand in the way of climate action or adaptation. From the writings of Rubow and Bird and others, it appears that theology and the grassroots faith of Pacific Christians have to change in order to meet the urgent activist needs of the present climate crisis. At worst, this means ignoring theology and lived Christianity and simply allowing oneself to be guided solely by science and politics, reducing Christianity in this field to advocacy for the climate justice, leaving one's faith commitments to the side. This vanguardist approach to doing theology risks leaving a lot of church members, particularly those in outer islands and villages far from the influence of urban liberalism, out of the conversation. Returning to the works of Martin Luther (and other theologians) can offer Christian teachings from a Christian worldview more compatible with the theologies and worldviews of grassroots Pacific Christians. They might not fully agree with Martin Luther's medieval theology in radically changed and more modern times, but his theology does offer some ways of thinking through climate change that can help the Christian Pacific adapt to a changing climate in relation to their faith commitments.

Conclusion

In this chapter I have introduced some insights from the Reformer Martin Luther on creation and divine justice that may help inform our understanding of how some Pacific Christians might react to climate change, sea-level rise, and severe weather events. Many Pacific Islanders rightly see themselves as victims of human created climate injustice, while at the same time seeing in weather events a working of divine justice against their sinfulness. I have shown how Luther's works might speak into current climate debates in the Pacific, by considering how he viewed the existential crises of his time.

As discussed in this chapter, this is done both negatively and positively. From the negative side, Luther can help Pacific Christians remove obstacles to climate adaptation. As shown above, reading the narrative of Noah differently can help to remove the important obstacle to climate inaction, being an unhelpful belief in God's protection through the Rainbow Covenant. Likewise, Luther promotes a heavenly focus for Christians that can assist in undermining worldly attachments to place that sometimes hinders village relocation. From a more positive angle, Luther's theology provides the Pacific churches reasons to supporting material adaptation spiritually, rather than see the material adaptation and God's protection as opposed. Luther also encourages seeing the interconnectedness of all creation, with important implications for human behavior toward other creatures, including viewing the land and sea as bothers and sisters worthy or love and protection.

Even though he lived 500 years ago, Martin Luther's voice can speak to current debates about how Christians might think about climate change. Of course, we must acknowledge that the world is a very different place in the twenty-first century from the sixteenth century. This is especially true when it comes to explaining weather and climate events, which is now done scientifically. While we can appreciate the valuable work of scientists in this work, it can leave people wondering where God is in all this. In adapting to climate change in the Pacific an acknowledgment must be made of the faith of the people on the ground and the importance this plays in their understanding of weather events and climate disasters. By doing this, theology can help climate change adaptation in the region by cutting with the grain of lived Christian faith. There remains, however, still much more work for the region's churches and theologians to do in relating Pacific cultures, Christianity, and climate change.

References

All biblical references are to the New Revised Standard Version

Bainton R (1950) Here I stand: a life of Martin Luther. Abingdon Press, New York

Barker HM (2008) The inequities of climate change and the small island experience. Counterpunch, Nov 4, 2008. https://www.counterpunch.org/2008/11/04/the-inequities-of-climate-change-and-the-small-island-experience/. Accessed 19 Nov 2020

Barnett J, Campbell J (2010) Climate change and small island states: power, knowledge and the South Pacific. Earthscan, London

Bhagwan J (2016) God and natural disasters. Fiji Times Wednesday, Mar 2, 2016. https://www.fijitimes.com/god-and-natural-disasters/. Accessed 19 Nov 2020

Cox J, Finau G, Kant R, Tarai J, Titifanue J (2018) Disaster, divine judgment, and original sin: Christian interpretations of tropical cyclone Winston and climate change in Fiji. Contemp Pac 30(2):380–410. https://doi.org/10.1353/cp.2018.0032

Cox J, Varea R, Finau G, Tarai J, Kant R, Titifanue J, Andreas Neef A (2019) Disaster preparedness and the abeyance of agency: Christian responses to tropical cyclone Winston in Fiji. Anthropol Forum, 1–16. https://doi.org/10.1080/00664677.2019.1647833
Davis R (2015) The rainbow covenant, climate change, and Noah's exile. Pac J Theol II 54:37–44
Deane-Drummond C, Bergmann S, Vogt M (eds) (2017) Religion in the anthropocene. Cascade Books, Eugene, OR
Edwards J (1741) Sinners in the hands of an angry God. a Sermon preached at enfield, July 8th, 1741 (ed: Smolinski R). Electronic Texts in American Studies. 54. https://digitalcommons.unl.edu/etas/54. Accessed 19 Nov 2020
Fair H (2018) Three stories of Noah: navigating religious climate change narratives in the Pacific Island region, Geo: Geography and Environment. 2018; e00068. https://doi.org/10.1002/geo2.68
Fair H (2019) From apathy to agency: exploring religious responses to climate change in the Pacific Island region. In: Klöck C, Fink M (eds) Dealing with climate change on small islands: towards effective and sustainable adaptation. Göttingen University Press, Göttingen, pp 175–194. https://doi.org/10.17875/gup2019-1216
Finau G, Cox J, Tarai J, Kant R, Varea R, Titifanue, J (2018) Social media and disaster communication: a case study of Cyclone Winston. Pac JIsm Rev: Te Koakoa 24(1):123–137. https://doi.org/10.24135/pjr.v24i1.400
Holmgaard SB (2019) The role of religion in local perceptions of disasters: the case of post-tsunami religious and social change in Samoa. Environ Hazards 18(4):311–325. https://doi.org/10.1080/17477891.2018.1546664
Kasulis TP (2002) Intimacy or integrity: philosophy and cultural difference. University of Hawai'i Press, Honolulu
Luther M (1909) Luther's Epistle Sermons, Trinity Sunday to Advent (trans: Lenker JN). The Luther Press, Minneapolis
LW: Luther's Works. Luther, M (1968–1986) Edited by Jaroslav Pelikan (vols 1–30), Helmut T. Lehmann (vols 31–54) and Joel W. Lundeen (vol 55). Concordia, St Louis; Fortress Press, Philadelphia
Mortreux C, Barnett J (2009) Climate change, migration and adaptation in Funafuti, Tuvalu. Glob Environ Chang 19(1):105–112. https://doi.org/10.1016/j.gloenvcha.2008.09.006
Nunn PD, McNamara KE (2019) Failing adaptation in island contexts: the growing need for transformational change. In: Klöck C, Fink M (eds) Dealing with climate change on small islands: towards effective and sustainable adaptation? Göttingen University Press, Göttingen, pp 19–44. https://doi.org/10.17875/gup2019-1210
Nunn PD, Mulgrew K, Scott-Parker B et al (2016) Spirituality and attitudes towards nature in the Pacific Islands: insights for enabling climate-change adaptation. Climatic Change 136:477–493. https://doi.org/10.1007/s10584-016-1646-9
Pacific Conference of Churches (2004) The Otin Taai declaration. World Council of Churches Office in the Pacific, Suva. https://archive.org/details/2004KiribatiTheOtinTaaiDeclaration. Accessed 19 Nov 2020
Parks B, Roberts J (2006) Globalization, vulnerability to climate change, and perceived injustice. Soc & Nat Resour 19(4):337–355. https://doi.org/10.1080/08941920500519255
Ripple WJ, Wolf C, Newsome TM, Barnard P, Moomaw WR (2020) World scientists' warning of a climate emergency. Biosci 70(1):8–12. https://doi.org/10.1093/biosci/biz088
Rubow C, Bird C (2016) Eco-theological responses to Climate change in Oceania. Worldviews: Glob Relig, Cult, Ecol 20(2):150–168. https://doi.org/10.1163/15685357-02002003
Rudiak-Gould P (2014) Climate change and accusation: global warming and local blame in a small island state. Curr Anthropol 55(4):365–386. https://doi.org/10.1086/676969
Taylor S, Kumar L (2016) Global climate change impacts on Pacific Islands terrestrial biodiversity: a review. Trop Conserv Sci 9(1):203–223. https://doi.org/10.1177/194008291600900111
Walsh KJE, McInnes KL, McBride JL (2012) Climate change impacts on tropical cyclones and extreme sea levels in the South Pacific—A regional assessment. Glob Planet Chang 80–81:149–164. https://doi.org/10.1016/j.gloplacha.2011.10.006

Part III
Religion as Culture, Religion for Wellbeing

Chapter 7
Cyclone Winston: Catholic Women's Faith and Agency in a Coastal Village in Fiji

Shirleen Sahai, Tammy Tabe, and Jacqueline Ryle

Abstract This paper concerns experiences of women in Tokou, a coastal village on the island of Ovalau, Fiji, during Tropical Cyclone Winston in 2016. It discusses how Tokou women, reflecting on how they acted during Cyclone Winston, see their Catholic faith as the source of their strength and survival. Their belief that God would help and sustain them enabled them to support their families emotionally and spiritually. Bringing families together for prayers and devotions, such as praying the Rosary to the Virgin Mary in homes and in the village evacuation centre, women recreated in the midst of chaos a familiar, ritualized and empowering devotional order. Women's central roles during the cyclone and after—caring for children and the elderly, cooking, cleaning the debris, cleaning homes, washing clothes, walking to town to get medicine, fishing—were all enacted in the belief that God was present during everything that happened and was guiding their actions. It is essential, we argue, that Disaster Risk Reduction studies and implementation take into account the hugely empowering efficacy of people's faith and spirituality, how it informs their decisions and actions, and women's core roles in disaster situations and post-disaster rebuilding of everyday life.

> It was the middle of the afternoon and Tropical Cyclone Winston was battering Tokou. The winds were powerful and heavy rain pelted down from a dark sky. The first tidal surge hit. From her kitchen window, Reijeli[1] saw branches and roofing iron flying and waves pouring across the road into the village. She couldn't bear to look and turned away from her window. Families whose homes had been struck by the waves had sought refuge in her home. One woman was particularly distraught. Reijeli wanted to pray to comfort her. Her husband said this was no time to sit down and pray but Reijeli went ahead anyway. Reijeli, her daughter with a child in her lap, and the woman sat on a bed together. Reijeli started to pray. She

[1]In agreement with the women of this research, all names in this chapter are pseudonyms.

S. Sahai · J. Ryle (✉)
School of Law and Social Sciences, The University of the South Pacific, Suva, Fiji
e-mail: jacquelineryle@usp.ac.fj

T. Tabe
Pacific Centre for Environment and Sustainable Development, The University of the South Pacific, Suva, Fiji

J. M. Luetz and P. D. Nunn (eds.), *Beyond Belief*, Climate Change Management,
https://doi.org/10.1007/978-3-030-67602-5_7

prayed all the prayers she could think of. She prayed the Divine Mercy, the Joyful Mysteries of the Rosary, the Salve Regina. She prayed in Fijian, in English, in Latin. The others sat with eyes closed; they did not respond but Reijeli knew they were praying with her. Flood water started to seep under the door but Reijeli continued praying. By the end of the final prayer the water had reached just below the bed. But Reijeli was no longer afraid.

Introduction

On 20th February 2016 Fiji was hit by the strongest cyclone ever to hit the Southern Hemisphere and the second strongest storm ever recorded globally (Gard and Veitayaki 2017). Tropical Cyclone Winston (TC Winston) impacted over 540,000 people, 62% of Fiji's population. It claimed the lives of 44 people and destroyed 495 schools, 88 medical facilities and 30,369 houses, internally displacing 186,195 people (Mansur et al. 2017; Gard and Veitayaki 2017; Government of Fiji 2016).

This paper is about how women in Tokou, a Catholic[2] village on the island of Ovalau, Lomaiviti Group, Central Fiji, experienced Tropical Cyclone Winston. Tokou, located on the coast, was in the direct path of the cyclone and was badly hit. It experienced three tidal surges that inundated the village; 60 of its 130 houses were damaged, 18 of them beyond repair. The primary school and Catholic church, the only church in the village, were severely damaged. Despite this, there were no fatalities or serious injuries among the 502 people living there at the time. The paper discusses how Tokou women's faith and faith practices as Catholics were a source of personal strength and survival during and after the cyclone, how this strength was embedded in social relations, and how it directed women's actions for the good of the community.

The research is based on two periods of ethnographic fieldwork in Tokou in 2019. These included participant observation, informal *talanoa* (chats), and 17 in-depth, recorded and transcribed interviews held in English with 10 women between the ages of 40 and 75. Each participant recommended other participants. Most were maternal relatives of the lead author. All participants were proficient in English. Additional information was gathered in subsequent face-to-face interviews in Suva and numerous phone conversations. In the following sections we introduce Tokou, describe the core role of Catholic faith and faith practices in the community especially in relation to women's roles; and discuss women's experiences before, during and after TC Winston. Finally, we argue for the necessity of integrating faith, faith practices and spirituality in climate change and Disaster Risk Reduction (DRR) strategies.

[2]The most recent available statistics on religious affiliation in Fiji, from the 2007 Census, note that 64.5% of the population is Christian, 28% Hindu and 6.3% Muslim (Fiji Bureau of Statistics 2007). The Methodist Church in Fiji and Rotuma counts 35% of the population, the Roman Catholic Church just under 10% (Ryle 2010, p. xxxii).

Understanding Faith

In 2014 an expatriate representative from an international organisation presented a study on women and climate change in Tuvalu at The University of the South Pacific. The overseas gender consultant who conducted the study showed a pie chart with brightly coloured segments indicating the percentage of time women spent on different aspects of their daily lives. One core element was missing. A participant pointed out that 'church and faith' were not included. The presenter agreed this was missing and, clearly unsure of how to categorize this, asked the audience, 'What would that come under? … hobbies?' There were audible intakes of air in the room. This example, far from unprecedented, demonstrates the conspicuous lack of understanding by many outsiders of the pervasive nature of religious and spiritual beliefs in every aspect of Pacific Islanders' everyday activities and decision-making (see also Nunn et al. 2016).

Although the time women in Pacific Island communities spend on church-related activities is part of the extensive, frequently unacknowledged, work they do, an approach that compartmentalizes women's everyday lives and agency into neat segments and then merely adds 'church and faith' as another segment, has entirely missed the point. For, as Yabaki (2006) notes, *iTaukei* (indigenous Fijian) women's religious values influence the way they make sense of the world. Similarly, Meo-Sewabu's (2016) study of women in Lau, Fiji, shows that their faith is a source of inner strength that informs every aspect of their daily lives.

There is a widespread lack of understanding of the importance of faith and faith practices throughout much of development, climate change and DRR engagement and intervention, especially by people raised in secular societies. This situation has major ramifications. For example, Paton and Fairbairn-Dunlop (2010) note that the institutions valued most by Tuvaluans, such as family, traditional structures, local government and church, are never taken into account in climate change initiatives by foreign donors. Fair (2018, p.12) stresses that it is crucial to include "multiple epistemologies of climate change" and understand the links between local perspectives and science in climate change interventions. As Nunn (2017, p. 3) argues "the one reason for the failure of external interventions for climate-change adaptation in Pacific Island communities is the wholly secular nature of their messages". And, as the above story shows, even when there is some openness to a religious dimension, faith and faith practices are at best seen as appendages, feel-good add-ons that provide individuals with emotional comfort. At worst, they are seen as obstructions to development (Nunn 2017).

Wæver (2011) emphasizes that 85% of the world's population hold religious beliefs and half of the world believes that religion should have more significance in the workings of a society. Yet, despite the importance of spirituality, faith and faith practices in developing countries, these core elements of people's life worlds are largely absent in climate change and DRR strategies. Fountain et al. (2004) and McGregor (2010) for example, writing on DRR interventions in Samoa and in Indonesia, point out that the role of religion received scant attention. In addition, women play key roles in disaster situations and post-disaster rebuilding of everyday life, yet only rarely is this built into DRR strategies and implementations.

It is essential to develop holistic conceptual frameworks (Luetz and Nunn 2020) that integrate women's roles and the hugely empowering efficacy of people's faith and spirituality, the "lived practice of everyday religiosity" (Haluza-DeLay 2014, p. 14), if more effective and sustainable strategies for climate-change adaptation and the development of disaster resilience are to be developed. Such holistic frameworks need to be carefully attuned to each specific sociocultural context as neither women's roles nor faith and spirituality can be fitted into one specific mould.

Faith, Prayer and Spiritual Agency

Prayer, writes Bandak (2017, p. 93) "expresses the collective concerns of the people praying" and is "a way of acting in the world but it is also a way of believing and of formulating this belief". Prayer, he emphasizes "needs to be repeated" but such repetition is not a repetition of sameness but a movement in time that can uphold or change specific contexts (Bandak 2017, p. 94–95). As Catholics, the Tokou women direct their prayers not only to Christ but also to the Virgin Mary, praying for her protection in their hour of need, as is part of the repeated words of the Rosary. The vignette at the start of this chapter shows how, in the midst of chaos, destruction, shock and fear, Reijeli re-created a familiar, ritualised devotional order through her prayers. Overriding her husband's point that now was not the time to sit down and pray, she continued. As host to the displaced families sheltering in her home, she felt she should provide comfort and nurture. Reijeli was acting out *loloma* (love, compassion and care for others), relational, core values of indigenous Fijian culture and of her beliefs as a Catholic.

But Reijeli's actions are more than comforting; her prayers change and empower. She draws the traumatized woman into a familiar ritual that simultaneously extends horizontally through the sociality of praying together in close bodily proximity, and vertically to the sacred power of God. The rhythmic sound of the repetitive, familiar words and phrases of the Rosary, prayed in the familiar order, create a regular, "pulsating heartbeat" (Bandak 2017, p. 93). Such faith language is, Bandak emphasises, "a form of action … Prayer as a ritualised form of action … rather than merely stabilising already existing patterns … is a matter of acting upon the social world" (2017, p. 94). Reijeli was "acting upon her social world". Her faith led her to pray, and the words she prayed and her act of praying effected change within herself and in those around her. When Reijeli had finished praying, she was no longer afraid. Through her prayer and the act of praying, a situation of trauma, shock and fear was converted into one that was manageable.

This is one of several examples of women's central role in Tokou during Cyclone Winston and after—all enacted through faith that God was present in everything that was happening to them and in their own responses to this. Women were tapping into an experienced spiritual order within and beyond the chaos. Through their faith, the women found meaning in even the most painful experiences. Though their reality was shattered by forces of nature more powerful than they had ever known before, their belief in God's love and care for them in the midst of the nightmare they were experiencing, gave them strength and courage to endure, empowered them and directed their actions.

Reijeli's actions express what Ryle (2011, p. 73–74) terms as 'spiritual agency', a powerful creative force that comes into being in the embodied experience of the sacred. 'Agency' emphasizes the creativity, intentionality and inner 'drive' that engenders this spiritual agency. To the person of faith, 'spiritual agency' is at the core of what makes up the whole person, an integral and active component of a person's life, the way they relate to others and their surroundings, the emotions they experience, the choices they make, the way they face life's challenges. It is such spiritual agency that is central to the women's experiences in Tokou and which directed their actions during Cyclone Winston, empowering them to maintain and recreate social and spiritual order and to effect change.

Women and Natural Disasters in the Pacific Islands

Women have crucial roles in society yet they are generally classified as disproportionately vulnerable. This is mainly due to traditional hierarchical systems dominated by men, especially in Pacific Islands communities where women are subjected to decisions made by men and often restricted from engaging in decision-making or articulating their viewpoints in communal spaces (Charan et al. 2016). Traditional gender norms in the Pacific also often constrain women from being able to make decisions for themselves or on behalf of their families. Studies reveal that the culturally gendered roles of women in Pacific Island countries usually leave them more exposed and vulnerable to disaster risks and higher mortality rates than men (Ear 2017; UNDP 2013).

In contrast, Clissold, Westoby and McNamara (2020) argue that while women may often be viewed as vulnerable, they are also resourceful actors during disasters and invariably play the critical role of ensuring that their families are safe and adequately looked after. Women have unique skills in DRR and management yet these are not usually acknowledged or sufficiently explored in many non-western contexts (Luetz 2008; UNDP 2013; UN Women 2014).

In countries like Samoa and Kiribati, the roles of women during disasters include the security of the family, and preparation and storage of food, water and other necessities required during and in the immediate aftermath of these events (SPDRP 2002). In Vanuatu after Cyclone Pam in 2015, women of Efate Island used their informal networks to exchange much-needed goods such as food and tools with women from different islands so as to optimize recovery (Clissold et al. 2020). In Malaita Province, Solomon Islands, as described by Asugeni et al. (2019), it was the women of two remote coastal villages who took the lead to build seawalls as an adaptation measure to reduce coastal erosion. In matrilineal Chuuk, Micronesia, women, considered to be "custodians of the environment", and women's groups promote sustainable activities such as the management of waste, back-yard gardening and mangrove planting (Kim 2020, p. 148). And in the aftermath of Cyclone Winston in 2016, women in rural communities of Fiji were significantly involved in the rebuilding of their homes, while also attending to the needs of families (Charan et al. 2016).

However, women are also exposed to high risks of gender-based violence during disasters. Families often spend lengthy periods in evacuation centres where living conditions are poor, providing no privacy for couples and families. A study conducted by UN Women in 2012 on gender and gender-based violence during two floods in the Western Division of Fiji underlined the significant roles women played during both events. It described how women's role in food preparation for their families in the mornings enabled them to observe and identify signs of flooding through the rise in the level of the river located next to their homes so they were able to warn their families ahead of time before the event took place. But the report also discussed the exacerbated gender-based violence many women experienced during these disasters. Women were for example forced into sex by their husbands, despite their reluctance due to overcrowding and lack of privacy in the evacuation centres (UN Women 2012).

Despite women's significant role in disaster management and recovery, there is still limited research on this topic. Charan et al. (2016) criticize the lack of research on gender and DRR and climate change in the Pacific, including perspectives and responses before, during, and after intense cyclones in Fiji, a point also made by Nakamura and Kanemasu (2020). Clissold et al. (2020) emphasize that detailed research is required into women's experiences of post-disaster recovery, reconstruction and resilience. Similarly, Kopf et al. (2020) point to the need for more research on the relations between gender roles and power dynamics in households and how they affect women's exposure to natural disasters.

In Fiji, in addition to their household chores, Fijian women work in gardens, fish for household consumption, and sell fish to contribute to household income (Yabaki 2006, p. 188; Ram-Bidesi 2015). Noting the concept of the "triple burden" of women's reproductive, productive and community roles, Meo-Sewabu (2016) emphasizes that managing these complex roles and responsibilities is part of women's everyday life. This contributes to the strategic needs of families and the community. It is therefore empowering. Women are not "passive victims" but rather "active agents of change". This is borne out by the following ethnography.

Tokou—A Catholic Village on Ovalau

Ovalau island in Lomaiviti Province, central Fiji, is accessible by ferry from the main island and has a small domestic airport; it lies 60 kilometres from the capital of Suva. It has 24 villages and a population of around 9,100 (Map 1).

Tokou is located along the east coast, five kilometres or a 10-minute drive from Levuka, the old capital of Fiji. The population of Tokou in the second quarter of 2019 was 448 people, comprising 217 females and 231 males in 76 households. At the

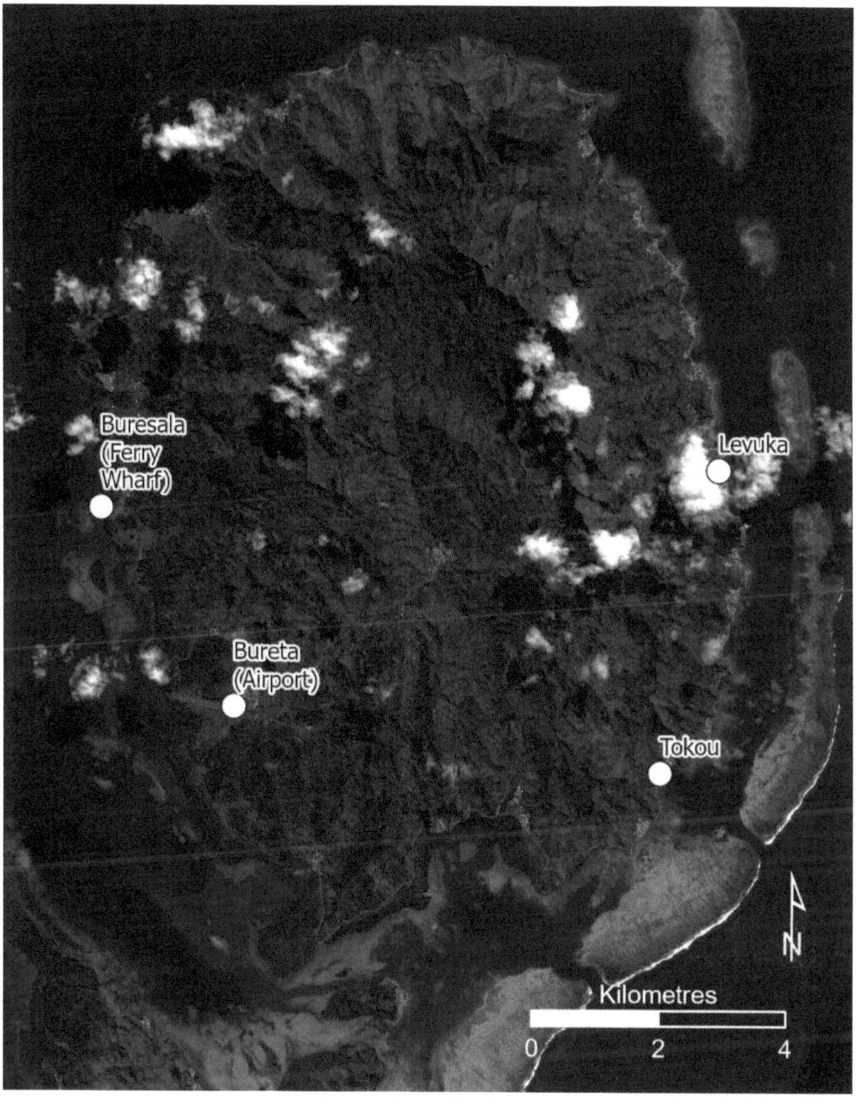

Map 1 Map of Ovalau island, Lomaiviti, Fiji. Courtesy of John Lowry, Massey University

time of Cyclone Winston, the resident population was 502. There were 33 cement houses, 16 wooden houses and 27 corrugated iron houses. 60 of these homes were damaged, 18 beyond repair, and the school and Catholic church, the only church in the village were also badly damaged. In the aftermath of TC Winston, Tokou was without electricity for six months (Mr Matareti Mateiwei, Conservation officer, Lomaiviti Provincial Council, pers comm, 12.11.19) (Map 2).

Map 2 Map of Tokou village, Ovalau island, Lomaiviti, Fiji. Courtesy of John Lowry, Massey University

Tokou's boundaries extend some eight hectares. The village is very close to the shoreline and much of it lies below high-tide level (Nunn 2000). This makes Tokou highly vulnerable to winds and tidal surges and significantly at risk and vulnerable to impacts of climate change especially sea-level rise and coastal erosion. A large river flanks the eastern boundary of the village, flooding the village during heavy rainfall. This added to the severity of the impact of the tidal surges that flooded the village.

The circuminsular road around Ovalau runs past Tokou. Before Cyclone Winston it was unsealed and Tokou had no seawall. In 2018, before the general elections, the government built a seawall and tar-sealed the part of the road that passes Tokou. Our Lady of Loreto Catholic church and Loreto Catholic primary school, about 300 metres from the village, are both within walking distance from the village along this road. All the village plantations are located beyond the village, extending to the hills behind. The first row of houses beside the roadside is 20 metres from the beach. The village *rara* (village green) and village hall are behind the first row of houses. The village hall is around 35 metres from the beach and is the designated Evacuation Centre as, aside from the church, it is the biggest and most solid building in the village, and can accommodate many people.

Villagers grow root crops and other vegetables in their clan plantations and collect fish, crabs and seaweed, and freshwater prawns from the river. As part of traditional marine conservation practices of the village, a *tabu* (taboo) was sometimes placed on a part of the traditional fishing grounds (*qoliqoli*) for a period of time. Yet in 2009, Tokou set up a 148 hectares Marine Protected Area (MPA) within the *qoliqoli* for 15 years in association with Seacology, a non-profit organisation, that in return pledged to construct a village hall (Sykes et al. 2018). In September 2018 Seacology funded repairs to the village hall on condition that Tokou extend the MPA for another decade. After Cyclone Winston, the shortage of fresh food in the village was perceived as an emergency so the villagers were permitted to fish from the protected area in front of the village for a month before the ban on the MPA was re-instated and villagers were forced further out to sea to fish.

In most households in Tokou at least one person is employed by the Pacific Fishing Company in Levuka (PAFCO) including three women who participated in this study. Five other women participants supplement household incomes by baking, sewing, and selling kerosene and cigarettes.

Tokou—Catholic Faith and Women

For indigenous Fijians *(iTaukei)* the relational term *vanua* is absolutely central to their understanding of life, their values and ways of living. It encompasses "various but related meanings: the people, the land, including living and [inanimate] creatures, and the ocean and marine life that belong to that land" (Kivi 2018, p. 1). *Vanua* is that which binds all this together, and with the adoption of Christianity, *vanua* and Christian faith became closely interwoven (Ryle 2010). Though Tahitian teachers first brought Christianity to Fiji in 1825 Christian mission did not really begin to grow until after 1835, following the arrival of English missionaries from Tonga. The first Roman Catholic missionaries of the French Society of Mary religious order (Marists) arrived from Uvea (Wallis) in 1844 in the eastern-most island group, Lau, moving from there to the island of Taveuni and then to Levuka, the capital at that time, on Ovalau.

In 1852 Tokou became the first village in Fiji to convert as a whole to Catholicism (Foi 2007). In 1885 the ancestors of Sawaniika *mataqali* (clan) of Tokou offered gave a 30-acre piece of land beside the village to the Catholic Church for the use of the Marists. It was named Loreto, after the shrine of the same name in Italy and devoted to Our Lady of Loreto (Knox 1997); Our Lady of Loreto Church and Loreto Catholic Primary School were later built there. From 1885 onwards, Loreto became the centre for the Marist mission on Ovalau and the surrounding Lomaiviti group of islands.

This history of a strong Catholic (Marist) presence since the mid-19th century forms a core part of Tokou villagers' sense of identity and is foundational to the central place of Catholic faith practices in the village. Furthermore, Tokou prides itself on having contributed several young men to the priesthood, most of whom joined the Marist order. Despite significant changes over the past 25 years to church adherence in Fiji, with Protestant Pentecostal churches gaining increased influence in urban and in rural areas (Ryle 2010; Morgain 2015), and many villages now housing more than one denomination, Tokou has remained Catholic and Marist.

Everyday life in Tokou is framed within a schedule of faithfully observed Catholic devotional practices. The week begins with the entire village attending Sunday Mass. Sunday School is conducted in the afternoon, followed by Rosary prayers. Weekday masses are held twice a week in the evening. Every day at 6 a.m., 12 noon and 6 p.m., the *vakavuvuli* (catechist) rings the Angelus bell and villagers pause their activities to pray. The Angelus is a prayer to the Virgin Mary recalling Mary's choice to accept the will of God and give herself to become the mother of the Messiah. Most people, including the children, pray the Angelus. Each day, except Sunday, is rounded off in all homes with Family Prayers from 7–8 p.m. which include hymns and praying the Rosary.

There are two women's church groups in Tokou: the Christian Mothers *(Marama tina va Kiristiano)* and the Catholic Women's League, and some women are also part of the charismatic group. When a young woman turns 16, she is automatically admitted to the *vanua* women's group, the *Soqosoqo Vakamarama* that organises projects for the village and the church. The women of this group assist the women's church groups in coordinating fundraising activities, cleaning the church and village hall and helping the men clean the village. The group ensures each household is equipped with household items and gardening equipment to carry out these functions efficiently, and members visit sick women and assist them with their household chores and gardening.

In Tokou it is first and foremost the women who, with their children, participate in devotions. This follows the pattern in other Catholic parishes (and in other denominations) that it is generally women who gather their family together for family prayers and who participate in communal prayer meetings. It is mothers and grandmothers who hand down the faith, teaching their children and grandchildren to pray, teaching them prayers and hymns. This gendered generational transmission of faith was also noted in villages in the Yasawa island group by Kline et al. (2013), in Tailevu Province by Ram-Bidesi (2015) and in Vanua Levu by Akanisi Tarabe and in the Lau group by Litea Meo-Sewabu (pers. comm. 2020) Such perpetuation of faith, described by Tarabe as "spiritual reproduction", is an integrated part of *iTaukei* women's lives and vital to community life, yet it is often unseen and unacknowledged.

In addition to the daily family and communal prayers, the women in Tokou also set time aside each day for their individual prayers. These can be conducted alone in the church or at home during the day when their families are out of the house, and some women pray the Rosary late at night when everyone else is asleep.

In Tokou the women are at the core of all the work for church fund-raising events and special Catholic Feast Day celebrations that occur many times throughout the year. On these occasions the women prepare and serve the food for priests, guests and parishioners as well as clean up afterwards. As hosts, the women will eat last.

According to Tarabe, *veiqaravi* is the Fijian term that best captures the agency of *iTaukei* women in family and community life described above and in the following sections in the disaster situation they faced. *Veiqaravi* is a relational term that means 'to serve'—though not 'to be subservient'—but to serve others in love, compassion and generosity, to give of oneself to others. These values are also expressively Christian. *iTaukei* women, especially married women, see their lives as that of service to their families, communities and the church. In so doing, the women act out of *loloma* (love, compassion and care for others) and see the maintenance of harmony and relationships as dependent on their actions. This is what makes women central figures in family and community life—their centredness is in their giving (Tarabe 2015; Meo-Sewabu 2015).

Tokou women value their families deeply and their responsibilities in their homes towards their husbands and family are their priority. Similarly, in Meo-Sewabu's study of rural women in Lau, most of the women "felt their role was to keep their husbands happy by doing everything to the best of their ability and many asked for spiritual help to remain humble and accept that they were there to serve" (Meo-Sewabu 2016, p. 105). One woman "felt that the happiness of her husband and her family is dependent on her. This sense of responsibility urges her to draw upon her spiritual strength to persevere with the marriage" (ibid, p. 108). So women are acting subjects and take pride in their indispensable roles in relation to the home and family. At the same time women's multiple contributions are often unseen, unrecognized, and taken for granted by men.

Though women are central to the management of households and the upbringing of children, men are recognized as head of the household and women are subject to their decisions on most matters. Then again, a woman can influence her husband and Meo-Sewabu points out (pers comm 2020) that if one wants to initiate change in a village, one goes through the women. They are skilled at finding appropriate ways to approach their husbands to broach a topic. The husband will then hopefully advance the issue among the men. But women do not generally have direct access to decision-making in Fijian villages.

As Tarabe points out, women "are usually placed at the lower end of the indigenous Fijian's social strata" (2018, p. 37). And she notes that "much of the major decision making at the traditional village institutions is carried out by men … as a patrilineal society, major decision making is done by men" (Tarabe 2020, p. 36). What access a woman may have, depends on her status: if she is married into the village, she has lower status and fewer rights than if she is of the village clan (Tarabe 2018). If she is of the chiefly clan, her status is highest. Even then, if a woman is present at a meeting, as a representative of say the Village Women's Group, she still would not necessarily freely engage in a discussion but would need to wait to (possibly) be asked to speak (Tarabe, pers comm 2020). It is not generally acceptable for women to speak out publicly in village community situations.

The high figures of Gender Based Violence in Fiji, as elsewhere in the Pacific, indicate serious structural gender inequalities that mean many women are subject to their husbands' coercion and control, as evidenced in the 2012 UN Report on GBV during the floods in western Fiji, discussed earlier. These different perspectives highlight the many layers, complexities and ambiguities in *iTaukei* women's status and roles in relation to men in community contexts. Therefore, the point we stress is that women are never passive victims but always acting subjects who consciously navigate and negotiate the options available to them.

Cyclone Winston

Tropical Cyclone Winston developed into a Category 5 cyclone on 19 February 2016 as it moved westwards from Tonga to Fiji, reaching peak intensity on Saturday 20 February with winds of 280 km/hour shortly before it hit the north coast of Viti Levu. It had been heading directly for Suva and Nadi, the most populated areas on the island, however it changed course, cutting a destructive path through the northern and eastern outlying islands of Taveuni, Vanua Levu, Koro and Ovalau, before it hit Viti Levu. This sudden change of path did not allow for advance warning (DFAT 2017).

Shortly after Cyclone Winston hit Fiji rumours starting circulating, among Protestant Pentecostal Christians in particular, about the sudden change of course of the cyclone. Why had it veered away from Suva at the last minute? Many people said that this was because of the intensity of people's prayer in Suva. Why was Koro island hit so terribly? A sustained explanation, still in circulation in 2020, was that it was God's punishment for "all the sorcery people do there" (see Cox et al. 2018 for similar discussions). Pacific Methodist theologian Rev Dr Cliff Bird sought to counter these "spiritualistic explanations" [that] "associate natural calamities and the resultant loss and destruction with divine rebuke and punishment", emphasizing that "We cannot say that God sent the Tropical Cyclone Winston back to Fiji to teach, rebuke or punish people … It is … freedom of nature that Tropical Cyclone Winston moved freely and erratically in the way that it did (Bird 2016).

No one in Tokou at the time or since saw the destruction Cyclone Winston wrought in the village as God's punishment. But in the following descriptions God is seen as present in all that happened.

Before

Most of the women heard the cyclone warning on the radio on Thursday and Friday, two days before it hit the village. One woman heard about the cyclone at PAFCO five days before and discussed the warning with her family, but at the time they did not perceive it then as a significant threat.

In Fijian villages, it is the traditional role of the *turaga ni koro* (village headman) to relay important messages or call villagers to meetings by walking around the village, calling out the information. On Friday, the day before the cyclone hit, at 8 am and 6 p.m. he walked about the village, warning villagers to prepare. His role is to advise the villagers, but it is up to each household whether and how to prepare for the cyclone or not. He told the villagers to prepare for a strong cyclone although he could not then have known its severity. There was reportedly a relaxed attitude among the villagers towards his warnings. Only a few families started their preparations in anticipation of the cyclone. Most started to prepare just hours before the cyclone reached Fiji on Saturday. They expected it to be similar to the many cyclones they had experienced in the past and prepared in the same way as for previous cyclones.

On Saturday morning just after 8 a.m., police officers from Levuka drove by the village, warning people about the severity of the imminent cyclone and alerted the *turaga ni koro*. Between 8 a.m. and 9 a.m. the *turaga ni koro* walked around the village again, calling out the upgraded cyclone warning. Tokou had no formal evacuation plan and the Evacuation Centre was not opened till around 12.30 p.m.

At about 9 a.m. Reijeli heard about the expected intensity of TC Winston directly from her niece in Australia. Since her niece lived overseas Reijeli felt this must be accurate information, so she started preparing for the storm. She stored rugs, mats and bedding in the wooden ceiling beams of her home. She packed away all the photos from her living room wall except for the Immaculate Heart of Mary picture which she kept there for protection. She prayed to Mary 'to protect Fiji and all its families'. Normally she has disaster essentials such as candles, batteries, bandages and paracetamol at home but that day she had run out of candles so she went to Levuka to stock up.

The *turaga ni koro's* warnings on Saturday morning prompted many of the villagers to start securing their homes and moving animals to higher ground. Women stored electrical items, bedding and mats on ceiling beams. The atmosphere in the village was not one of panic but of the usual preparations for any cyclone. Even two hours before the cyclone struck, one woman was trying to finish her sewing, knowing that a power cut was inevitable. Some of the women continued cooking lunch for their families so that they would have meals to take with them to the Evacuation Centre. Others made urgent trips to Levuka town to buy cyclone supplies such as kerosene, candles, flour and cooking oil.

During

How people responded to the situation depended on their experience of the intensity of the cyclone. This was dependent on where in the village their homes were located. So even though the *turaga ni koro* opened the village hall to evacuees around 12.30 p.m., most people stayed in their homes until the first tidal surge had swept through the village.

By early afternoon the winds were very strong. It was raining heavily and the tide was also high. Seini, a disabled 75-year-old woman, refused to be evacuated when her son came to take her to the Evacuation Centre. She was convinced her house was strong enough to withstand the storm and therefore wanted to remain in her home. But her son had helped in the construction of the village hall and knew that it was stronger than their home so he persuaded her to be taken to the Evacuation Centre where it would be safe.

Mid-afternoon the first tidal surge flooded the village. A woman recalled that "It started to rain, the wind was so strong…and very soon we can [sic] see the water coming in … it was like the ocean".

Some women took their bedding and kerosene stoves to the Evacuation Centre and continued cooking their lunch there. Others went with only the clothes they were wearing or just one change of clothes. Mothers and children took food and bags of clothes with them while many of the men and their sons insisted on staying in their homes to secure them.

Leilani, who was five months pregnant, was frightened as she watched the tidal surge flooding the village. The water was rising. She put blankets, pillows, mats and clothes in garbage bags and placed them in a laundry basket on a bed raised above the floor. While the rest of the family were at home, her husband had gone to a neighbour's house, thinking that the flooding was just the usual flooding caused by the river in heavy rain.

When Leilani heard calls outside that a "tsunami was coming', she hurriedly left for the Evacuation Centre, holding her two-year old son in her arms, but the water was already waist-deep, and she had to be assisted by young men. Even though Leilani's pregnancy made her vulnerable, she made sure to take a thermos with hot water to last her till the next morning so she could warm milk for her son. Leilani's response to the flooding of placing bedding, mats and clothes on the bed proved effective as the dry clothes, blankets and sleeping bag were used by her family that night in the Evacuation Centre.

Another woman who tried to make her way to the Evacuation Centre described the scene: "When we wanted to cross there was like … a whirlwind or something inside the village … we couldn't see anyone".

One woman and her family, who continued with a birthday party celebration despite the cyclone warnings, had to evacuate their home when it was flooded by the tidal surge. By this time it was too late to get to the Evacuation Centre. They tried to wade through the water to get to their neighbour's house, but it was too difficult for them to move through the water on their own because of the strong currents and the now chest-deep water. Young men had to assist the women, children and the elderly, while others used a *bilibili* (a hand-made raft of bamboo poles) that had floated into the village from the river.

Struggling to get to her husband through the chest-deep current surging through the village, Asinate was assisted by two young men who helped her to safety. As she moved from house to house through the flooded village she screamed in English "Lord, please give me another chance", and in Fijian, "*Isa, Turaga, rawa ni vukei au*".

By 4 p.m. three tidal surges had washed through Tokou. A large part of the village was flooded and remained submerged for about seven hours. The tidal surges reached as far as 120 metres inland, well beyond the Evacuation Centre which was located 35 metres from the shoreline and was flooded ankle high with water.

Faith and Prayer in the Midst of Chaos

All the women consistently said their faith in God kept them safe during the cyclone and that they felt strengthened and empowered by their faith and prayers. Placed close to the entrance of the Evacuation Centre in her wheelchair, Seini watched as the tidal surge brought water right into the building. She was unable to help herself but remembered that she had a bottle of holy water in her handbag beside her so she took it out and sprinkled the holy water around her, sure that God would help her.

From the verandah of the Evacuation Centre villagers watched the tidal surges sweep through the village. One woman watching her home being destroyed, prayed, "Lord, I love my family and everyone but you love us more and you won't let anything happen to us".

Elderly Akanisi and her husband watched as an electricity post fell on their house and the waves washed most of it away. Akanisi whispered to her husband to comfort him, "That's God's will to us … God brings it, God takes it, he'll give it again". Her words link to the Old Testament Book of Job in which the Devil wagers with God that he can make Job lose his faith. Job loses everything he ever owned and all those he loved, yet he never wavers in his faith, holding fast to the belief that all he had was God-given and that God could take it all from him. Job lost everything but was subsequently richly rewarded for his strong faith. Akanisi's words express the belief shared by all the women that everything in life is a gift from God; that God was present in all that was happening; and that they should accept what they were going through, trusting that God would take care of them.

One woman said her faith helped her get through it all. She prayed the Rosary at home during the flooding, trusting that Mary would help her. "We just pray… God helps in everything", she said. Ruci, a single mother, said that praying the Rosary every day was what helped her most during and after the cyclone; she knew her faith in God would get her through the difficult times and that "everything was going to work out". Reijeli said she knew God was listening to her when she prayed because "what was asked for, was given to me". For Arieta, her daily prayers were her major support that made her strong for her family and children:

> The one thing that helped me [was] I had to pray every day, I believe that, with my health being okay and with my prayers, we'll get through everything together … My faith kept me strong and kept me going to get the family going and be there for everyone.

By 10.30 p.m. the cyclone had passed and the wind had died down. There had been no electricity from around midday so the only source of light in Reijeli's home was a torch and a solar light.

While everyone was cleaning the house, the Eucharistic Minister from the church came to ask the household to look after the ciborium. The church had been damaged and flooded and he was searching for a safe place to keep it. This was an unprecedented situation. The ciborium contains hosts (the wafers consumed at Holy Communion) that have been 'consecrated' during Mass. To Catholic faithful, the ciborium is understood as containing Christ's actual Body under the appearance of the hosts. It is kept in what is known as the tabernacle that is locked and placed in the most holy area of the church, the sanctuary, to which only the priest and sacristan, who looks after the church, or Eucharistic ministers, have access.

To Reijeli and her family, it was not merely an extraordinary situation that the ciborium had been removed from the sanctuary in the church. It was as if Christ himself had been brought into their home. They reverently placed the ciborium on a piece of cloth, high up on a shelf and the only candle in the house was lit and placed next to it. That night they all sensed a spiritual presence in the house:

> It was amazing because He was there … everybody … noticed that He was there … so they were not frightened … we were doing our work in the dark with one torch and one candle … but the Light was there … inside our house … a different Light.

Late at night, when the floor of the living room had been cleaned and dried, the tarpaulin, rug and mats that Reijeli had stored away safely before the cyclone were laid out and the families sat down. Reijeli wanted to give thanks for being alive. Though a few young men wanted to explore the situation in the village, Reijeli asked if they could all first sit down and pray. She felt she had to be in control as it was her house and she felt a responsibility for the eight families who had taken shelter there, so she started conducting her usual evening prayers. She could hear the sobs of the women sitting around her. Although no one was praying the responses, she continued. She ended with the much-loved Fijian hymn to the Virgin Mary, '*i Maria na Tinaqu'* (Oh Mary, my Mother) which everyone joined in singing. "I did not feel down", Reijeli later reflected, "I knew we were all saved".

In the Evacuation Centre, people also prayed their own personal prayers. Akanisi prayed for her family and for the restoration of her home that had almost all been destroyed by the waves. Lisi felt sad for everyone's loss but was grateful to be alive. Her prayers, as she lay down to sleep that night, were for her family, the villagers, and for God's intervention in relation to the loss of her family's home and belongings.

After

As day broke on Sunday, the villagers ventured outside to assess the damage. They walked through the ravaged village in silence, trying to comprehend the devastation. Rooftops of houses had been blown away, doors and windows washed away. Driftwood, thick layers of seaweed and debris lay all around, inside homes and across the road. Clothes, fridges and stoves had been swept out of homes and tossed about the village. A fiberglass boat had been swept through the village as far as the Evacuation Centre. The road was cracked and damaged. The church and school were badly damaged. Never before had Tokou been so badly hit by a cyclone. All was different; "as if everything was dead", one woman said.

The first supplies to arrive, on Day Three after the cyclone, were from Fiji Red Cross: one package for each family, containing blankets and non-perishable essential household items. After this, government aid packages with food rations arrived.

Meanwhile, many families remained in the Evacuation Centre as they cleaned and repaired their homes. Those whose homes had been totally destroyed stayed in the Evacuation Centre for months. One family stayed there over a year. People shared their food and assisted one another in cleaning up their homes and the village. The women cooked, washed and sunned clothes and bedding, while caring for the children. Both the men and women cleared debris from the village, fished and went to the plantations to recover crops. The men patched the damaged roofs, walls and windows.

In the Evacuation Centre the women organised and rationed the food supplies and made decisions on what meals to prepare for those sheltering there, based on the availability of food stocks. They created a roster to ensure there was enough help to prepare and serve meals on time. They woke around 5 am to prepare breakfast. Before breakfast, there was morning prayer. Everyone sat on the floor, facing the makeshift altar in the front of the hall, a table covered with a white cloth with a statue of Our Lady of Loreto placed on it and a lit candle. Each day all families included the standard Tokou morning prayer in their devotions.

After breakfast, some women prepared lunch while others continued cleaning up the village and homes. Some women went fishing with their children, others helped the men gather cassava roots and firewood from the plantations. The women served lunch at 12 p.m. In the afternoon villagers continued clearing debris and cleaning up; women started preparing dinner around 3 p.m. and everyone had dinner around 6 p.m. Every evening at 7 p.m. they prayed the Rosary and sang hymns. After prayers, they relaxed in the hall before sleeping.

Other women were caring for relatives and neighbours in their homes. When Reijeli later wondered how she had managed to care for the eight families staying with her and provide meals for them every day, her daughter told her that this was made possible because of her daily prayers.

Lessons Learnt

Many different factors contributed to the severe impact of TC Winston in Tokou:

- Tokou's location on the coast exposes it to extreme events and climate change impacts such as sea-level rise and coastal erosion. The lack of natural protection such as mangroves and a seawall placed the village at high risk to exceptionally strong winds and the unusual level of tidal surges.
- There was no evacuation plan for the village. Had such measures been in place before, people would have been better prepared.
- Although there was a Red Cross Committee in the village, it was inactive at the time of TC Winston. A few months after TC Winston the Red Cross Committee was reactivated. It has since run regular evacuation drills in the village and has established a flag warning system.

- The designated evacuation centre (the village hall) is too close to the shoreline. A safer evacuation space further from the shoreline and on higher ground needs to be constructed.
- According to the villagers, the urgency and level of necessary preparedness was not clearly conveyed by the responsible authorities; warnings by the police were given only a few hours before the cyclone. This was because the path of the cyclone shifted suddenly only hours before.
- The lack of preparation meant that Tokou did not have adequate medical supplies for the injured. Since villagers could not access the health facilities by vehicle, many had to walk to the hospital in Levuka after the cyclone had passed to receive health assistance. Women had to walk to Levuka to buy medical supplies for their sick children.
- No one had ever previously experienced a cyclone of this magnitude so many of the villagers did not take the cyclone warning seriously. They did not follow the early warnings on the radio and warnings by relatives and work colleagues who better understood the likely intensity of the cyclone. They neglected the village headman's warnings the day before the cyclone hit. They expected a cyclone similar to previous ones, so they boarded up windows, secured rooftops with cement bricks; trimmed plants and trees that might cause harm; gathered root-crops and other fresh produce from the gardens as food supplies. This was what they would do in preparation for cyclones they had experienced in the past.
- According to villagers there had never previously been a tidal surge in Tokou. The women did not expect the tidal surges to flood the village so badly, "We did not expect this thing from the sea …we were just told about the strong winds and the rain …that's why I think most of the families stayed in their home [preparing] the [cyclone] shutters".
- Most people did not evacuate early enough to the Evacuation Centre. For several, when they tried, the water was too high and the current too strong for them to get there.

The extensive damage caused by TC Winston in Tokou effected a change of attitude among the villagers. They recognize the need to better understand the nature of natural hazards and climate change impacts they will face in the future; the need to develop and comply with an effective evacuation plan to improve disaster preparedness; and the importance of heeding early warnings.

Lack of preparedness, not heeding warnings, and not taking the situation seriously was common across Fiji during Cyclone Winston and was the main causes of fatalities. Changes in attitude to disaster preparedness were evident across the islands when TC Gita hit in 2018. Fiji Red Cross noted then that the lack of casualties in the outer islands of Ono-i-Lau and Votua was due to increased disaster preparedness knowledge and disaster evacuation plans, resulting in swift moves to evacuation centres (Tuilevuka 2018). This is a good illustration of the kinds of culturally-grounded autonomous processes for coping with external (environmental) stresses that continue to characterize the resilience of peripheral communities in Fiji and elsewhere (Korovulavula et al. 2019; Nunn and Kumar 2019).

Conclusion

Our ethnography shows how central *iTaukei* women were to community coping before, during and after TC Winston. We argue that the key roles women play in practical coping with disaster and in passing on their faith are complementary and mutually reinforcing. Our study shows women's organisational strengths in different ritualized everyday actions, such as caring for their families' and neighbours' well-being. It shows how women are the holders of faith, the ones who ensure that faith and faith practices are handed down through the generations in families and maintained in everyday life. And it shows how women had core roles through their faith and prayer in maintaining a sense of spiritual order during the crisis of the cyclone which gave hope and comfort to people and empowered them to act in the midst of an unprecedented situation.

We conclude that a holistic DRR conceptual framework is needed that is malleable enough to integrate within it from inception to completion a gendered approach to the specific socio-cultural, spiritual and faith complexities of the local context. A one-size-fits-all approach based on non-local conceptions and assumptions of what is needed is not effective. Such a holistic framework needs to integrate within it time and space for local dialogue. It needs to be open to listening to and integrating local voices, local knowledge, local experience and local wisdom. In particular, it needs to be tailored towards learning from women and their experience, as well as partnering with women on how best to strengthen community resilience. Strengthening women's options for agency and resilience empowers not only women but the whole community.

Acknowledgements We thank Margie Sahai, Ruth Sahai, Akata Conoa Baranisavu, Alekisio Kubunavanua, Fr. Donato Kivi SM, Lomaiviti Provincial Council, and Katherine Foi for their time and assistance. To the Tokou women in this study and the lead author's family in Tokou, *vinaka vakalevu* for sharing your faith and experiences. And we thank Dr John Lowry, Massey University for providing the maps.

References

Asugeni R, Redman-MacLaren M, Asugeni J, Esau T, Timothy F, Masse P, MacLaren D (2019) A community builds a "bridge": an example of community-led adaptation to sea-level rise in East Kwaio, Solomon Islands. Clim Dev 11(1):91–96

Bandak A (2017) Repeated prayers: saying the rosary in contemporary, Syria. Relig 47(1):92–110

Bird C (2016) God and natural disasters, 1 March 2016. www.methodistji.org

Charan D, Kaur M, Singh P (2016) Indigenous Fijian women's role in disaster risk management and climate change adaptation. Pac Asia Inq 7(1):106–122

Clissold R, Westoby R, McNamara KE (2020) Women as recovery enablers in the face of disasters in Vanuatu. Geoforum 113:101–110

Cox J, Finau G, Kant R, Tarai J, Titifanue J (2018) Disaster, divine judgment, and original sin: Christian interpretations of tropical cyclone winston and climate change in Fiji. Contemp Pac 30(2):380–411

Australian Government Department of Foreign Affairs and Trade (2017) Tropical cyclone winston education response evaluation. Retrieved from https://www.dfat.gov.au/sites/default/files/tropical-cyclone-winston-education-response-evaluation.pdf
Ear J (2017) Women's role in disaster management and implications for national security. Asia-Pacific Center for Security Studies Honolulu United States
Fair H (2018) Three stories of Noah: navigating religious climate change narratives in the Pacific Island region. Geo: Geogr Environ 5(2)
Fiji Bureau of Statistics (2007) Population by religion. Fiji Bureau of Statistics - Fiji Government, Suva
Foi K (2007) Cawaci 1890–2007: a microcosm of Catholic education in Fiji. Forum Novum 8(9):166–182
Fountain PM, Kindon SL, Murray WE (2004) Christianity, calamity, and culture: the involvement of Christian churches in the 1998 Aitape tsunami disaster relief. Contemp Pac 16(2):321–355
Gard AR, Veitayaki J (2017) In the wake of winston-climate change, mobility and resiliency In Fiji. Int J Saf Secur Eng 7(2):157–168
Government of Fiji (2016) Post-disaster needs assessment: tropical cyclone Winston, February 20, 2016
Haluza-DeLay R (2014) Religion and climate change: varieties in viewpoints and practices. Wiley Interdiscip Rev: Clim Chang 5(2):261–279
Kim MM (2020) Nesor Annim, Niteikapar (Good Morning, Cardinal Honeyeater): indigenous reflections on micronesian women and the environment. Contemp Pac 32(1):147–163
Kline MA, Boyd R, Henrich J (2013). Teaching and the life history of cultural transmission in Fijian villages. Human Nature 24(4):351–374
Kopf A, Fink M, Weber E (2020) Gender vulnerability to climate change and natural hazards: the case of tropical cyclone winston, Fiji. In: Amin SN, Watson D, Girard C (eds) Mapping security in the pacific: A focus on context, gender and organisational culture. Routledge Studies in Crime and Justice in Asia and the Global South Routledge, London, pp 119–132
Korovulavula I, Nunn PD, Kumar R, Fong T (2019) Peripherality as key to understanding opportunities and needs for effective and sustainable climate-change adaptation: a case study from Viti Levu Island, Fiji. Climate and Development
Knox M (1997) Voyage of faith. the Story of the first 100 years of catholic missionary endeavour in Fiji and Rotuma. Archdiocese of Suva, Suva
Kivi D (2018) Towards a marian ecological spirituality for the formation and re-evangelization of the Vanua: the people and the land of Fiji. Unpublished PhD thesis, Universitatem S. Thoma in URBA
Luetz JM (2008) Planet prepare—preparing coastal communities in Asia for future catastrophes: Asia Pacific disaster report, World Vision International
Luetz JM, Nunn PD (2020) Climate change adaptation in the Pacific islands: a review of faith-engaged approaches and opportunities. In: Filho WL (ed) Managing climate change adaptation in the Pacific region. Springer Nature, Cham, Switzerland, pp 293–311
Mansur A, Doyle J, Ivaschenko O (2017) Social protection and humanitarian assistance nexus for disaster response: lessons learnt from Fiji's Tropical Cyclone Winston, World Bank
McGregor A (2010) Geographies of religion and development: rebuilding sacred spaces in Aceh, Indonesia, after the tsunami. Environ Plan A 42(3):729–746
McMichael C, Katonivualiku M (2020) Thick temporalities of planned relocation in Fiji. Geoforum 108:286–294
Meo-Sewabu LD (2015) 'Tu ga na inima ka luvu na waqa':(The bail to get water out of the boat is in the boat yet the boat sinks): the cultural constructs of health and wellbeing amongst *Marama iTaukei* in a Fijian village in Lau and in a transnational Fijian community in Whanganui, Aotearoa. Unpublished PhD thesis, Massey University
Meo-Sewabu L (2016) Na Marama iTaukei Kei Na Vanua: culturally embedded agency of indigenous Fijian women-opportunities and constraints. New Zealand Sociol 31(2):96–122

Morgain R (2015) 'Break down these walls': space, relations, and hierarchy in Fijian evangelical christianity. Ocean 85(1):105–118

Nakamura N, Kanemasu Y (2020) Traditional knowledge, social capital, and community response to a disaster: resilience of remote communities in Fiji after a severe climatic event. Reg Environ Change 20(1):1–14

Nunn PD (2000) Coastal changes over the past 200 years around Ovalau and Moturiki Islands, Fiji: implications for coastal zone management. Aust Geogr 31(1):21–39

Nunn PD, Mulgrew K, Scott-Parker B, Hine DW, Marks ADG, Mahar D, Maebuta J (2016) Spirituality and attitudes towards nature in the Pacific islands: insights for enabling climate-change adaptation. Clim Change 136(3–4):477–493

Nunn P (2017) Sidelining god: why secular climate projects in the Pacific islands are failing. *The Conversation*

Nunn PD, Kumar R (2019) Measuring peripherality as a proxy for autonomous community coping capacity: a case study from Bua Province, Fiji Islands, for improving climate change adaptation. Soc Sci 8(225):1–26

Paton K, Fairbairn-Dunlop P (2010) Listening to local voices: Tuvaluans respond to climate change. Local Environ 15(7):687–698

Ram-Bidesi V (2015) Recognizing the role of women in supporting marine stewardship in the Pacific islands. Mar Policy 59:1–8

Ryle J (2010) My God, my land—interwoven paths of Christianity and tradition in Fiji. Ashgate, Farnham

Ryle J (2011) Laying our sins and sorrows on the altar: ritualizing Catholic charismatic reconciliation and healing in Fiji. In: Lindhardt M (ed) Practicing the faith: the ritual life of Pentecostal-Charismatic Christians, New York & Oxford, Berghahn, pp 68–97

Sykes H, Mangubhai S, Manley M (2018) Contribution of marine conservation agreements to biodiversity protection, fisheries management and sustainable financing in Fiji. Report No. 02/18. Wildlife Conservation Society, Suva, Fiji

South Pacific Disaster Reduction Program (2002) Gender, households, community and disaster risk management: case studies from the Pacific Islands

Tarabe A (2015) Re-assessing gender roles: a study of indigenous Fijian women's roles in funerals. Unpublished MA Thesis, The University of the South Pacific

Tarabe A (2018) Fiji: navigating a dual system and REDD + . cstomary tenure systems and REDD + : ensuring benefits for indigenous peoples. Tebtebba Foundation, Baguio City, Philippines, pp 23–41

Tarabe A (2020) Fiji: inseparable indigenous knowledge system and forest. Forest policies and indigenous peoples' traditional knowledge and practices on sustainable forest management: Fiji, Cambodia and Thailand. Tebtebba Foundation, Baguio City, Philippines, pp 21–48

Tuilevuka N (2018) Islanders' readiness comes under praise. Fiji Sun, 19 Feb

Wæver O (2011) Velkommen til Flertallets Verden—den religiøse [Welcome to the World of the Majority—the Religious World]. Udenrigs [Foreign Affairs] 2011: 1 'Religion i international politik' [Religion in International Politics], 8–18

United Nations Development Programme (2013) Gender and disaster risk reduction. UNDP, New York

United Nations Entity for Gender Equality and the Empowerment of Women (2012) The 2012 Fiji floods, gender sensitivity in disaster management: a qualitative review of gender and protection issues in disaster response, Suva

United Nations Entity for Gender Equality and the Empowerment of Women (2014) Putting women at the forefront of climate change and disaster response in the Pacific, UN Women

Yabaki T (2006) Women's life in a Fijian village. Unpublished doctoral dissertation, University of Canberra

Chapter 8
Signs of "The End Times": Perspectives on Climate Change Among Market Sellers in Madang, Papua New Guinea

Kylie McKenna and Lorelle Tekopiri Yakam

Abstract This paper describes understandings of climate change among market sellers in Madang Province, Papua New Guinea (PNG). Market sellers in PNG are likely to experience negative impacts associated with climate change due to their dependence on ecosystem services for their livelihood. Based on qualitative interviews at the Madang Town Market, this paper highlights an acute awareness of environmental change among market sellers such as, rising temperatures, changes in rainfall patterns and unpredictable seasons. Yet the interviews reveal unique interpretations of the cause of these changes. In contrast to anthropogenic greenhouse gas emissions emphasized in international climate change discourse, interviewees refer to biblical notions of "the End Times". This appears to give rise to a fatalistic view of climate change, as "God alone" is seen to be in control. Despite this apparent resignation, there is eagerness among market sellers for more information on climate change. These findings stress the need for sustainability practitioners working in PNG to consider the integration of environmental information with Christian beliefs. It is argued that this can be done in ways that respect spiritual understandings, while also promoting practical mitigation and adaptation strategies.

Introduction

Papua New Guinea (PNG) is the largest country in the Pacific region with by far the largest number of inhabitants. It encompasses the eastern half of the island of New Guinea and a number of island provinces to the north and east (Manus, New Britain, New Ireland and Bougainville). Administratively, the nation is divided into 22 provinces (including one Autonomous region and one National Capital District)

K. McKenna (✉) · L. T. Yakam
Divine Word University, Madang, Papua New Guinea
e-mail: kmckenna@dwu.ac.pg

L. T. Yakam
e-mail: lyakam@dwu.ac.pg

J. M. Luetz and P. D. Nunn (eds.), *Beyond Belief*, Climate Change Management,
https://doi.org/10.1007/978-3-030-67602-5_8

across four geographically distinct regions: Islands, Momase, Southern, and Highlands. PNG is one of the most culturally diverse countries in the world with over 800 indigenous languages. Madang Province, where this study was based, is the country's most linguistically diverse (Sullivan et al. 2003).

For some decades, PNG has been experiencing numerous environmental variations attributable to a changing climate. Since 1943 and 1962 respectively, annual and half-year air temperatures at Port Moresby and Kavieng have been warming (BOM and CSIRO 2014). Since 1993, the sea level around the islands of PNG has risen by approximately 7 mm per year, around double the global average (Becker et al. 2012). Due to sea level rise and other climate related pressures, communities in small atolls, such as in the Carteret Islands of Bougainville, have commenced out-migration (Luetz and Havea 2018). Climate model projections for PNG also suggest the continuation of El Niño and La Niña events, rising temperatures, changes in rainfall patterns, ocean acidification and deoxygenation, and sea-level rise (BOM and CSIRO 2014).

In spite of increased international attention to the ways in which climate change is affecting Pacific Island Countries (Leal Filho et al. 2020), critics rightly argue that "indigenous peoples are either scarcely mentioned or completely absent from any discussions regarding their perspectives and knowledge of climate change" (Jacka 2009, p. 197). One consequence of this exclusion is the side-lining of ontological frameworks that differ from western scientific knowledge systems (Foale 2004) which "has given rise to a plethora of ineffective and unsustainable adaptation responses…" (Luetz and Nunn 2020, p. 297). Ethnographic studies among specific language groups in PNG, for example, illustrate variations in conceptions of the environment, as well as explanations for environmental change that differ from those regarded as global/universal in many western contexts (Jacka 2009, 2010; Lipset 2011). Similar conclusions have come from studies looking at culturally-grounded concepts of climate change in Micronesia (Rudiak-Gould 2014).

Less is known about interpretations of climate change in urban centres in Pacific Island nations, places that can bring together communities with contrasting worldviews in one space, such as marketplaces. This chapter reports the results of an innovative study that attempted to record and analyse syncretic views of this kind, motivated by the likelihood that such views will increasingly become those that dominate in the future in many Pacific Island Countries. In an attempt to engage with perspectives and understand knowledge of climate change in a town setting in PNG, the researchers conducted qualitative interviews with market sellers at the Madang Town Market, Madang Province. Market sellers are prioritized in this analysis as they represent a segment of the population likely to experience negative impacts of climate change arising from their dependence on ecosystem services for their livelihood.

Three research questions guided the analysis:

- What changes have market sellers at the Madang Town Market observed in the environment?
- How do market sellers believe these changes are impacting fruit and vegetable production?

- What do market sellers consider to be the cause of these changes?

Answers to these questions were obtained through 14 qualitative interviews at the Madang Town Market over a two-month period. While the research team intended to conduct a larger number of interviews, the research coincided with the enforcement of a nation-wide State of Emergency in response to the COVID-19 pandemic that resulted in the temporary closure of the market. Yet given the dearth of research engagement with culturally-grounded understandings of climate change and its causes in PNG, the authors believe that even this comparatively small dataset will be of interest to sustainability practitioners working in the PNG context and similar elsewhere.

This study found a realistic and nuanced awareness among market sellers of environmental changes perceived to be impacting on fruit and vegetable produce. These changes are plausibly attributable to rising temperatures, changes in rainfall patterns and less-predictable seasons. Interviews also reveal different views about the causes of these changes compared to those that are central to western climate science. Rather than an emphasis on anthropogenic greenhouse gas emissions as the driver of global warming and climate change, explanations offered by market sellers invariably refer to biblical notions of the End Times. This referend appears to give rise to a fatalistic view of climate change and environmental degradation, one in which "God alone" is seen to be in control. Yet the interviews also reveal an eagerness among market sellers for more information about climate change and potential mitigation and adaptation strategies.

Such findings stress the need for sustainability practitioners working in PNG (and comparable contexts) to consider the integration of environmental information with Christian beliefs (Nyawo 2017; Schuman et al. 2018). This echoes calls concerning other issues in PNG (e.g. health messaging), that dissemination of scientific explanations are contextualized and complemented with Christian principles in recognition of the "varying viewpoints or frameworks of ideas and beliefs through which one perceives and interprets reality" (Gibbs 2020, p. 67). It is argued that this can be done in ways that respect spiritual understandings, while also promoting practical mitigation and adaptation strategies.

The paper begins with an outline of PNG's key climate related risks and hazards. The paper then explains the ontological differences between Melanesian (PNG) and Western conceptions of the environment. Ethnographic studies conducted in PNG are then referenced to highlight the ways in which worldviews in PNG can also be aligned with Christian beliefs and principles. Following an introduction to the study area of Madang, interview findings are presented alongside the key themes identified.

Climate Related Risks and Hazards Facing PNG's Communities

85% of the PNG population live in rural areas (Ramakrishna and Bang 2018), where subsistence agriculture is central to the daily lives of most communities (Rudiak-Gould 2014). More than three-quarters of food consumed in PNG is locally grown (Schmidt et al. 2019). Common food crops are sweet potato, banana (both sweet and cooking varieties), yam and taro (Ramakrishna and Bang 2018; Schmidt et al. 2019). Most households in PNG sell some of the crops they produce in addition to cash crops like cocoa, betel nut and vanilla, but this varies by area (Schmidt et al. 2019). Crops for food production (e.g. roots, vegetables and fruits) are sold mostly to consumers at local markets (Schmidt et al. 2019), providing a source of income that can be more stable and reliable than export tree crops (Sharp 2019).

PNG has a young and growing population and is expected to increase from 7.6 million to over 10 million by 2030 (Hayward-Jones 2016). Concerns have been raised about future land shortages for food production (Koczberski et al. 2018), alongside continued human development challenges. Although rich in natural resources, the Asian Development Bank (2020) estimates that 37.5% of the PNG population live below the National Poverty Line. Incomes and poverty levels vary across the country and within provinces (Gibson et al. 2005), influenced by factors such as access to schools (Anderson 2008), road infrastructure (Wiegand et al. 2017), electricity provision, locations of health centres, and economic development opportunities (CCDA 2017). Climate-related hazards could amplify existing challenges and add new ones (OCCD 2014). Key risks include (OCCD 2014):

- Coastal flooding and sea-level rise
- Worsened effects and spread of malaria
- Inland flooding, driven by heavy irregular rainfalls
- Sea temperature increase and acidification
- Landslides, triggered by increased rainfall intensity and land use changes
- Variability in agricultural yields

The impacts of climate change are likely to be experienced differently within PNG owing to the country's environmental diversity. Yet general nationwide concerns arise from changes in temperature and rainfall patterns and their impacts on agricultural yields and even the type of crops that can be grown (Ramakrishna and Bang 2018). Impacts on agricultural production arguably receive less international attention compared to those linked to sea-level rise which is affecting PNG's coastal and low-island communities (Connell 2016). This seems surprising given the importance of domestic/subsistence agriculture to the welfare of the majority of Papua New Guineans (Ramakrishna and Bang 2018).

Analysts also argue that the focus on averages typically depicted in climate change models could be a distraction from the greater risks to PNG posed to by climatic extremes of drought/flood cycles (Ramakrishna and Bang 2018; Jacka 2009). In 2015/2016 for example, much of rural PNG suffered a severe El Niño associated drought and severe frosts in the Highlands (Bourke et al. 2016). The drought and frost led to widespread crop failure and affected water availability (Bourke et al. 2016), causing a food crisis and loss of income (Gwatirisa et al. 2017). Interspersed with El Niño induced drought, many parts of PNG have experienced extreme rainfall events linked to La Niña events contributing to landslides and flooding (Cobon et al. 2016).

Understandings of Environmental Change in PNG

As mentioned above, PNG is extraordinarily diverse, linguistically, culturally, and environmentally. Critics of international climate change discourse argue that engagement with this diversity, especially with cultural diversity, is lacking (Jacka 2009; Bryant-Tokalau 2018). This can marginalise ontological frameworks that differ from western scientific knowledge systems (Jupiter 2017; Foale 2004).

A key ontological difference noted in the literature on Melanesian conceptions of the environment, is "the Melanesian perception of humans as part of nature versus western views of people apart from nature, termed the nature-culture dualism" (Jupiter 2017, p. 139), something noted in other studies (Nunn et al. 2016). In contrast to "this dualistic opposition, which allows people to see themselves as separate from and thereby able to control (and protect) the natural world" (Foale 2006, p. 131), the Melanesian worldview is described "as holistic…in relationship between the natural world, living creatures and persons; with illness seen as a disruption of the harmony of these relationships…" (Vallance 2012, p. 4). While climate change discourse "makes claims about the consequences of industrial capitalism on processes of various kinds in nature" (Lipset 2011, p. 20), traditional belief systems in Melanesian societies are more likely to attribute "the health and abundance of natural resources to the spirit realm" (Foale 2004, pp. 2–3). For example, an ethnographic study among Porgerans in the highlands of PNG reveal understandings of the cause of climate change as "not due to their own impacts on the environment, or the global increase of greenhouse gases, but rather as a societal breakdown between themselves and the rituals oriented toward powerful spirits that control the cosmos" (Jacka 2009, p. 206). Differences in worldview can thus lead to dissimilar emphases on the contribution of human activities to environmental degradation, as well as the efficacy of human interventions (Foale 2004; Nunn and Kumar 2019).

The most recent PNG census data available shows that the majority of citizens identify as Christians[1] (96%) (National Statistics Office 2011). Given the influence of religious narratives in contemporary PNG society (Gibbs 2005), a number of anthropologists have sought to document the ways in which Christianity is now "integrating with traditional resource management practices" (Jacka 2010, p. 24). For example, in an ethnographic study among the Urapmin in the West Sepik of

[1]Major denominations include, Roman Catholic (26%), Evangelical Lutheran (18%), and Seventh Day Adventist (National Statistical Office 2011, p. 33).

PNG, it was found that a key influence of Christianity on Urapmin ecology is that it has provided:

> … a rationale for doing away with almost all of the taboos that formerly interdicted food consumption and land use. The basis of the abrogation of these taboos is found in an Urapmin cliché: "God," they say, "made everything and He has left it for us to use." This is a clear statement of human dominion over the earth and its fruits, the familiar Christian idea of the human domination of nature." (Robbins 1995, p. 215)

A key point of interest in this literature appears to be the degree to which Christianity might erode indigenous ecological frameworks, and/or give rise to hybrid forms. For example, an ethnographic study on changing environmental values in Porgera found that the introduction of Christianity "has not resulted in the demise of beliefs in spirits, in general, although it has altered the beliefs in certain kinds of spirits and their powers" (Jacka 2010, p. 40). Furthermore, and akin to the interview data discussed in this paper, it was learned that many Porgerans understand environmental changes (e.g. climatic extremes of drought/flood cycles) "as somehow linked to the impending return of Jesus" (Jacka 2009, p. 205). These studies imply that conceptions of the environment and environmental change in PNG are overlaid with Christian belief systems, albeit to differing degrees, yet a strong emphasis appears to remain on the spirituality-environment nexus. Similar research in other Pacific Island countries has reached comparable conclusions (Nunn et al. 2016; Rudiak-Gould 2014).

Methods

Study Area

Madang Province (Momase Region) is located on the northern coast of the PNG mainland. In 2011 (most recent census data), the population of Madang Province comprised 493,906 people (6.8% of the PNG total population) (National Statistics Office 2011). Madang Province is environmentally diverse, extending from the central mountain ranges on the island of New Guinea to the offshore islands of Manam, Karkar, Bagabag, and Long Island (Bourke et al. 1994). Madang has a tropical climate. Temperatures are highest on average in May and most precipitation falls in April (average 418 mm).

There are considerable variations in economic development across Madang Province. Those located closest to the town centre have access to modern services and cash income. Other communities are based in remote areas that are more dependent on subsistence agriculture (Bourke et al. 2016). Madang is home to Divine Word University; a number of hotels; a nearby Chinese operated nickel cobalt mine (MCC Ramu Nico Ltd); and the Philippines-owned RD Tuna fishing company. Many people travel to Madang from other provinces in search of small business and employment opportunities and to access services (Sumb 2015).

Numerous conservation programs have been implemented by non-governmental organisations in Madang (Benson 2012). Recent research engagement has also been conducted with communities in Madang Province on the impacts of climate change, current and projected (Jacobs et al. 2020). The lead author of this chapter, for example, was involved in a project involving three communities (Riwo, Ohu and Bundi) in Madang province that revealed the ways in which these communities are experiencing threats to biodiversity that they, in part, attribute to a changing climate. Changes in rainfall and rising temperatures were considered among the three communities to be the most important climate changes. An increasing number of hot days is also limiting the number of hours that women could work in food gardens, something also reported from a study of rural Fiji communities (Nunn and Kumar 2019). Further, for coastal communities such as Riwo, rising sea levels are a major concern. These changes have flow-on effects for food security, human health, and ecosystem health and function (Foale 2006). Although Madang has established a Provincial Climate Change Committee, it has no resources and currently is actively supported only by the agriculture agency.

The Madang Town Market

The current site of the Madang Town Market was rehabilitated and re-opened in 2016 through funding provided by the Japanese International Cooperation Agency (JICA). The redevelopment followed concerns raised by the Government of PNG regarding damaged roofs, lack of space, and unhygienic conditions of the previous facility. From a sample of 210 respondents, a survey conducted by JICA (2013) found that 81% were female. The majority were residents of a range of Districts across Madang Province and 28% from outside of the Province (JICA 2013). The daily quantity of products including vegetables and fruits is more than 8 tons and the sales amount of each retailer per day mostly falls between 50 and 150 PNG Kina (USD 43.21) (JICA 2013).

In Madang and elsewhere in PNG, it is mostly women who produce and sell fruit and vegetables at local marketplaces (Busse and Sharp 2019; Gwatirisa et al. 2017; Anderson 2008). Selling fruit and vegetables at the market can be more profitable, flexible and often safer for women in PNG compared to 'formal sector' employment at foreign-owned companies. As Havice and Reed (2012) write:

> …women selling produce or fish at local markets earn incomes similar to those of tuna processors [between PNG Kina 150–200, roughly US$66–88], but work fewer hours…They market their wares 3–5 days per week, balancing market time with child care, farming, fishing, household duties, or extra wage work such as mending clothes. (p. 426)

In addition to family livelihoods, Busse and Sharp (2019) write that marketplace incomes:

> …contribute to rural food security, through improved nutrition from the purchase of energy dense foods, and by enabling people to purchase food when food crops fail. (p. 127)

Beyond selling and buying produce, marketplaces in PNG are:

> …important meeting places where people gather to see friends, hear the latest news, attend court cases, play cards and be entertained…While marketplaces are places of trade, trade is only part of what goes on in marketplaces. (Busse and Sharp 2019, pp. 126–7)

Places categorised under the term *maket* in Tok Pisin, the common language of PNG, range in size and scale and include large urban marketplaces like that at Madang, small clusters of sellers on roadsides and street corners (*kona maket*), and small stores in the front of individual houses (*haus market*) (Busse and Sharp 2019). There are many marketplaces in the Madang town centre, selling food, betel nut, tobacco, *bilums* (string bags) and clothing. This paper focuses on the largest market in Madang, the Madang Town Market.

Qualitative Interviews

14 market sellers participated in this research, 10 women and 4 men. Interviews were conducted at market stalls while vendors sold produce and interacted with their customers. Interviews were conducted by two researchers (one Australian and one PNG national) together in English and translated synchronously into Tok Pisin.

Participants involved in this research came to the Market from a range of locations within Madang Province, as well as from the Western and Eastern Highlands Provinces. Highlanders travel to Madang in search of higher prices for produce grown in cooler climates, especially broccoli, cauliflower, potatoes and sweet potato. Two main categories of sellers participated in this research: producers and re-sellers. Re-sellers buy produce from locations outside of Madang and/or purchase from bulk sellers at the market gate. With the exception of those selling peanuts, most market sellers trade a range of vegetables and fruits at their stalls.

The amount of income generated by market sellers varies according to types of produce being sold and the level of demand. The lowest average reported income was 5–6 Kina (USD 1.44–1.73) per day by a woman selling coconuts only (Participant 1). In contrast, the highest reported income was 500–700 Kina (USD 144.04–201.66) per day by a man from Goroka (Eastern Highlands) who sells a wider range of products including onion, carrot, ginger, beans, cabbage, pineapple, tomato, radish and lettuce (Participant 11). Most participants reported incomes averaging 50–150 Kina (USD 14.40–43.21) per day. Broccoli and fruits (especially melon and rambutan) are the most profitable items; "cabbage is also good, and it can be stored to sell for 2–3 days" (Participant 4).

To analyse the interview data, the authors followed the four basic steps of thematic data analysis outlined by Green et al. (2007): data emersion, coding, creating categories and identifying themes. As an emphasis on Christian beliefs emerged inductively from the data, the authors then reflected on existing literature related to faith-based approaches to sustainability to inform the discussion.

Results

Observed Changes in Fruit and Vegetable Produce

The majority of participants have observed long-term changes in their produce such as, size, quality, quantity, taste, colour and smell. These changes were attributed mainly to rising temperatures by 36% of respondents (Participants 1, 3, 7, 12 & 13). As one participant said:

> …market vendors talk about it together. The quality of the food has changed, and we discuss this. It's really hot. (Participant 1)

Another two explained:

> …even in the highlands it is becoming hot. It is not affecting cabbage, but it is affecting broccoli and cauliflower and potato. (Participants 12 & 13)

Warmer temperatures are said to be impeding the growth of taro, greens, bananas and bok choy in particular. Crops previously only able to be grown in PNG's coastal areas (e.g. bananas and peanuts) are now also growing in the Highlands (Participant 5, 12, 13, 11), something attributed to recent warming. Some participants report spending less time in their gardens due to the increasing daytime heat (Participant 4).

Compounding the impacts of increasing temperatures, some participants have observed less rainfall (Participants 1, 3, 5), commonly referred to as a "longer dry season". As the following quotes illustrate:

> …before we had a dry season and then rain. Now there is one day hot but then no rain. We can't do gardening because it is too hot. It is a dry season for a whole week and the soil will break. (Participant 4)

> …there is a longer dry season and the sun is really hot. In the dry season, sometimes the water dries out and it makes life hard. We have to look for water to wash and have to fetch water for the garden, especially the greens. (Participant 5)

> …there is little rainfall and the heat is much more than before. It's like burning. It was alright before but now little rain. (Participant 9 & 10)

While less rainfall is thought to be inhibiting the growth of produce in the Madang low country such as bok choy (Participant 3), two participants from the Highlands are experiencing too much rain, or continuous rain and then a period of no rain (Participants 11, 14), which they adjudge a problem. As the following market sellers explain,

> …before there was equal balance between rain and sun and [now] more rain is affecting the produce. There is more leaf than fruit on the *kaukau* [sweet potato] and yam. (Participant 11)

> …the continuous rain makes it so we don't harvest properly. It affects how we make gardens… [There is] not enough sunlight. Continuous rain and then [a] period of no rain. (Participant 14)

Three participants reported no observed changes in produce including, peanuts (Participant 6), carrots, onion, ginger (Participant 5), and sago (Participant 8). One re-seller also explained that while warmer temperatures are damaging vegetables such as bok choy, the quality of ginger and carrots is now actually better because it has become warmer for longer (Participant 3).

Interpretations of the Cause of Change

When participants were asked about the cause of these changes in the climate, a common theme quickly emerged in most interviews referring to Christian notions of "the Last Days" or "the End Times". The following quotes illustrate the point:

> … These things are because we're living in the Last Days. So, there is not much we can do about it. That's why these things are happening. (Participant 11)

> … the Last Days in biblical terms – that's the only thing and we have to go to church and start changing our ways. It's God who is doing this. (Participant 1)

> … I'm thinking it is the end of the world and God will come. *Big man ba come [big man [God or Jesus] will return]*. The world is ending – that is why the climate is changing. The only person with control of this is God. (Participant 2)

> …thinking the End of Days is why this is happening. Before it was never happening. The end of times. According to the Bible, in Noah's time, in the Old Testament, water covered the whole earth. Brought new changes. We will go through the same. Fire will consume the earth. This heat could be that end of world is coming. Before we used to be living in cold weather. (Participants 12 & 13)

Participants also associated the healthy growth and cultivation of their garden produce with their faith in God. This is clearly expressed by the following participant;

> …food is growing well because God blesses the garden. If you have a good relationship with God, it will grow well. (Participant 6)

The participants identify with a range of Christian denominations including Seventh Day Adventist, Lutheran and Catholic and report learning about the End Times at church and through reading the Bible.

There are two exceptions on the cause of observed changes amongst interviewees. One participant made reference to global warming which she had learned about at school "but I stayed out of school to help my family and [have now] forgotten about it" (Participants 12 & 13). Another participant linked climate change to the testing of nuclear weapons:

> …It's a global problem because bigger countries are building nuclear weapons and testing bombs. America, North Korea, Russia – all the big countries. (Participant 11)

Perspectives on Climate-Change Mitigation and Adaptation

When asked what might be done to adapt to and minimize the changes observed by the participants, the interview data highlight the limited agency of humans when God is seen to be the sole authority:

> …it's the Creator himself that brings these changes. I can try my best to change it but I can't. Everything happens according to God's will. (Participant 5)

> …we are fighting with nature up in heaven. It is uncontrollable, I think. We can talk about climate change, but we can't change it. (Participant 4)

Based on this understanding, most interviewees consider that the primary response of humans should be church attendance and repentance:

> …we need to share the word of God. We need to repent and go to church before the end of the world. (Participant 2)

> …to avoid it, we should continue to go to church and worship God – if we do that we will go to heaven. (Participants 12 & 13)

One participant noted that governments across the world also have an important role to play in mitigating climate change:

> …the Governments of other countries should shut down big factories and use less trucks and use bicycles. Small countries like PNG and Pacific countries have to do awareness on bushfires… The Government here in Madang is cutting trees down. The people are meant to be caring about greenhouse gas. (Participant 4)

Yet the same participant expressed little confidence in governments to act on climate change, particularly in PNG, largely due to concerns about corruption:

> …the leaders don't see this one. They give money to their friends. Nothing is done. (Participant 4)

Awareness and Understanding of the Term 'Climate Change'

The majority of participants report having heard of the term 'climate change', primarily through conversations on the PMV (Public Motor Vehicle) while transporting their produce to market (Participants 7, 9, 10), via the radio (Participants 7 & 14), from family members (Participants 9 & 10) and pastors (Participant 1). Descriptions of climate change include:

> …climate change is referring to strong winds, rainfall, sea-level rise, earthquakes. (Participant 11)

> …I have heard of climate change on the radio. That's why the seasons are changing. (Participant 2)

> …climate change is a change in weather pattern. (Participant 3)

Two of the 14 interviewees had never heard the term ‘climate change’ (Participants 1 & 6). One young woman, for example, explained that she does not have access to the media as she lives in the bush (Participant 6). Others said that although they have heard of the term, their knowledge about its meaning is limited:

> …it feels like it is not a good thing. I don’t really understand what it is – I’m not informed about it. (Participant 7)

> …I only hear about it on the radio. But I don’t understand it. (Participant 14)

> …I don’t know why the climate has changed. (Participant 9)

Most participants expressed a desire for more information about climate change as existing knowledge is mostly “hearsay – conversations with other people” (Participant 10). One participant suggested this information be distributed via TV and radio (Participant 3). Yet another participant expressed doubts about the value of increasing awareness because of his doomsday beliefs. He stated:

> …media people talk about it. I don’t think anyone on earth can change this. We try to create awareness, but the decision has been made. We can talk about it, but we are getting to the end of the world. (Participant 4)

Discussion

The findings of this study support arguments that holistic worldviews require holistic approaches to sustainability (Moyer et al. 2012; Havea et al. 2017). This is particularly relevant for Melanesian societies such as PNG where conceptions of the environment fuse “the social, cultural, political, economic, and environmental landscapes” (Banks 2002, p. 41).

This chapter has highlighted the influence of spirituality, specifically Christianity, on interpretations of the causes of observed environmental change. Changes in the environment such as rising temperatures and changes in rainfall patterns were believed by most participants to be affecting the quality of their produce. This was primarily interpreted with reference to the “Last Days”, “the End Times”, and the second coming of Christ. These beliefs (albeit, from a very small sample size) mirror research on the spirituality-climate change nexus conducted in other Pacific Island Countries such as Tonga, where it is said that “because God made this universe, He is the only one who can reverse these impacts of climate change and natural hazards from these coastal areas and communities” (Havea et al. 2017, p. 10). Similar beliefs have been encountered in Tuvalu and Fiji and in the Pacific Islands region more generally (Mortreux and Barnett 2009; Janif et al. 2016; Nunn et al. 2016). These beliefs are shared by a proportion of the market sellers interviewed for this study, whose primary response to changes observed in the environment is to attend church and reflect upon their personal relationship with God. If they have a good relationship with God, it is thought that their garden will be blessed and result in a good healthy harvest. Yet in apparent contradiction to this, most participants expressed interest in increasing their knowledge about climate change and how it might affect their future livelihoods and wellbeing.

Previous research has found that information about climate science can be delivered in ways that are respectful to Christian (and other) belief systems (Moyer et al. 2012). In response to fatalistic views towards climate change mitigation for example, Luetz et al. (2018) state that information on current and future climate impacts can be supplemented with messaging on stewardship and 'creation care'

> …to denote the idea that humankind is chiefly responsible for the sustenance, maintenance, or simply the 'care' of God's creation. (p. 54)

Further, drawing from the experience of faith-based organisations in Kenya, Moyer et al. (2012) write that;

> …people appreciate learning about their environment, their relationship to the land and their everyday activities from a biblical perspective, having never been shown these aspects of the Bible before. (p. 81)

Opportunities for Future (Participatory) Research

In the PNG context, it is possible to imagine Christian communities playing an active role in initiating activities on environmental change, but they may need assistance and initiative from an external organisation. One practical example might be the beginning of a church outreach program designed for market vendors, which starts by exploring the notion of stewardship and then introduces "up-to-date environmental information" (Hitzhusen 2012, p. 40). In Madang, which houses Divine Word University's main campus and also a conservation not-for-profit organisation (New Guinea Binatang Research Centre), initiating dialogue between climate science, Christian communities and market vendors would seem both practical and achievable. This could take the form of an action research project coordinated by researchers at Divine Word University, with the participation of church leaders, community members/leaders, representatives of the Madang Provincial Administration, and climate experts from organisations such as, New Guinea Binatang Research Centre, the PNG National Research Institute (NRI), Conservation International (CI) PNG, and the PNG Climate Change Development Authority. The project would be centred on finding a contextualized and common understanding of the causes and effects of climate change, considering the different worldviews present; and discussion of community-level solutions to address identified climate related risks and hazards. The findings of this chapter indicate that to be effective, the project would need to incorporate not only climate science knowledge, but also Christianity and local, indigenous knowledge. Themes and questions that may be beneficial to incorporate into project activities include:

- The changing climate: is it really happening? How can we tell? What has caused these changes?

- Christianity and climate change: what does Christianity say about a changing climate? What can or cannot be done? Are there visible connections between climate science and Christianity? If so, can they be integrated to formulate a common understanding?
- Traditional practices and beliefs of the environment: What traditional/cultural practices or beliefs were used in the past that protected the environment and encouraged conservation? Can these practices be applied today? Why? Why not?
- Middle-ground: Where can we find common ground? How can we work together to mitigate and adapt to climate change?
- Moving forward: What practical approaches to addressing climate change could be implemented at the community level? Who should be involved in these projects or programs? Who should take lead?

A further recommendation is that sustainability practitioners working in PNG ensure the involvement of religious leaders in the design, implementation, monitoring and evaluation of conservation programs. This might necessitate an initial dialogue phase whereby practitioners and religious leaders share their worldviews concerning the environment and climate change, which may require external facilitation. As Moyer et al. (2012) write,

> …while there are technical, scientific, and political elements to addressing environmental problems, at the root of the issues are profound ethical decisions and worldview orientations which fit more into the realm of faith than they do into the realm of science, which has been a primary realm in which these issues have been engaged. (p. 985)

Efforts to establish new collaborations between researchers, non-government organisations and church congregations, can also draw on guidance materials developed for the international development sector, such as Research for Development Impact Network's (2018) 'How to collaborate with Pacific Churches for development research' handbook.

Conclusion

Market sellers in Madang are observing changes in the environment which are perceived to be impacting on the quality and quantity of fruit and vegetable produce. The principal causes identified are rising temperatures and changes in rainfall patterns. The root cause of these changes is interpreted with reference to Biblical notions of the 'End Times', which appears linked to defeatist responses to climate change mitigation and adaptation. Despite this belief, there is eagerness for more information on climate change among market sellers. It has been argued that this can be achieved in ways that respect Christian beliefs. Two examples are increasing engagement with notions of stewardship and creation care within Church communities and improving and increasing dialogue between climate scientists and religious communities through participatory and collaborative research processes.

Acknowledgements We would like to acknowledge the contributions of Calista Hamadi and Samuel Ray to the development of this chapter. We are also grateful to Johannes M. Luetz and Patrick D. Nunn for their generous and thoughtful feedback on earlier drafts of this chapter.

References

Anderson T (2008) Women roadside sellers in Madang. Pac Econ Bull 23(1):59–73

Asian Development Bank (2020) Poverty data: Papua New Guinea. https://www.adb.org/countries/papua-new-guinea/poverty

Banks G (2002) Mining and the environment in Melanesia: contemporary debates reviewed. *The Contemporary Pacific* 39–67. https://doi.org/10.1353/cp.2002.0002

Becker M, Meyssignac B, Letetrel C, Llovel W, Cazenave A, Delcroix T (2012) Sea level variations at tropical Pacific islands since 1950. Glob Planet Chang 80–81:85–98

Benson C (2012) Conservation NGOs in Madang, Papua New Guinea: understanding community and donor expectations. Soc & Nat Resour 25(1):71–86. https://doi.org/10.1080/08941920.2011.603141

Bureau of Meteorology (BOM) & Commonwealth Scientific and Industrial Research Organization (CSIRO) (2014) Climate variability, extremes and change in the Western Tropical Pacific: new science and updated country reports. *Pacific-Australia climate change science and adaptation planning program technical report.* https://www.pacificclimatechangescience.org/wp-content/uploads/2014/07/PACCSAP_CountryReports2014_WEB_140710.pdf

Bourke RM, Allen B, Lowe M (2016) Estimated impact of drought and frost on food supply in rural PNG in 2015. *Development Policy Centre Policy Brief* 11

Bourke RM, Allen BJ, Hide RL (1994) Notes on smallholder agriculture, Eastern Highlands Province. Department of Human Geography, RSPAS, Australian National University, Papua New Guinea

Bryant-Tokalau J (2018) Indigenous Pacific approaches to climate change. Springer, Berlin Heidelberg

Busse M, Sharp TL (2019) Marketplaces and morality in Papua New Guinea: place, personhood and exchange. Ocean 89(2):126–153. https://doi.org/10.1002/ocea.5218

Climate Change Development Authority (CCDA) (2017) Papua New Guinea national REDD + strategy 2017–2027. https://redd.unfccc.int/uploads/4838_1_papua_new_guinea_national_redd_2B_strategy.pdf

Cobon DH, Ewai M, Inape K, Bourke RM (2016) Food shortages are associated with droughts, floods, frosts and ENSO in Papua New Guinea. Agric Syst 145:150–164

Connell J (2016) Last days in the Carteret Islands? climate change, livelihoods and migration on coral atolls. Asia Pac Viewp 57(1):3–15. https://doi.org/10.1111/apv.12118

Foale S (2004) Commensurability of scientific and indigenous ecological knowledge in coastal Melanesia: implications for contemporary marine resource management strategies. *Resource management in Asia-Pacific Working Paper* 38. https://openresearch-repository.anu.edu.au/bitstream/1885/40986/3/rmap_wp38.pdf

Foale S (2006) The intersection of scientific and indigenous ecological knowledge in coastal Melanesia: implications for contemporary marine resource management. Int Soc Sci J 58(187):129–137. https://doi.org/10.1111/j.1468-2451.2006.00607.x

Gibbs P (2020) Witchcraft accusations and hybrid formulations in Papua New Guinea. On Knowing Humlty J 4(1):65–68

Gibbs P (2005) Political discourse and religious narratives of church and state in Papua New Guinea. *SSGM Working Papers* 1. https://openresearch-repository.anu.edu.au/bitstream/1885/43157/2/05_01wp_Gibbs.pdf

Gibson J, Datt G, Allen B, Hwang V, Parajuli D, Bourke M (2005) Mapping poverty in rural Papua New Guinea. Pac Econ Bull 20(1):27–43

Green J, Willis K, Hughes E, Small R, Welch N, Gibbs L, Daly J (2007) Generating best evidence from qualitative research: the role of data analysis. Aust N Z J Public Health 31(6):545–550. https://doi.org/10.1111/j.1753-6405.2007.00141.x

Gwatirisa PR, Pamphilon B, Mikhailovich K (2017) Coping with drought in rural Papua New Guinea: a Western Highlands case study. Ecol Food Nutr 56(5):393–410. https://doi.org/10.1080/03670244.2017.1352504

Havea PH, Hemstock SL, Des Combes HJ, Luetz JM (2017) "God and tonga are my inheritance!"—climate change impact on perceived spirituality, adaptation and lessons learnt from Kanokupolu, 'Ahau, Tukutonga, Popua and Manuka in Tongatapu, Tonga. In: Leal Filho W (ed) Climate change impacts and adaptation strategies for coastal communities. Springer Nature, Cham, pp 167–186. https://doi.org/10.1007/978-3-319-70703-7_9

Havice E, Reed K (2012) Fishing for development? Tuna resource access and industrial change in Papua New Guinea. J Agrar Chang 12(2–3):413–435

Hayward-Jones J (2016) The future of Papua New Guinea: old challenges for new leaders. Lowy Institute for International Policy. https://www.files.ethz.ch/isn/196414/future-of-papua-new-guinea1.pdf

Hitzhusen GE (2012) Going green and renewing life: environmental education in faith communities. New Dir Adult Contin Educ 2012(133):35–44. https://doi.org/10.1002/ace.20005

Jacka J (2010) The spirits of conservation: ecology, Christianity, and resource management in highlands Papua New Guinea. J Study Religion Nat Cult 4(1):24–47

Jacka J (2009) Global averages, local extremes: the subtleties and complexities of climate change in Papua New Guinea. In: Susan A Crate, Mark Nuttall (eds) Anthropology and climate change: from encounters to actions. Routledge, Oxon, pp 197–208

Jacobs B, McKenna K, Boronyak L, Dem F, Sui S, Pomoh K, Jimbudo M, Maraia H (2020) Engaging communities and government in biodiversity conservation and climate change adaptation in Papua New Guinea. In W Leal Filho (ed) Managing climate change adaptation in the Pacific region, climate change management series. Springer, Switzerland, pp 213–230

Janif S, Nunn PD, Geraghty P, Aalbersberg W, Thomas FR, Camailakeba M (2016) Value of traditional oral narratives in building climate-change resilience: insights from rural communities in Fiji. Ecol Soc 21(2). http://dx.doi.org/10.5751/ES-08100-210207

Japan International Cooperation Agency (JICA) (2013) Preparatory survey report on the project for rehabilitation of Madang Town Market in the Independent State of Papua New Guinea. Unpublished document

Jupiter S (2017) Culture, kastom and conservation in Melanesia: what happens when worldviews collide? Pac Conserv Biol 23(2):139–145. https://doi.org/10.1071/PC16031

Koczberski G, Curry GN, Bue V, Germis E, Nake S, Tilden GM (2018) Diffusing risk and building resilience through innovation: reciprocal exchange relationships, livelihood vulnerability and food security amongst smallholder farmers in Papua New Guinea. Hum Ecol 46(6):801–814

Kumar V, Melet A, Meyssignac B, Ganachaud A, Kessler WS, Singh A, Aucan J (2018) Reconstruction of local sea levels at south west pacific islands: a multiple linear regression approach (1988–2014). J Geophys Res-Ocean 123:1502–1518

Leal Filho W, Ha'apio MO, Luetz JM, Li C (2020) Climate change adaptation as a development challenge to small Island states: a case study from the Solomon Islands. Environ Sci & Policy 107:179–187. https://doi.org/10.1016/j.envsci.2020.03.008

Lipset D (2011) The tides: masculinity and climate change in coastal Papua New Guinea. J Roy Anthropol Inst 17(1):20–43. https://doi.org/10.1111/j.1467-9655.2010.01667.x

Luetz JM, Nunn PD (2020) Climate change adaptation in the Pacific Islands: a review of faith-engaged approaches and opportunities. In: Leal Filho W (ed) Managing climate change adaptation in the Pacific region. Springer, Cham, pp 293–311

Luetz JM, Havea PH (2018) "We're not refugees, we'll stay here until we die!"—climate change adaptation and migration experiences gathered from the Tulun and Nissan atolls of Bougainville,

Papua New Guinea. In: Leal Filho W (ed) Climate change impacts and adaptation strategies for coastal communities. Springer Nature, Switzerland, pp 3–29. https://doi.org/10.1007/978-3-319-70703-7_1

Luetz JM, Buxton G, Bangert K (2018) Christian theological, hermeneutical and eschatological perspectives on environmental sustainability and creation care—the role of holistic education. In: Luetz JM, Dowden T, Norsworthy B (eds) Reimagining Christian education, Springer, Singapore, pp 51–73. http://dx.doi.org/10.1007/978-981-13-0851-2_4

Mortreux C, Barnett J (2009) Climate change, migration and adaptation in Funafuti, Tuvalu. Glob Environ Change 19:105–112

Moyer JM, Sinclair AJ, Spaling H (2012) Working for God and sustainability: the activities of faith-based organizations in Kenya. Volunt: Int J Volunt Nonprofit Organ 23(4):959–992

National Statistics Office (2011) Papua New Guinea 2011 National Report. Prepared by the National Statistics Office, Port Moresby, Papua New Guinea. https://png-data.sprep.org/system/files/2011%20Census%20National%20Report.pdf

Nunn PD, Kumar R (2019) Measuring peripherality as a proxy for autonomous community coping capacity: a case study from Bua Province, Fiji Islands, for improving climate change adaptation. Soc Sci 8:1–26

Nunn PD, Mulgrew K, Scott-Parker B, Hine DW, Marks ADG, Mahar D, Maebuta J (2016) Spirituality and attitudes towards Nature in the Pacific Islands: insights for enabling climate-change adaptation. Clim Change 136:477–493

Nyawo S (2017) Are prayers a Panacea for climate uncertainties?: an African traditional perspective from Swaziland. EcumAl Rev 69:362–374

Office of Climate Change & Development (OCCD) (2014) National climate compatible development management policy. https://www.pacificclimatechange.net/sites/default/files/documents/National_Climate_Change_Policy1.pdf

Ramakrishna A, Bang S (2018) The impacts of climate change on food security and management in Papua New Guinea. In: Information Resources Management Association (ed) Climate change and environmental concerns: breakthroughs in research and practice. IGI Global, Hershey, PA, pp 219–241

Robbins J (1995) Dispossessing the spirits: Christian transformations of desire and ecology among the Urapmin of Papua New Guinea. Ethnol 34(3):211–224. https://doi.org/10.2307/3773824

Rudiak-Gould P (2014) The influence of science communication on indigenous climate change perception: theoretical and practical implications. Hum Ecol 42:75–86

Schmidt E, Gilbert R, Holtemeyer B, Rosenbach G, Benson T (2019) Papua New Guinea survey report: Rural household survey on food systems (Vol. 1801). International Food Policy Research Institute

Schuman S, Dokken JV, Van Niekerk D, Loubser RA (2018) Religious beliefs and climate change adaptation: a study of three rural South African communities. Jamba-J Disaster Risk Stud 10

Sharp TL (2019) Haggling highlanders: marketplaces, middlemen and moral economy in the Papua New Guinean betel nut Trade. Ocean 89(2):182–204. https://doi.org/10.1002/ocea.5221

Sullivan N, Warr T, Rainbubu J, Kunoko J, Akauna J, Angasa M, Wenda Y (2003) Tinpis maror: a social impact study of proposed RD Tuna Cannery at Vidar Wharf, Madang

Sumb A (2015) Analysing business opportunities in the Madang Lagoon area. Contemp PNG Stud 22:119–131

Vallance RJ (2012) A Melanesian research methodology. Paper presented at the Joint AARE/APERA Conference, Sydney. https://bit.ly/2VWtsuA

Wiegand M, Koomen E, Pradhan M, Edmonds C (2017) The impact of road development on household welfare in rural Papua New Guinea. Tinbergen Institute Discussion Paper, No 17-076/V. https://www.econstor.eu/bitstream/10419/177644/1/17076.pdf

Chapter 9
Climate Change and Eschatology

Kurt Bangert

Abstract Global Warming and other environmental hazards have conjured up among Bible readers an impending end-time apocalypse. Based on a previous paper offering three biblical eschatological scenarios (the Rainbow Model, the Nineveh Model and the Apocalypse Model), this paper expands on the three scenarios, giving more theological and, if any, historical background. While the Rainbow Model (Noah's flood story) assumes that another worldwide destruction will never again take place and the Apocalypse Model (Jesus' end-time expectation) assumes the ultimate end-time world annihilation, the Nineveh Model (Jonah's doomsday message) assumes that the prospect of an envisioned end-time annihilation can be avoided through repentance and a change of behavior. For each scenario, this paper looks at possible historical roots and the theological intent. The result of the study suggests that each of the three narratives (Noah, Jesus, Jonah) is characterized by an inherent *component of contingency,* explicit or not, and that none of them is unequivocally unconditional. Each narrative assumes the basic tenet of God's compassion and mercy in light of man's willingness to change heart. Explicit or not, all biblical predictions entail contingencies that make it impossible for man to exactly predict the future. While the biblical covenants put man into a special relationship with God, God remains at all times sovereign and supreme to extend His mercy and compassion to whomever He wishes if and whenever man confesses his sins and changes his ways.

Three End-Time Scenarios Revisited

Global warming, environmental degradation and the extinction of many species at times conjure up doomsday scenarios of an impending end-time apocalypse. In light of humanity's relentless exploitation of nature, some orthodox people interpret such apocalyptic end-time scenarios as the fulfilment of biblical eschatological prophecies about the end of the world. Without any doubt, there are such biblical end-time prophecies: "Heaven and earth will pass away." (Mt 24:35 ESV) Or: "We are looking

K. Bangert (✉)
Bad Nauheim, Germany
e-mail: kontakt@kurtbangert.de

J. M. Luetz and P. D. Nunn (eds.), *Beyond Belief*, Climate Change Management,
https://doi.org/10.1007/978-3-030-67602-5_9

forward to a new heaven and a new earth, where righteousness dwells." (2Pet 3:13 NIV; see also: Is 65:17 or Rev 21:1). Hence, for today's Biblicist end-time soothsayers the end of the world is imminent, doomsday is just around the corner, and God is about to establish His own kingdom (see Grey 2013). Climate change—no longer questioned today (IPCC 2015)—seems to confirm the expectation of the world's annihilation and the second coming of Christ. Should it be true that mankind is meddling with our climate (anthropogenic causes of global warming—which some are still doubting), then man's hubris and sinfulness only serves to speed up the coming of the final judgment and the "restoration of all things, which God has spoken by the mouth of all His holy prophets since the world began" (Acts 3:21 NKJV).

There is ample evidence of climate denial due to religious reasoning. Referring to climate change, the Republican Representative Tim Walberg (Michigan) once told his constituents: "As a Christian, I believe that there is a creator God who is much bigger than us"; so he was quoted by the *Washington Post*, "And I'm confident that, if there's a real problem, he can take care of it." (Washington Post 2017) The popular Christian American Hal Lindsey, referring to God's covenants with this earth, expressed his belief "that as long as this old earth is floating in space, there will be no man-made climate change" (Lindsey 2015). Lisa Vox, author of the book *Existential Threats: American Apocalyptic Beliefs in the Technological Era,* came to the conclusion: "We must accept that a number of conservative evangelicals, especially from the older generations, will never support significant action on climate change, especially if it means signing a global treaty" (Washington Post 2017).

Climate denial is prevalent also in the very countries and regions most affected by global warming, such as in the Pacific Islands. Teburoro Tito, a former President of Kiribati, was quoted as saying that "God is not so silly to allow people to perish just like that" (Fair 2018, p. 5). Some authors have belittled religious thought as being "barriers" in the fight against global warming. On the other hand, though, while there is evidence "that the Church has significant potential as a climate change actor" there is also, according to Hannah Fair, a need "to consider the power and potency of religious ideas themselves" (Fair 2018, p. 3) which may serve to support an appropriate climate response in tune with scientific evidence.

A minority of Marshall Islanders, for instance, have combined scientific knowledge and religious faith as a springboard for action (Rudiak-Gould 2009). According to Fair, "while there is both the potential and prerogative for spiritualising climate change, it must be recognised that not all religious groups approach the morals and ethics of climate change in the same way" (Fair 2018, p. 4). Fair strongly advocates a "balancing" of religious and scientific knowledge which she terms *tufala save* (meaning "double knowledge"), but she also recognizes that Pacific Islanders while seeking to balance religious and scientific knowledge, most often give priority to their religious beliefs. One Presbyterian pastor was quoted as saying: "Yeah, we can by scientific research say something will happen but if God says there will be rain, there'll be rain. ... God is sovereign" (Fair 2018, p. 7).

Vital Questions Regarding Eschatology

In light of the foregoing, Luetz, Buxton and Bangert have asked these vital questions:

> Does climate change herald the end of the world? Does environmental degradation constitute the beginning of the end? Is God about to create a new heaven and a new earth because we are on the verge of destroying the one He entrusted to us in the first place? Must the world really pass away before God's kingdom can come? Did God expect humans to destroy creation so that He would replace it? What did Jesus actually mean when he spoke of the 'Kingdom of God'? Did he mean a future kingdom, or did he interpret the 'Kingdom of God' as something already begun in the here and now and [as something] that was and is to expand through his disciples? Has God set a time when the world would come to an end? Or is He leaving it up to humans to determine the time when the Earth becomes uninhabitable? Does the progressive anthropogenic destruction of God's creation in any way relieve humanity of the originally entrusted creational stewardship responsibility? Further, could it be that some Bible-believing Christians have adopted a sort of apocalyptic fatalism? Could it be that doomsday Christians, although not necessarily yearning for the end of the world, still reckon with it, as it were, in a somewhat placid and unperturbed fatalistic expectation, hoping that when the end comes, God will then be able to finally establish His kingdom? – according to the dictum: 'When the last tree has come down, Christ will come'.[1] (Luetz et al. 2018, p. 61 f.)

The authors responded to those questions by unfolding three possible end-time scenarios (or *models*) which they thought could help in dealing responsibly with the Bible's eschatological language, especially in light of what they called the "creation crisis". They hoped that these three models would help clarify the crucial question of whether it made sense for Christians to engage in fighting global warming, considering that—according to Biblicist end-time prophets—the imminent end of the world cannot be avoided anyhow.

Three End-Time Models

The first end-time scenario was called the *Rainbow Model*. It was described as follows:

> After Noah and his family survived the flood (seemingly as the only family), he offered a burnt offering in gratitude for their escape. And God promised: 'Never again will I destroy all living creatures, as I have done. As long as the earth endures, seedtime and harvest, cold and heat, summer and winter, day and night will never cease.' (Gen 8:21–22 NIV). As a sign of this promise God reportedly created the rainbow: 'Whenever the rainbow appears in the clouds, I will see it and remember the everlasting covenant between God and all living creatures of every kind on the earth.' (Gen 9:16 NIV). According to the Rainbow Model, there will not be another worldwide apocalyptic catastrophe extinguishing virtually all life, because God has promised it. Relatedly and importantly, according to this model, there is

[1] The dictum is an adaptation of the Native American saying: "When the last tree is cut down, the last fish eaten, and the last stream poisoned, you will realize that you cannot eat money." J. Simpson and J. Speake (Eds): *The Oxford Dictionary of Proverbs*, 5th edition, Oxford University Press: Oxford 2009.

> no immediate need to take action on climate change and creation care because the earth, all living creatures and the climate are essentially predestined to endure unharmed in perpetuity. (Luetz et al. 2018, p. 63)

Another scenario was the *Nineveh Model* which is based on the prophet Jonah who, according to the narrative in the Book of Jonah, was sent by God to the city of Nineveh to preach doomsday judgment: "Go to the great city of Nineveh and preach against it, because its wickedness has come up before me." (Jon 1:2 NIV) Because Jonah had little inclination to be a doomsday preacher, "it took some dramatic detours before he eventually gave in and went to Nineveh to preach the message God had instructed him to deliver: 'Yet forty days, and Nineveh will be overthrown.' (Jon 3:4 KJV)." (Luetz et al., p. 65) But then came the surprising turnabout: "The Ninevites believed God. A fast was proclaimed, and all of them, from the greatest to the least, put on sackcloth." (Verse 5 NIV). Even the king "took off his royal robes, covered himself with sackcloth and sat down in the dust" (Verse 6 NIV). And when God saw how they turned away from their evil ways and repented, he decided not to destroy Nineveh (Verse 10). Luetz et al. commented this narrative:

> The lesson of the Nineveh Model is that apocalyptic announcements do not of necessity come to pass in every case but constitute first and foremost a call to repentance and behavioral transformation. According to the Nineveh Model, God could be giving humans – as the appointed stewards of His creation – another chance to repent and amend their ways. (Luetz et al. p. 65)

The third scenario was called the *Apocalypse Model*. It worked from the opposite assumption of the Rainbow model (Noah's flood) in "that there actually will be such a worldwide apocalyptic doomsday one day – regardless and in spite of the Rainbow motif." (Luetz et al., p. 64) This scenario was based on the end-time sermons of Jesus as reported in the New Testament. Jesus in fact drew a parallel to Noah's flood: "For in the days before the flood, people were eating and drinking, marrying and giving in marriage, up to the day Noah entered the ark; and they knew nothing about what would happen until the flood came and took them all away. That is how it will be at the coming of the Son of Man." (Mt 24:38–39 NIV) At the time of the final judgment which Jesus was here imagining, some would be saved, many others would be destroyed, leaving the rainbow promise aside. Luetz et al. commented:

> According to this model, there is no need to take action on climate change because the earth is destined to be doomed *anyway*, and all living creatures will ultimately be terminated by inescapable end time disaster. This model is alive and well, whenever politicians either deny or make light of the enormous damage unmitigated anthropogenic greenhouse gas emissions are inflicting on God's creation. (Luetz et al., p. 64)

For many who cling to the Apocalypse Model, "it is also too late to save the Planet, too hard and too futile to decarbonize the global economy, and perhaps even too 'unspiritual' to take action on climate change, because the Apocalypse is the preordained 'Divine design' in respect of the prophesied coming end" (Luetz et al., p. 64).

In synthesizing the three end-time models, Luetz et al. came to the conclusion—in light of the global warming and the general threat that humanity poses to the equilibrium of creation – that while they could not categorically exclude the possibility of an end-time Armageddon (Apocalypse model) as a contingency, they strongly favored the Nineveh Model: "It invites us to take very seriously the warnings of today's prophets, namely the thousands of scientific experts who advocate urgent change [...], begin to cover ourselves (figurately speaking) with sackcloths, and organize a dramatic turnabout of our thinking and behavior" (Luetz et al., p. 66).

The authors also reminded their readers of the oft-cited verse from 2 Chr 7:14: "If my people, which are called by my name, shall humble themselves, and pray, and seek my face, and *turn from their wicked ways;* then will I hear from heaven, and will forgive their sin, and will *heal their land.*" (NIV, emphasis by Luetz et al.) And the authors asked: "Could it be that such heartfelt human repentance may prompt God to hear, forgive and *heal* the land?" (Luetz et al., p. 65).

To summarize: While the *Rainbow Model* assumes that another worldwide destruction will never take place again (the rainbow being its warranty) and the *Apocalypse Model* presumes an unequivocal and ultimate end-time world annihilation, the *Nineveh Model* advocates that the prospect of an envisioned end-time annihilation can still be avoided if humanity repent from their destructive behavior and amend their ways to avoid the ultimate eschatological disaster. In other words: The detrimental (and potentially lethal) effects of climate change must not of necessity constitute the "end of the world" that Biblicist conservatives anticipate—IF humankind chooses, and successfully manages, to reverse its destructive course.

In this paper, I shall now make an attempt to give more background to the three models by looking at the respective *theological and*—if any—*historical* contexts of the Noah story, the Jonah story, and the apocalyptic scenarios which were so abundantly prevalent during the time of Jesus. In terms of the *Noah narrative*, I shall not only look at the theological intent of the story but also on what we today can know about the historical roots of the flood account. In terms of the *Jonah narrative*, I shall not only investigate the intent of the Book of Jonah and its plot, but also its later usage by Jesus (the "sign of Jonah") and its possible application to today's global challenges. In terms of the *apocalyptic end-time scenarios*, I shall investigate the nature and intent of apocalyptic literature in general as well as the intent of the apocalyptic passage in Matthew 24–25 where Jesus apparently announced the imminent end of the world (which German theologians have called *Naherwartung*). How must we interpret this imminent end-time expectation (*Naherwartung*) today—in light of it being thwarted and postponed again and again for some 2000 years?

The result of this paper will show that all of the three narratives (Rainbow, Nineveh, Apocalyptic scenario) are characterized by an inherent *component of contingency*. That is to say, none of them are unequivocally unconditional, but each contains an inherent conditionality or contingency. And the question is to be raised what this conditionality means in light of the current climate crisis.

The Rainbow Model (Noah's Message)

In the following section, I wish to first look at how the flood narrative has been utilized by the Pacific islanders, especially in relation to the climate crisis, before going on to consider the *theological* implication and possibly *historical* background of the *Rainbow Model*. From that, I shall draw some conclusions with regard to an eschatological perspective.

Noah's Flood as Primordial Archetype

In the Pacific Islands, Noah's flood narrative has been cited as "a reoccurring feature of discussions with pastors and parishioners" and thus "held meaning and relevance" for many islanders who were questioned by Hannah Fair. The biblical story is also a good example for what Fair calls "epistemological entanglements" as it highlights "the polysemic potential of scripture". Hence, Fair used the narrative "as a means to explore the balance, enmeshment and tensions between different epistemologies". There were three discursive manifestations of the flood narrative in the Pacific Islands, namely: (1) the rainbow covenant as denial, (2) Noah as example for preparation, and (3) Islanders being unjustly left outside of the ark (Fair 2018, p. 8).

While Fair did not directly encounter the first manifestation with her interlocutors, her interviewees did complain, however, that Noah's covenant was often cited by elderly people as a reason for inaction and for a religious counternarrative to the threat pictured by science. To them, the rainbow covenant serves as an argument to profess the ongoing continuity of the world, as was shown also by Rubow and Bird (2016). But most pastors who were questioned on the subject, explicitly rejected that interpretation.

The second manifestation of the Noah narrative was encountered in connection with cyclone disasters that may (have) hit the islanders. Pastors evoked Noah as an example for timely preparedness—not only in case of a cyclone hitting the islands but even in terms of climate adaptation. "Thus, here the story of Noah is not a vehicle for doubt, but is explicitly mobilised as a weapon against it …" (Fair 2018, p. 9) (Similarly, the biblical story of Joseph being prepared for a famine was invoked by some participants as an example of preparation.)

The third manifestation was encountered with Ezekiel, a church-based Tuvaluan climate advocate who emphasized the need for human action. According to this manifestation Noah is not the hero of the story but is counted among those who have safely sought refuge in the ark, while many others (including many animals) perished outside the ark. "Those outside the ark need to be liberated and I think God is with those who are outside the ark. God is struggling with them, trying to alleviate them while Noah is enjoying the luxury life, you know", said Ezekiel. Comments Hannah Fair: "While this reading is theologically unorthodox, it is highly pertinent to the question of how climate change is framed" (Fair 2018, p. 11).

I shall now review the theological and historical background of the Noah narrative.

Theological Background

According to the biblical flood narrative, God saw that the earth "was filled with violence" and that "all flesh had corrupted their way on earth" which caused God to decide "to make an end of all flesh" (Gen 6:12–13). "Behold, I will destroy them with the earth", He said (verse 13). But because He also saw that "Noah was a righteous man, blameless in his generation", He spared Noah the fate of the destruction and ordered him to build the ark so that he, his family, and "two of every sort" of animal would go into the ark and survive the flood (verse 19). Through the flood, God "blotted out every living thing that was on the face of the ground, man and animals and creeping things and birds of the heavens. They were blotted out from the earth. Only Noah was left, and those who were with him in the ark." (Gen 7:23 ESV) After Noah survived the flood, he built an altar to the Lord, and when the Lord smelled the pleasing aroma, the Lord said in his heart, "I will never again curse the ground because of man, for the intention of man's heart is evil from his youth. Neither will I ever again strike down every living creature as I have done. While the earth remains, seedtime and harvest, cold and heat, summer and winter, day and night, shall not cease." (Gen 8, 21–22 ESV) And God made a covenant between him and the earth. The rainbow in the sky was to be the sign of this covenant. "This is the sign of the covenant that I have established between me and all flesh that is on the earth" (Gen 9:17 ESV).

It is to be noted that God actually made two covenants with Noah. The first one was before the flood: "I shall establish my covenant with you, and you shall come into the ark …" (Gen 6: 17–19 ESV). This was a covenant that promised salvation from disaster. The second covenant was concluded after the rescue: "I establish my covenant with you, that never again shall all flesh be cut off by the waters of the flood …" (Gen 9:11 ESV). The first covenant was of a salvific nature and limited to a select group (Noah and his family and the animals inside the ark). The second covenant was of a preventive nature and a promise given to the whole world.

The theme of the *covenant* pervades the whole Bible. According to the biblical narrative, God made a covenant with Noah, later with Abraham, then with the people whom Moses and Joshua led from Egypt into the promised land. In the New Testament the "old covenant" is replaced by a new one (Hebr 8:13), according to which God "will put my laws in their hearts, and I will write them on their minds" (Hebr 10,16 NIV). In the Old Testament, the covenant (Hebrew *berit*) is often connected with the word *shalom* which not only means peace but also implies wholeness, wholesomeness and complete harmony (von Rad Bd 1, 1978, p. 144).

If Scripture is taken at face value, the covenants widen from Noah to the whole world, but then are narrowed down to Abraham, Isaac and Jacob (alias Israel) and their offspring, the people of Israel. Abraham is promised that his many descendants will inherit a vast land (from the Nile to the Euphrates). However, when the Bible is read from an historical-critical perspective, the covenant theology ought to be reversed: Based on the covenant consciousness of the Jews during the time of their exile, the covenant idea was projected back into Israel's earlier history (cf von Rad Bd 1, 1978, pp. 147–149.). For instance, when in Leviticus and Deuteronomy, it is repeatedly stated that God will scatter his people amongst the nations in response to their disobedience (Lev. 26:27–35; Deut. 4:25–31, 40; 8:19–20; 28:36–37, 62–68), then these references may be considered to be back projections by which the covenant idea is extended to the time of the exodus, then to the time of the patriarchs and finally even to the time of Noah. The covenant given to Abraham may have served to reassure the exiles not only of their identity as Abraham's offspring but also of their right to the holy land.

We must also recognize a contingency factor in the covenant idea. The fact that Jerusalem and its temple were actually destroyed and the Jews were indeed scattered amongst the nations suggests that God had, at least for the time being, abandoned his "eternal" covenant because His people had broken it repeatedly: "[T]hey did not obey your voice or walk in your law. They did nothing of all you commanded them to do. Therefore you have made all this disaster come upon them." (Jer 32:23 ESV) What was originally meant to be an *everlasting* covenant (cf. Gen 17:13; Num 18:19; 1Chr 16:17; Ps 105:10; Is 55:3; Jer 50:5; Ez 16:6 etc.) was cut short and—at least for the time being—terminated due to Israel's disobedience. Hence, despite God's intention to abide by his eternal covenant, the contingency of that covenant cannot be overlooked.

One of the important theological lessons to be learned from the flood story is that man is inherently depraved ("man's heart is evil from his youth": Gen 8:21), and that violence and evil will therefore always be an existing reality on earth. The destruction of the earth could not and would not prevent evil from rising up again, even amongst Noah's own descendants. Man's depravity is a given in this world. Nevertheless, it is not only desirable but also possible that some people (like Noah and others after him) can and will be righteous "in their generation". The continuation of the Genesis story (and the whole biblical account) offers ample evidence for man's depravity while at the same time giving many examples of men and women adhering to the covenant with God and living a righteous life.

The flood story, then, is a narrative that encourages man to live righteously and, hence, receive the blessings of a merciful and benevolent God who will consider man's righteousness as a "pleasing aroma". The story, of course, also assumes very strongly that there will not be another world-wide disaster destroying life on earth. That promise—vouched for by the rainbow in the sky—was a comforting and reassuring pledge that was to take away from humankind any fear of another such global disaster.

Historical Background

I will now turn to the historical aspect of the flood story. There are two questions over which Christians have battled for years. The first is: Is the flood story a myth or has it an historical basis? The second question is: If there is some historical basis to the flood story, was it a world-wide flood or a regional event? I will address both questions here.

As for the first question and the proposed alternative between *myth* and *history*, this is—in my opinion—a false dichotomy. For the Greek word *mythos* (μύθος) does not connote "unhistorical", but rather implies an event which transcends, in significance and meaning, the realm of time and space. In other words, a biblical event may very well have historical roots which in the course of time gained a mythical implication for man's existence. The myths of an ethnic group or a nation say much about that group's or nation's identity and self-understanding. What matters is not so much whether these myths are "true" in a strict historical sense but whether they are true in an existential sense. Myths are rooted in reality, but grow and blossom in the collective subconsciousness of an ethnic group which draws from it its identity, the meaning of its existence, and the interpretation of the lives of its members. Myths are the presuppositions upon which people base their thinking, reasoning, and *weltanschauung*.[2]

As for Noah's flood, there has come to the fore in recent years a theory which commands and deserves attention when we speak of the biblical flood narrative. In 1998 two geologists, Walter Pitman and William Ryan, published their book *Noah's Flood. The New Scientific Discoveries About the Event That Changed History* (Pitman and Ryan 1998). In it, the authors describe their theory about the Black Sea Flood which occurred some 7500 years ago (ca. 5500 B.C.) when the rising waters of the Mediterranean Sea spilled over the Bosporus to flood the Black Sea which up to that time had been a freshwater lake some 100 meters (more than 300 ft) below the level of the Mediterranean. The spilled-over salty seawater flooded much of the coastal areas of the Black Sea, especially in the north where the coastal regions were rather flat, turning the once freshwater lake into a deadly ("black") saltwater sea and irrevocably submerging large swaths of land never to be retrieved again. The evidence for the Black Sea Flooding was impressive, and a Bulgarian research team around Petko Dimitrov substantiated much of the evidence previously collected by Pitman and Ryan and other teams (Dimitrov 2004). While there are still controversies regarding some details of the Black Sea Flood, the actual flooding appears to be factual. Culture and language expert Harald Haarmann has described the cultural and linguistic repercussions of the Black Sea Flood, consolidating and corroborating

---

[2]For a fuller discussion of my understanding of myth, see my chapter „Schöpfungsmythen" in: Kurt Bangert, *Und sie dreht sich doch! 50 Antworten auf die Frage, wie alles begann,* Theiss: Darmstadt 2015, S. 14–41. For further discussion on the concept of *myth* see also: Norbert Bischof: *Das Kraftfeld der Mythen. Signale aus der Zeit, in der wir die Welt erschaffen haben,* Piper: München 1996/[3]2004; Christian Danz and Werner Schuessler: *Die Macht des Mythos. Das Mythosverständnis Paul Tillichs im Kontext,* de Gruyter/ Kindle Edition 2014; Mircea Eliade: *Myth and Reality,* Waveland Press/HarperRow: Prospect Hights, Ill. 1963.

much of the evidence. (Haarmann 2003) Christian and Siegfried Schoppe have even connected the Black Sea Flood to Plato's Atlantis Saga (Schoppe and Schoppe 2004). The scenario was also supported by two Indian geographers (Roy and Mukhopadhyay 2014). And various media have reported on these findings (for an example, see Axel Bojanowski 2010).

What remains a matter of conjecture though is whether the Black Sea Flood constitutes the historical basis for the biblical flood narrative. What speaks in favor of this connection are the flood stories that sprang up around the Black Sea. While it is true that flood stories have been discovered all around the world, the most prominent and detailed ones, however, were found in the vicinity of the Black Sea, such as the biblical narrative, the Gilgamesh Epic or the Greek Deucalion legend. If we assume that indeed there was a near-miraculous rescue story of one family having luckily survived the Black Sea Flood on a houseboat that eventually landed on the eastern shores of today's Black Sea, then it takes no phantasy to assume that this story was told all around the Near East and even beyond.

This brings us to the question of whether the biblical flood was a worldwide event or a regional occurrence. While there can be no doubt that those who transmitted Noah's story and included it in the Genesis account considered the flood to be worldwide, it is also true that no serious geologist alive today would speak of any evidence for a worldwide flood. Nevertheless: What are we to make of the numerous flood stories that have sprung up all around the world? There are some 500 of them. Biblicists claim that the mere existence of these flood stories prove beyond a reasonable doubt that there must have been a worldwide flood.

What these biblical literalists overlook, though, is that each of the worldwide flood stories tells of a handful of survivors: either a married couple, a pair of siblings or one large family. If indeed there was a worldwide flood and there were surviving families all around the world, then this would obviously belie the biblical narrative according to which only one single family survived—namely Noah's family in the Near East. The only alternative explanation, then, for the many flood stories with numerous survivors would be to assume that the story of the rescued family actually travelled from one place to all corners of the earth, thereby being transformed and adapted to the respective local environment (for instance, the mountain upon which the boat or the raft was supposed to have landed, usually assumed the name of a local mountain). If that is the case, then the numerous flood stories are anything but a proof for a worldwide flood. A review of the literature on the world's flood stories suggests that these accounts are likely to have travelled from the Near East to the rest of the world (cf. Isaak 1996–2002) (For an extensive discussion of the effects of the Black Sea Flood and its cultural, linguistic and theological implication for the interpretation of the biblical flood story, see Bangert 2015, pp. 60–136).

Eschatological Implications

The biblical flood narrative offers the comforting assurance that there would never again be another catastrophe that would destroy the whole earth. But that optimistic promise was relativized by none other than Jesus who in his forewarnings, on the Mount of Olives, of the time when "heaven and earth will pass away", made explicit reference to Noah's time by saying:

> As it was in the days of Noah, so it will be at the coming of the Son of Man. For in the days before the flood, people were eating and drinking, marrying and giving in marriage, up to the day Noah entered the ark; and they knew nothing about what would happen until the flood came and took them all away. *That is how it will be at the coming of the Son of Man.* ... Therefore keep watch, because you do not know on what day your Lord will come. (Mt 24:37–39.42 NIV)

Hence, according to Jesus, the rainbow in the sky was no guarantee that there would not be another global and final destruction of the earth. Jesus, therefore, assumed the contingency of the Rainbow Model. As already shown above in the discussion of the covenant, a covenant promise is generally hinged upon the conditions which both parties of the covenant must adhere to in order for the covenant to hold. What was true for the covenant with Israel from which God retracted after Israel had failed to fulfill her part (hence the destruction of Jerusalem and the scattering into the diaspora), may also be true for the covenant that God made with all of mankind after the flood: God can retract from it, if mankind fails to abide by the covenant. One could argue, though, that in the case of the rainbow covenant God had made only a one-sided covenant without detailing any requirements to which man must adhere in order for the covenant to hold. Nevertheless, I would suggest two implicit conditionals for the rainbow promise: For one, it was Noah's thank offering after the rescue that prompted God to grant the covenant in the first place; hence, man's gratitude and faithfulness to God was reciprocated by God's promise and constituted man's inherent part in the deal. For another, if mankind works against the very protection that God had promised to provide, then this covenant too becomes null and void. Why should God protect the earth if man undermines God's effort in that regard? It would be tantamount to man calling off the covenant.

The Nineveh Model (Jonah's Message)

Literary and Theological Background

Although the Book of Jonah is enlisted among the "twelve minor prophets" (Dodekapropheton), it does not, contrary to the other eleven books, contain any prophetic message of the prophet that gave the book its name—except for the one message God told him to deliver to the Ninevites: "Forty more days and Nineveh will be overthrown." (Jon 3:4 NIV) Hence, the book is not the record of a prophet's message, but a narrative telling the convoluted story of someone whom God had ordered to fulfil a task which he took great pains to avoid. Neither does the book contain an *explicit* message directed at the people of Israel, as Jonah was sent exclusively to the Ninevites. The book may, however, contain an *implicit* message to the Jews. But which one?

The general theme of the book of Jonah is debatable. Several possibilities have been offered, some of which were discussed by Martin Roth (2005). According to one alternative the book seems to suggest that when God announces a judgment, He does not necessarily follow through with it, as He leaves open a back door for a rescue. And for this, Nineveh is but an example (p. 111). Another alternative assumes a fundamental Jewish debate over whether or not sin must of necessity be atoned for through punishment or whether it can be atoned for simply by turning away from sin. According to this alternative, the book of Jonah aimed its message at those who had insisted on the necessity of punishment (p. 111 f.). Another alternative presupposes the absolute sovereignty of God's freedom who may extend his mercy and compassion to whomever he wishes (Cooper 1993, p. 144). Finally, one can assume that the citation of the psalm in Jon 2:3–10 refers to the destruction of the temple in Jerusalem and the ensuing exile. Based on that assumption one can see the book of Jonah in the context of the Dodekapropheton (twelve minor prophets) and consider Jonah to be a key to an historical interpretation: Jonah was punished for his disobedience as the people of Israel had been punished for their disobedience (Roth, p. 114 f.). In this sense, Jonah becomes the epitome for Israel.

Martin Roth concludes that any book is rarely written for one purpose only, but that most books seek to convey several messages at once. So in this case, too (Roth, ibid.; likewise also Schmid 2019, p. 394). One can generalize that the book was written in an historical context in which Nineveh and the surrounding nations were seen very negatively by many Jews. And the book of Jonah takes a stand against this prejudiced viewpoint (Roth, p. 116).

One could also assume, I might add, that the book of Jonah was used as an aetiology around the psalm of Jon 2. The language there of the psalmist being engulfed by the depth of the sea—originally meant to describe the destruction of Jerusalem and the exile in Babylon—may have been taken literally to describe the wreckage of Jonah's vessel and his wonderful rescue amidst the heathen. Incidentally, it has been widely debated whether the psalm of Jon 2 was a later addition or not. Apart from this debate, the book is widely considered to be a literary unit (Schmid 2019, p. 393).

As the language of the book points to a time *after* the Babylonian exile (there are words in the book reflecting late Hebrew as well as some Aramaisms), the origin of the narrative is usually dated after the fifth century B.C. Some have even dated it as late as between 400 and 200 B.C. (cf. The Interpreter's Bible VI, p. 873). Based on language and allusions to Greek mythology, the book appears to have been written during the time of the Ptolomaeans (third century B.C.) (Schmid 2019, p. 393).

As is well known, the Jews, during the exile in Babylon and Persia, had come to the recognition that their God (Yahweh) was not just a tribal god of their small ethnic group, but the one and only creator of the world who was the God of all peoples, the Assyrians, Babylonians and Persians included. The implication was that God's mercy was not limited and reserved for the Jewish people, but applicable to all peoples everywhere.

The Jews also gained the conviction that it was their holy task to tell the world about this (monotheistic) Creator God who was merciful and forgiving to those who would believe in Him, repent and turn away from their evil ways. In Isaiah 66:18–19 it says: "The time is coming to gather all nations and tongues. And they shall come and shall see my glory. I will set a sign among them, and I will send some of those who survive to the nations …" (Is 66:18–19 ESV/NIV) Hence, the Jewish people were to be a *sign* for the surrounding nations. But that missionary notion may have given way to "a spirit of bitterness and vengefulness toward other lands" after the exile (The Interpreters Bible VI, p. 872). That is the probable theological backdrop against which one must read the Jonah narrative, the intention of which seems to be to reawaken in the nation of Israel the missionary destiny to which they once felt called by their God.

The mercifulness of God is obviously an important theme of the book and is exemplified by the story of Jonah whose doomsday message to the Ninevites was thwarted by their contrition and repentance on the one hand and by God's mercy and forgiveness on the other. When Jonah grew angry after seeing that his prediction had not come to pass, he told God with a reproachful undertone: "I knew that you are a gracious God and merciful, slow to anger and abounding in steadfast love, and relenting from disaster. Therefore now, O Lord, please take my life from me, for it is better for me to die than to live." And God's answer to him was: "Do you do well to be angry? … Should not I pity Nineveh, that great city, in which there are more than 120.000 persons …?" (Jon 4:2–3.11 ESV). The main message of the Book of Jonah appears to be that God's mercy is for all to grasp: the Assyrians, the Babylonians and the Persians. God is for all, and his mercy is for all. In this way, the book of Jonah modifies the dooming judgment messages of the Israelite prophets against their surrounding nations (see, for instance, Joel 4) by setting a counterpoint. Commenting on the lesson of the Jonah narrative and applying it to the here and now, the German theologian Helmut Fischer writes: "We are being challenged to clarify and re-arrange our relationship to those who have a different world view, a different faith and a different value system than we have" (Fischer 2019, p. 99).

Historical-Critical Considerations

While it is true that some biblical literalists still consider the Book of Jonah to have a historical basis, most modern exegetes consider it to be a fictional narrative with an important lesson. (For a discussion of the pros and cons of the historical option, see, for instance, *The Interpreter's Bible,* Vol. VI, p. 871 f.) "The popular modern view is that Jonah was written as an imaginative tale either as an allegory of the exile and mission of Israel (*cf.* Je. 51:34), or to teach the truth of God's universal love by way of a parable." (NBC 1970, p. 746) Fischer suggests that the Jonah text "does not wish to be read as reporting an actual event." (Fischer 2019, p. 96)

We know that Jesus himself told many allegorical stories (called "parables") which were to transmit important truths. It appears that we have allegorical books in the Old Testament, such as Job, Esther or Jonah. The question to be asked is not whether they are true in an historical sense but whether they are true in a theological and existential sense. And there can be little doubt that Jonah's narrative delivers messages about God's relationship to us as individuals (Jonah) and to us as collective communities (Ninevites). It has also been suggested that the prophet Jonah is the embodiment of Israel in her relation to the surrounding heathen nations. For instance, according to The Interpreter's Bible, "the nation, insofar as it rejects its commission to be a light to the Gentiles, is identical with Jonah" (The Interpreters Bible VI, p. 873).

Jonah and Jesus

Not only Noah's flood story was featured in the teachings of Jesus, but also Jonah's story. When the Pharisees demanded that he legitimize his role by performing a "sign", Jesus retorted, according to one synoptic version: "A wicked and adulterous generation asks for a sign! But none will be given it except the sign of the prophet Jonah." (Mt 12:39) And he continued: "The men of Nineveh will stand up at the judgment with this generation and condemn it; for they repented at the preaching of Jonah" (Verse 41).

There are altogether five synoptic versions of the demand for a sign (Mt 12:38–42; Mt 16: 1 + 4; Mark 8:11–12; Luke 11:29–32 and John 6:30). In only one of the five version is there a reference to the *three days and nights* which Jonah was said to have spent in the belly of a fish and which were then taken as an allegory to the "three days and three nights" which Jesus spent in the "heart of the earth" (Mt 12:40). Apart from the problem, that Jesus could not have spent three days and three nights in the earth, if we accept the common narrative according to which Jesus was crucified on a Friday and resurrected on a Sunday, an historical-critical comparison of the five different synoptic versions (two of them in Matthew) leads to the inescapable conclusion that the reference to the *three days and nights* must be a later interpolation that actually distorts the message of that tradition. But what is the message?

The gist of the narrative is best reflected in the Mark version in which the response of Jesus actually makes no reference to Jonah at all: "Why does this generation seek a sign? Truly, I say to you, no sign will be given to this generation." (Mark 8:12) The message is: Jesus will not give them a sign (i.e. a miracle). Fullstop.

The reference to the "sign of Jonah" is found only in the two Matthew versions and in the Luke version. In Matthew 16 Jesus simply says: "An evil and adulterous generation seeks for a sign, but no sign will be given to it *except the sign of Jonah.*" (Verse 4) Then Jesus left them without further ado. That version does not give an explanation of what the "sign of Jonah" was supposed to be. That explanation is only given in the versions of Mt 12 and Luke11. In both places the explanation is as follows: "The men of Nineveh will rise up at the judgment with this generation and condemn it, for they repented at the preaching of Jonah, and behold, something greater than Jonah is here." (Luke 11:32). Mt 12 offers the same text. The meaning in both places is clear: Jesus has no intention whatsoever of giving his critics and doubters a sign—except the sign of Jonah: It was merely *at the preaching of Jonah* that the Ninevites repented, and it should simply be *at the preaching of Jesus* that the Jews in his days should repent and mend their ways. The "sign of Jonah", then, stands for the preacher himself (Jonah and Jesus) and his preaching. In other words: Believe my words, Jesus is saying, that should be sufficient. No sign, no miracle is needed; just the preaching and the listening and the believing and the repenting. It only takes listening and a change of heart to experience God's grace and mercy.

Hence, not only in the Jonah narrative, but also in the preaching of Jesus, is there an inherent contingency in the message: Don't look for miracles or signs in the sky but simply listen to the message of the messenger and change your ways, and God will show His mercy and compassion to you. God's prophecies always carry an intrinsic contingency in them that is largely dependent on man's attitude.

The Apocalypse Model (Jesus' End-Time Message)

We now turn to the third scenario, the Apocalypse Model. We are referring here especially to Jesus' prophecies and warnings regarding the coming of the Son of Man and the final judgment as described in Mt 24 (and, partly, in 25). This is not to deny texts of an apocalyptic nature in the other three gospels or other New Testament books. (For a more extensive discussion of NT apocalyptic cf. Becker and Öhler 2006.)

The Phenomenon of Apocalypticism

The passage of Mt 24–25 is a typical apocalyptic text. The phenomenon of *apocalyticism*—the name is derived from Revelation 1:1—is characterized by a *specific literary genre* on the one hand and a particular *eschatological world view*, on the other, which is preoccupied with the end of the world and the consummation of history. "Generally, where religions are occupied with the problem of the end of the world and the close of history, we meet apocalyptic thoughts and their literary deposits, especially in the Orient, and particularly in Iranian religion, but also in Hellenistic syncretism and among the ancient Germans. Apocalypticism developed most profusely on the soil of Judaism and primitive Christianity." (Kümmel 1966, p. 317)

The term *Apocalypticism* (as a terminus technicus for a particular literary genre) can be traced back to Friedrich Lücke (1791–1855) who used the word *Apokalyptik* in his Introduction to the Apocalypse of John and other related apocalyptic literature (1852). The original meaning of the word ("uncovering") has little to do with the term as it is today used for apocalyptic literature which implies both a particular form and content. While the term is nowadays applied exclusively in the modern scientific discourse, it was not used in ancient times to describe apocalyptic literature.

Apocalyptic literature proceeded from what is considered an apocalyptic movement (Tilly 2012, p. 12) that is generally characterized by certain basic principles such as: (1) the experience of the presence as being fraught with suffering, threat, injustice, hopelessness and suppression; (2) a dualistic world view of opposites such as light and darkness, good and evil, life and death, truth and deception etc.; (3) the idea of several aeons ("Weltzeitalter") such as the idealized past, the disorderly presence, the Armageddon-type end-time aeon, and the time of deliverance; and (4) the expectation of a final judgement. Typical for apocalyptic literature is that the future and the hereafter become the projection of today's hopes and fears. The apocalyptic paradigm pretends to have received revelations about God's salvific plans even though that plan is nowhere to be verified in today's reality (Tilly, p. 15).

Apocalyptic literature (which generally deals with eschatological events) is to be distinguished from *prophetic* literature which is altogether a different genre and historically precedes the apocalyptic type. According to Israel's prophets the presence allows for divine intervention at any time, while the apocalyptic viewpoint does not reckon with any divine intervention during the present era but expects God to put a radical end to the presence at some point in the future. Hence, apocalyptic thought is an expression of the disappointment over God's virtual non-intervention in times past. Another difference between the prophetic and the apocalyptic view is that while the prophets reckoned with God's judgmental intervention as a consequence of man's sin during the current history, the apocalyptic view places God's retribution at the eschatological final judgment (Tilly, p. 22 f.).

Apocalyptic literature has flourished since the time of Ezekiel and Deutero-Isaiah when—amongst the diaspora Jews—the expectation of the imminent end of the world and the coming of the kingdom of God (led by the *Messiah*) came more and more to the fore. While Mt 24–25 constitutes only a portion of the Gospel of Matthew, "the true Jewish apocalypses present complete books of prophecy in pseudonymous form" (Kümmel 1966, p. 317). One of the most prominent—and fully preserved—Jewish apocalyptic books is the *Book of Daniel* which, according to most modern scholars, was probably written at the time of the Maccabean revolt. There are also apocalyptic passages in other Old Testament books, notably Is 24–27; 33–35; Jer 33; Ez 39–39; Joel 3; and Zech 12–14.

There are also numerous non-canonical apocalyptic books many of which were written during the Jewish era. Jewish apocalyptic books were, for instance, the *Book of Enoch*, the *Assumption of Moses*, *Second and Fourth Ezra*, the *Jewish Sibyllines*, the *Syrian Apocalypse of Baruch* etc. But there were also apocalyptic books authored in the early Christian era. The *Ascension of Isaiah* is believed to have originated during the Judaeo-Christian era, as it contains Jewish and Christian elements. The early Christians either revised some of the Jewish apocalyptic literature, such as *Ezra*, the *Testament of the Twelve Patriarchs* and the *Christian Sibyllines*, or they wrote altogether new books of the same genre, such as the *Apocalypse of John* (known as Revelation), the *Apocalypse of Peter*, *Shepherd of Hermas,* etc. most of which were not included in the New Testament canon. Some apocalyptic narratives have even surfaced in the Quran—such as the *Christian Legend Concerning Alexander* or the *Apocalypse of Abraham.* (For further discussion on Quranic apocalyptic literature, see Bangert 2016, pp. 426–450).

According to Tilly, Jewish apocalyptic literature emerged as a result of the "Makkabaean crisis" when Greek and Jewish elites attempted to turn Jerusalem into an Hellenistic metropolis. The Makkabaens succeeded in staving off that Hellenistic challenge and in reaffirming the conservative Jewish tradition. The apocalyptic revision did no longer assume the wholesale salvation of the Jewish people as a nation but reduced the salvific promise to a pious remnant amongst the Jews who remained faithful to the traditional law (Tilly 2012, p. 39).

One of the important features of apocalyptic literature is that the authors "do not dare to step forth personally, but conceal themselves behind the authority of some ancient worthy, such as Enoch, Noah, Abraham, the twelve patriarchs, Moses, Elijah, Daniel, Baruch and Ezra." (Kümmel 1966, p. 317) The nominal would-be author usually epitomizes an idealistic authority and represents a normative past to which the content of the apocalyptic texts is linked (Tilly 2012, p. 49 f.).

Another characteristic, albeit connected with the previous one, is that the history portrayed in apocalyptic literature usually runs from the time of the pretended would-be author through the time of the unnamed actual author to a distant end-time future. By purporting an ancient author, the book appears to accurately "predict" the events of the future—at least up to a certain point in time—and therefore solicits trust also for the final events of its anticipated eschatology. "They write history in prophetic form from the time of the alleged author until the end of the world. About the time of the actual author, which is treated with special accuracy, there often occurs a break between representation of past history and construction of future history, which gives a clue to the time of composition." (Kümmel 1966, p. 317) Such *vaticinia ex eventu* ("prophecy from the event") represent the fictitious construction of a future of which the first part has already come to pass and thereby instills confidence in the would-be author to also predict accurately the remaining events (cf. also Tilly 2012, p. 51).

Apocalyptic literature—Jewish and Christian alike—is characterized by an "uncompromising cosmic dualism" (Tilly 2012, p. 51), contrasting antagonistic powers, principalities and world aeons. Opposing each other are the evil world and the future salvation, the present impurity and the future purity, the light and the darkness, the unjust and the pious, God's adversary and God Himself (ibid.). Apocalypticism distinguishes particularly between the evil present aeon and the time of the end, when God "will again reveal himself mightily and will finally prevail. The world lies in wickedness, but God will help. This aeon of the godless, earthly kingdom is coming to an end; the new aeon of God's transcendental kingdom stands at the door." (Kümmel 1966, p. 318) The end is portrayed as being close at hand, as in the preaching of John the Baptist and Jesus of Nazareth, and it behooves man to prepare for it by repentance and a change of heart. The end is usually preceded by cataclysmic cosmic events and a chain of catastrophes on earth which are the forebodings of the ultimate consummation of history and the final judgment at which man's fate will be decided. "The seriousness of the threat of judgment and the glow of hope in imminent salvation through God's wonderful rule exercised strong influence upon Jewish piety in the centuries around the turn of our era." (Kümmel 1966, p. 318) The apocalyptic dualism of the Jews (and later, the Christians) is most likely to have been influenced by the Persian dualism which at times has also been identified as Gnostic philosophy (For a discussion on Persian dualism, see Widengren 1965; for Gnostic dualism, see Jonas 1999).

According to Tilly, there is a subtle difference between Jewish and Christian apocalypticism, although both are historically closely related. The dissimilarity lies in the availability of salvation. While the Jewish apocalyptic expects salvation only at the very end of time, the Christian apocalyptic believes that salvation was made accessible already through the Christ event (Tilly 2012, p. 88). While the Jewish apocalyptic has given up on the current wicked aeon, the Christian apocalyptic still believes in some sort of divine intervention during the present era. Says Tilly: "The most important difference between the early Christian kerygma and the Jewish apocalyptic consists in the Christian assumption of God being operative in this world and in history."[3] (Tilly 2012, p. 90) God's salvific intervention is seen in the Christ event (the crucifixion and resurrection of Jesus) by which the eschatological aeon has already been introduced vicariously. The presence is not entirely godless and hopeless, as it is seen by the Jewish apocalyptic, but is a time when God makes His salvation available to those who believe (Tilly 2012, p. 91). God's salvation has been mediated through the life, death and resurrection of Christ, and it is in man's relation to Christ that the final judgment becomes effective even now. "The decisive judgement over all men and women takes place, as it were, already in their encounter with the earthly Jesus ..."[4] (Tilly 2012, p. 104).

The Expectation of Jesus

As a Jew, Jesus apparently had more or less adopted the eschatological/apocalyptic world view of his time and his people, and it should be no surprise to note that his preaching centered around the coming of the eschatological "kingdom of God" and man's preparation for it. One will also have to reckon with the possibility (if not the fact) that the thinking of the early (Judaeo-) Christian church influenced the synoptic tradition which also found its way into the book of Matthew. It is quite likely, for instance, when Jesus predicted of the temple in Jerusalem that "there will not be left here one stone upon another" (Mt 24:2), that this was a back projection by the early church which was already fully aware of the fate of the Jewish temple which the Romans destroyed in 70 A.D. (The book of Matthew is believed to date from 80 to 100 A.D.).

---

[3]Original German: "Die wichtigste Differenz zwischen dem frühchristlichen Kerygma und der jüdischen Apokalyptik besteht nämlich in der christlichen Annahme des Wirkens Gottes in der Welt und in der Geschichte.".

[4]Original German: "Das Entscheidungsgericht über alle Menschen vollzieht sich vielmehr bereits aktuell in ihrer Begegnung mit dem irdischen Jesus ...".

In his apocalyptic sermon of Mt 24–25, Jesus announced the coming of the Son of Man, i.e. the coming of the Messiah. When "the sun will be darkened, and the moon will not give its light, and the stars will fall from heaven, and the powers of the heavens will be shaken, then will appear in heaven the sign of the Son of Man" (Mt 24:29–30 ESV). Then he said: "When you see all these things, you know that he is near, at the very gates. Truly, I say to you, this generation will not pass away until all these things take place" (Verses 33 and 34).

It has been a puzzle for many theologians that Jesus predicted the coming of the Son of Man to be happening during the lifetime of those who were contemporaries of Jesus. But did Jesus actually utter those words? Or were they put into his mouth by Christians who had seen the destruction of the Jewish temple? Some theologians have argued that Jesus erred when assuming the end would come during the life time of his contemporaries. Werner Zager, for instance, wrote: "The historical Jesus was filled with the urgent expectation of an imminent kingdom of God and the final judgment. And he erred in that expectation." (Zager 1996) But was it really the expectation of Jesus? Or was it the hope of those having seen the destruction of Jerusalem who anticipated the nearby end of the world and who placed their expectation into the mouth of Jesus, as it were?

Be that as it may, we have in any case, in the subsequent verses, the warnings of Jesus who is also reported to have said: "But concerning that day and hour no one knows, not even the angels of heaven, nor the Son, but the Father only." (Verse 36) According to this verse, not even Jesus, the Son of God, could have known the time of the coming of the Son of Man and the end of the world. Even if this verse were considered a later interpolation by the early church—in light of the predicted events not coming to pass as early as anticipated—one cannot insinuate that Jesus, even while expecting the Kingdom of God to come soon, pretended to know the exact time and date of its coming. First on his mind was not the timeline of the eschatological events, but whether or not people would be ready for them. Earlier in the chapter, Jesus warned: "You will hear of wars and rumors of wars. See that you are not alarmed, for this must take place, but the end is not yet." (Verse 6 ESV) So there seems to be an ambiguity about what Jesus may or may not have claimed to know about the coming of the Son of Man. But he definitely wanted his listeners to stay alert and prepare for whatever was coming whenever it was coming. "Therefore, stay awake, for you do not know on what day your Lord is coming" (Verse 42).

From the apocalyptic perspective of John the Baptist, the fate of the present aeon was already sealed: it would be utterly destroyed and replaced by a future Kingdom of God at the end of time; and the final judgement would decide who may enter that kingdom and who may not. From the perspective of Jesus, however, the Kingdom of God was already in the making; it was "all but coming"—to use a famous phrase of Walter Rauschenbusch (1907, p. 309). God's Kingdom is already a present reality—however feeble and unassertive it may yet be. It is not entirely deferred to the time of the end. In fact, the time of the end has already begun, as it were,—as seen in the "finger of God" through which Jesus was healing the bodies and souls of his contemporaries (Lk 11:20: "If it is by the finger of God that I cast out demons, then the kingdom of God has come upon you.") Perhaps Jesus considered himself less an *apocalyptic* (like John the Baptist) but rather a *prophet*, for like the prophets of old Jesus, too, still seemed to believe in the reshaping and reforming of present history. That is not to deny the fact that the final and consummative establishment of God's Kingdom remained elusive and was to be expected at the time of the end by God's doing.

Whatever we may think of Jesus' predictions, it will be expedient to assume that there has always been a contingent element in virtually all prophetic and apocalyptic predictions of the Bible; for—as Jonah had to learn for himself—God was "a gracious God and merciful, slow to anger and abounding in steadfast love, and relenting from disaster" (Jonah 4:2), and "behold, something greater than Jonah is here" (Math 12:41). No one in this world should claim for him- or herself to know the mind of God and to be able to predict the future. As the adage goes: "It is difficult to make predictions, especially about the future."

Today, after 2000 years of futile expectations of the end of the world (despite many dates set), we do not know for sure if and when doomsday comes. With Jesus, one may ask: "Who then is the faithful and wise servant?" (Verse 45) What matters is not *when* we face our day of reckoning (which may be at our death) but whether or not *we are ready* for it.

And assuming there will actually be a final apocalyptic event extinguishing the human race, the question may well be asked, from today's perspective, whether that final end-time destruction will be brought about by a God who may once again "regret that I have made them" (Gen 6:7) or whether that event is precipitated not by a vengeful God but by man's own carelessness and recklessness. If the latter, it should give us pause to amend our ways. If the former, i.e. if God were to have set a date in his mind for the world's final hour, then He may not be very pleased if we were to beat him to the task. Should man be guilty of jeopardizing life on earth because he chooses to ignore the warning signs nature is giving him and because he confuses his own sinful doings with the acts of God?

The Contingency of Prophetic Predictions

It may now be worthwhile, in this connection, to take a closer look at the possible contingencies of prophetic predictions from a general biblical viewpoint. Let's first look at the Old Testament prophetic literature. When was a prophetic prediction conditional and when was it not?

The correctness of biblical predictions ranks among the basic presuppositions of many Christians. Especially evangelicals have maintained that virtually all of the Old Testament predictions about the coming of the Messiah were fulfilled at the time of Jesus (while liberal theologians would question at least some of these "fulfillments", arguing that they were sometimes artificially construed on the basis of the Greek wordings in the Septuaginta). Some prominent evangelical literalists have also claimed that it is possible to predict the events of the future simply by carefully reading the Bible. Hal Lindsay (1970) and John Walvoord (1974) are examples of those who claimed to know exactly what is going to happen during the time of the end. Richard Pratt spoke of a "monomania in the interpretation of biblical prophecy" (Pratt 1993, n.p.) and set out to challenge this widespread hermeneutical orientation by exploring the role of Old Testament predictions and investigating their respective conditionality and fulfillment (Pratt 1993, n.p.). He observed that sometimes future events conformed to a prophet's words, but at other times they did not. How is this to be interpreted?

As a Reformed theologian, Pratt takes seriously the Calvinistic tradition according to which "God from all eternity did, by the most wise and holy counsel of his own will, freely and unchangeably ordain whatever comes to pass" (Westminster Confession of Faith of 1647). Hence, what God has once ordained, remains immutable. But the Reformed tradition of Calvin not only spoke of the *immutability* of God's plan but also of *divine providence*: "Although in relation to the decree of God, the first cause, all things come to pass immutably and infallibly, yet by the same providence he often orders them to fall out, according to the nature of second causes." (Westminster Confession 5.2) Calvin balanced the immutability of God's decrees with an acknowledgement of God's complex involvement in the progression of history. "In this sense", writes Pratt, "belief in God's immutability does not negate the importance of historical contingencies, especially human choice. ... For this reason, we should not be surprised to find that intervening historical contingencies, especially human reactions, had significant effects on the way predictions were realized."

Pratt goes on to distinguish between three different kinds of prophetic predictions: (1) predictions qualified by conditions, (2) predictions qualified by assurances, and (3) predictions without qualifications.

An example of the *first* type (predictions qualified by conditions) is Isa 1:19–20: "If you are ready and obey, you will eat the best produce of the land; but if you resist and rebel, you will be eaten by the sword." The conditionality of such a prophecy is bipolar and goes into two directions. There are also conditional predictions which are only unipolar such as Isa 7:9: "If you are not faithful, then you will not stand at all." The prediction does not spell out what would happen if Israel is faithful, hence, it aims at one direction only.

As for the *second* type (predictions qualified by assurances), these prophecies forbid the people to intervene or pray to God as this would be futile and would not lead to Him changing his mind. "And they may cry to me, but I will not listen to them" is an example of this (Jer 11:11b). Or God takes a solemn oath (in the form of "As I live ..." or "As Yahweh lives ...") in order to add weight to his prediction and to emphasize that the prediction is unequivocal. However, the reason why God forbade the people to pray or for the prophet to intervene on their behalf, seems to be, so Pratt, that such interventions had been proven to actually be effective. "Yahweh forbade prayers in response to some oracles precisely because prayer usually had the potential of effecting outcomes (Jer 26:19; Jonah 3:10; Amos 7:1–9). Similarly, Yahweh declared that he would not 'turn back' or 'relent' from some courses of action because he normally left those options open (Joel 2:14; Amos 7:3, 6 Jon 3:9)." So the prohibition to pray or intervene paradoxically implies the tacit and unexpressed conditionality of these very prophecies. Such conditionality is corroborated by a passage in Jeremiah 18 where the prophet visits a potter's house and where the people of Israel are likened to the clay in the potter's hands (i.e. in Yahweh's hands) and which the potter may mold into whatever pleases him. Then we read:

> If at any time I declare concerning a nation or a kingdom, that I will pluck up and break down and destroy it, and if that nation, concerning which I have spoken, turns from its evil, I will relent of the disaster that I intended to do to it. And if at any time I declare concerning a nation or a kingdom that I will build and plant it, and if it does evil in my sight, not listening to my voice, then I will relent of the good that I had intended to do to it. (Jer 18,7–10 NEV)

The lesson to be learned from this passage is that not only the qualified predictions but also the other prophecies were subject to implicit conditionalities. Sincere repentance as well as flagrant disobedience could effect the outcome of predictions which was the reason why i n some cases God advised not to intervene. And Pratt's survey of Scripture reveals "that the descriptions of God's reactions in Jeremiah 18 are only representative" in that God reacted to human responses in many different ways.

One might think that the *third* type (predictions without qualifications) would be unequivocal and definite. However, argues Pratt, the Old Testament abounds with examples of unqualified predictions of events that did not take place, Jonah's announcement to the city of Nineveh being probably the most prominent one. But there are other such predictions as well in which the predicted future simply did not come to pass. What causes these turns of event? "Each text explicitly sights human responses as the grounds for the deviation." The people repented or prayed upon hearing the prophetic word. "These passages indicate that the fulfillment of at least some unqualified predictions were subject to the contingency of human response. Conditions did not have to be stated explicitly to be operative."

But unfulfilled predictions also pose a problem. When a prediction fails to come to pass, this is considered a false prophecy, at least according to Deut 18:22 where it states: "When a prophet speaks in the name of the Lord, if the word does not come to pass or come true, that is a word that the Lord has not spoken; the prophet has spoken it presumptuously." (ESV) Hence, the fulfillment of a prediction was to be a test and a hallmark of a prophet's genuineness and truthfulness. However, even that seemingly straightforward test of true prophecy doesn't stand the Old Testament evidence, as "it does not account for the many predictions from canonical (and thus true) prophets that were not realized", so Pratt. Hence, he concludes that "intervening historical contingencies had a bearing on all three major types of prophetic predictions".

One question to be raised in light of such prophetic contingencies is this: How could the listeners be certain of the future or the truthfulness of the prophetic predictions if there was always an implicit conditionality in them? For an answer, Pratt points to an Old Testament parameter to which God had bound himself. "To be more specific, the prophets looked to Yahweh's covenants to guide their expectations of what the future held." In other words, the prophets always operated within the framework of God's covenants. We already discussed above the significance of the covenants within the context of Biblical theology. The prophetic corpus explicitly mentions the covenant with Noah (Isa 54:9), Abraham (e.g. Isa 41:8 51:2Jer 33:26 Mic 7:20), Moses (e.g. Isa 63:11, 12 Dan 9:11, 13 Mic 6:4; Mal 4:4) and David (e.g.

Isa 9:7; Jer 30:9; Hos 3:5 et al.). In principle, divine covenants were not thought to be subject to revisions. On the contrary, they reflected the immutability of God's character. Hence, "Yahweh would never react to historical contingencies in ways that transgressed his covenants." The recipients of Old Testament prophecies normally could rest assured that God would abide by his covenant promises, "but no particular prophecy was completely free from the potential influence of intervening historical contingencies. In this sense, those who heard and read the prophets faced a future whose precise contours remained hidden."

Pratt concludes that Old Testament prophets "did not speak of what had to be, but of what might be". Even the few predictions that guaranteed fulfillment did not address their timing or manner of realization. "Old Testament prophets did not speak of a fixed but potential future." As a consequence, Pratt warns that "we should not speculate as to how current events fit within a fixed future". The future is certain only to God. But even that statement may be questioned as God has repeatedly subjected himself to the ways human beings responded to His warnings.

It appears, then, that on the basis of Pratt's investigations, the conditionality or contingency of Old Testament predictions has been well established. And that has very practical consequences for what the future holds in store for us. "The way we handle biblical predictions will greatly effect how they are fulfilled."

One could, at this juncture, raise the question of whether these insights with regard to the Old Testament prophecies also apply to the apocalyptic literature. Could it be that apocalyptic predictions are of a different nature than prophetic prediction? And could it not be that while prophetic predictions entail explicit or implicit conditionalities, apocalyptic texts ought to be read as unequivocal and final predictions without any contingencies?

Stephen O'Leary has looked specifically at apocalyptic literature (1998) and identified a paradigm that explains not only the motivation which drove the authors of the ancient apocalyptic texts but which also illuminates the impetus of modern end-time scenarios propagated by today's soothsayers. He describes the paradigm as follows: "The eschatological understanding [of apocalypticism] can be seen in the age-old lament that describes the decline of morals in society: young people no longer respect their elders, while war and all kinds of immorality increase in direct correlation with the growth of humanity's knowledge and technical skills." (p. 5) O'Leary goes on to show that this apocalyptic paradigm assumes an ideal world at the beginning of time, right after God had created the world, followed by a subsequent deterioration of morals due to God's non-interference, until a time when things get totally out of hand so that God, at the very end of time, is prompted to once again intervene in order set things straight and restore the old order. Interestingly, O'Leary offers a prominent example for this pattern from a prominent secular author, Plato, who wrote:

> It is from God's act when He sets [the universe] in its order that it has received all the virtues it possesses, while it is from its primal chaotic condition that all the wrongs and evils arise in it—evils which it engenders in turn in the living creatures within it. When it is guided by the Divine Pilot, it produces much good and but little evil in the creatures it raises and sustains. When it must travel on without God, things go well enough in the years immediately after He abandons control, but as time goes on and forgetfulness of God arises in it, the ancient

> condition of chaos also begins to assert its sway. At last, as this cosmic era draws to a close, this disorder comes to a head. The few good things it produces it corrupts with so gross a taint of evil that it hovers on the very brink of destruction, both of itself and of the creatures in it. – The God looks upon it again, He who first set it in order. Beholding it in its troubles, and anxious for it lest it sink racked by storms and confusion. He takes control of the helm once more. Its former sickness he heals; what was disrupted in its former revolution under its own impulse He brings back into the way of regularity, and, so ordering and correcting it, He achieves for it its agelessness and deathlessness. (Plato's Statesman, 273c–273e, p. 152)

O-Leary sees parallels between Plato and Paul who, in light of the general deterioration of the world at his time (see, for instance, Rom 1:18–28) anticipates a restoration of creation (Rom 8:18–22; 11:25–27). "The hope or expectation of God's regaining control over the helm of the universe that Saint Paul held in common with Plato testifies to the enduring power of eschatological myth as a way of understanding the problem of evil." (p. 5) Apocalyptic myths offer a "temporal and teleological framework" for understanding the outworkings of evil in this world by claiming that evil must grow in power in order to reveal its full nature before an appointed time of the end, when God will once again reinstate his own power and order.

Apocalypticism, then, is the attempt to make sense of the deteriorating development of evil in a once perfect world that is being corrupted to a point when only God can put an end to it. Such a message seems to have been especially appealing to those groups who suffered most under the weight of persecution, injustice and marginalization. "The early Christians who responded favorably to the book of Revelation were, by most historical accounts, subject to intense persecution that included execution and public torture." (p. 11) Hence, they were most receptive to the promises and predictions of the apocalyptics who appeared to offer a comforting time-line with redemption close at hand.

Unfortunately (or fortunately), most of the apocalyptic predictions over the past 2000 years have failed. There exists a long list of apocalyptic predictions over the centuries which did not come to pass. Whether it were the Millerites in the 19th century who proclaimed the end of the world in 1844, or the false prophets who anticipated the coming of the Christ or the Antichrist in the year 2000, or even the present-day soothsayers around the QAnon movement; they have all built their case on the dissatisfaction of marginalized peoples in order to mobilize their followers into an awakening for a better future. Take this modern case, for an example:

In the anonymous book *QAnon: An Invitation to the Great Awakening* we read: "The world collapsed into darkness … Destroyed factories, declining job numbers, sicker people, opioids, destruction of Iraq, Syria and Yemen with pointless wars, displacement of people into Europe, Isis, terrorism, collapsed governments, poverty and genocide. TOTAL MISERY" (p. 6). But the QAnon fellowship offers a cure: "Our job is to generate a future that reclaims the reins of power from the forces that would abuse our children and enslave us. Our intention is to take them back and restore power to 'We the People'. We are committed to achieving that for all of us, you included." (p. 2) Those who sense a secret conspiracy of the powerful ("Deep State") have themselves created a conspiracy of which they want each of us to be a part.

But so much seems clear: In light of the many failed predictions, apocalyptic texts also have built-in contingencies that are not made explicit because, if they were, they would undermine the very objectives which their authors aim to achieve.

Towards an Hermeneutic of the Three Models

This survey of the three biblical narratives has yielded a number of hermeneutical insights with regard to how God may be seen as dealing with mankind. I shall enumerate a number of such discernments:

The Rainbow Narrative

1. If we accept the modern discoveries about the Black Sea flood some 7500 years ago as the basis for the flood myths, then we must also accept the fact that there was never a worldwide flood which extinguished virtually all life on earth.
2. Be that as it may, the flood narrative was meant to convey a comforting message: that as long as the rainbow would be seen in the sky, there would not cease day and night, summer and winter, seedtime and harvest; life will always go on!
3. The narrative also testified to the general depravity and sinfulness of man. But despite that depravity God will abide by his eternal covenant with mankind and not punish mankind again as He appears to have done at the time of the flood.
4. The story also encouraged the listeners and readers to live a righteous life as Noah had lived, who set himself apart from the majority of his contemporaries who did not heed his warnings.
5. The flood narrative was also about making appropriate preparation for a time when calamity does strike. Entering the ark thus becomes a symbol for disaster preparedness which is today as needful as it ever was.
6. In light of Jesus comparing the sinfulness of Noah's contemporaries with the sinfulness of the people at the time of the end (Mt 24:37–39), the rainbow promise (of there never again being worldwide destruction) emerges as being contingent.

The Nineveh Narrative

1. The Jonah narrative exemplifies, first and foremost, the contingency inherent in all prophetic predictions which may or may not contain explicit (or implicit) conditionalities.
2. The narrative also witnesses to the sublime sovereignty of a God who may reward or punish whom he wishes, whether it be His own people or the Gentiles. For God is no longer a tribal god, but the sovereign Creator of the universe.

3. The story shows that God allows himself to revoke any predictions of impending disaster whenever man repents and changes his or her way of life. In other words: man's sin can be atoned for by man's confession and by turning from his (or her) evil ways.
4. Conversely, God can also punish His own, despite the overarching covenant relationship (be that an individual such as Jonah or the people of Israel), if they choose to disregard His commandments.
5. But God can also, after demonstrating disapproval with His own (Jonah's near disaster or Israel's near destruction), rescue them and restore His covenant relationship with them. If He is free to show mercy even to the Ninevites, how much more will He show compassion to His own covenant people, if they repent and turn from their evil ways.
6. Finally, the story also shows that man cannot run away from God or from what man is destined to be and to do—regardless of the ultimate outcome.

The Apocalyptic Narrative

1. The apocalyptic narratives suggest that there will be an eventual end to history and to life on earth as we know it—be that through an external catastrophe or through man's self-inflicted calamities.
2. Apocalypticism presumes the principle non-interference of God into the affairs of man and hence assumes man's responsibility for history and for the well-being of the planet.
3. Apocalyptic end-time scenarios should not serve as a rationale or excuse to continue ad infinitum man's squandering of earth's resources such as water, air, non-renewable energies, virgin forests, or endangered species, by claiming that it doesn't matter as the final judgment cannot be avoided anyway.
4. Even when assuming, on the basis of apocalyptic texts, that the destruction of the earth and the consummation of man's history lie in the hands of an unfathomable sovereign God, that assumption must not and cannot relieve us of the responsibility of working diligently for the effectuation of the principles of God's kingdom in the here and now, as Jesus taught.
5. Whatever we may do to implement these divine principles and to change the world for the better in order to preserve the earth as a livable environment for all, we as human beings will never be able to re-create a perfect world or an earthly paradise; hence, there always remains a *promise surplus* (*Verheißungsüberschuss*) that we hope for, as God remains the final answer to man's history and to the fate of the earth.
6. In other words, although apocalyptic myths convey the message of a final restoration through the sole act of God (apocalypticism according to John the Baptist), they implicitly also call for man's commitment to righteousness, peace, liberty and justice in this world (apocalypticism according to Jesus).

A New Understanding of God and of Man's Role?

In light of the principal contingencies of biblical prophecies, well established above, and also in view of the contemporary climate crisis, the question has been raised in what way the traditional view of God ought to be revised.

In October 2008, the Department for Theology and Studies of the Lutheran World Federation (LWF) convened a consultation of biblical scholars, theologians and ethicists active in the area of climate change. They authored the Publication *God, Creation and Climate Change* (Bloomquist 2009). In it, the group reflected, among other things, on the question of God in light of climate change, and how that view also affects the self-understanding of man.

"When people think about 'God', the authors stated, "they often refer to a supreme being who reigns over and above the world as an almighty ruler or monarch (almost always as 'he'). When something goes wrong in nature, such as occurs under climate change, it is then immediately assumed that this is caused by 'God'—as an almighty actor standing outside of and controlling all that occurs on earth." (p. 17). They then described what they considered a "shift" toward an understanding of God based on biblical research. According to this shift, God is "not an all-controlling monarch who punishes even the innocent, but God revealed yet hidden throughout creation. God's grace and love are ultimately more crucial than might and power. God is intimately related with humans and the rest of creation, present in the midst of vulnerability and suffering" (p. 18).

While that shift was a much needed revision in Protestant theology, today in light of climate change and other developments, "a similar shift is called for in how we imagine or think of God", said the authors. "God is neither to be seen nor sought behind creation nor inferred from it, but only recognized in and through it." (p. 18 f.) "God is the source of all being rather than one who intervenes from the outside." And referring to the writings of Sallie McFague, they come to a yet newer understanding of God: "Rather than assuming God to be like a will or intellect ordering and controlling the world, God is the breath that enlivens and energizes the living breathing planet. God permeates, suffers with and energizes the innermost aspect of all that is created, in ways known and unknowable, in ways that are both intimate and transcendent" (McFague 1993, pp. 145, 147).

That new shift toward a postmodern understanding of God during the Anthropocene of climate change leads, of necessity, also to a new understanding of man's role within creation. The Lutheran scholars lamented that "[s]ome human beings have acted as if they were demigods who can order and control, for their own self-interests, the land, trees, air, water and other creatures, including vulnerable human communities. This often occurs in the name of 'development' or 'progress'." And in consequence, "the delicate interrelationships within creation have been upset. Creation's protest is now being experienced through climate change" (Bloomquist 2009, p. 21). The assumption that human beings are separate from or above nature, has ensnared man to "escape our creaturehood" (p. 22).

But given the Kairos moment[5] of climate change today, there is a need for repentance and conversion that involves also a shift in the way man understands himself. That shift ought to involve a shift from human independence to human interdependence with the rest of creation; from creation understood only as the backdrop for human worship, to creation pulsating with life, pathos and worship of God; from an exclusive focus on God active in human history, to God being active in, with and through the spatial realities of the whole creation, in which humans participate; from an understanding of man's sin only as a broken relationship between humans and God, to sin being understood as a broken relationship with all creation; from an obsession with progress and development as measured in economic terms, to what will result in more sustainable life for all of creation; from a focus on technological or market-based fixes, to the healing of creation (p. 23 f.).

The Lutheran scholars summarized their thoughts by suggesting that we "must move beyond narrow anthropocentric views of life, and embrace more interconnected views in which God, human beings and the rest of creation are intimately related" (p. 26).

Conclusion

Global Warming and other environmental hazards have conjured up among Bible readers an impending end-time apocalypse. Based on a previous paper offering three biblical eschatological scenarios (the Rainbow Model, the Nineveh Model and the Apocalypse Model), this paper expanded on the three scenarios, giving more theological and historical background and drawing some hermeneutical conclusions. While the Rainbow Model (Noah's flood story) assumed that another worldwide destruction would never again take place, and the Apocalypse Model (Jesus' end-time expectation) assumed the ultimate annihilation of the world, the Nineveh Model (Jonah's doomsday message) assumed that the prospect of an envisaged end-time annihilation can be still avoided through repentance and a change of behavior.

For each of the three scenarios, this paper looked at possible historical roots and the theological intent, but also discussed the contingency aspects. The result of the study suggests that each of the three narratives (Noah, Jonah, Jesus) is characterized by an inherent *component of contingency,* be it explicit or not, and that none of them are unequivocally unconditional. Each narrative (and, in principle, any biblical prophecy) assumes the basic tenet of God's compassion and mercy in light of man's willingness to change heart. Explicit or not, all biblical predictions entail contingencies that make it virtually impossible for man to exactly predict the future. Even with God at the helm of history, history remains open-ended as God has subjected Himself to the way humans will relate to God. While the biblical covenants place man into a special

[5] A „*Kairos*-Moment" (from Greek καιρός implying a moment in time, in contrast to *chronos*/χρόνος which stands for a period of time) connotes a crucial time requiring decisions of vital consequences.

relationship with God, God remains at all times sovereign and supreme to extend His mercy and compassion to whomever He wishes—especially if and when man confesses his sins and changes his ways. Hence, God's predictions mediated by His prophets are contingent.

The lesson to be learned from the contingency of biblical prophecies, as demonstrated in this paper, should caution us to assume that apocalyptic end-time scenarios, as they have been depicted in some biblical and extra-biblical texts, are final and unequivocal and that supposedly, as a consequence of this, any effort by man to preserve the natural environment would remain futile as the final outcome has supposedly already been fixed. On the contrary, man as a steward for God's creation (see Gen 1:28) has a holy responsibility to care for our planet, for the environment and for all life. Man is to adhere to the divine principle of "Reverence for life" (*Ehrfurcht vor dem Leben*) a s propagated by Albert Schweitzer.

The lesson of prophetic contingencies may, in all likelihood, also require a shift in our understanding of God, away from God as an almighty actor standing outside of creation and controlling everything that occurs on earth, to a view of God as the source of all creation and who suffers with it and whose spirit permeates it in a way that is both immanent and transcendent.

But above all, the climate change problem requires a shift in man's behavior. Barbara Rossing, an American New Testament scholar in Chicago who regularly speaks against the apocalypticism that some have associated with climate change, found the Nineveh narrative "the most intriguing biblical model" as it "can serve as a parable for how the greatest 'empire' on earth today can shift course and avoid disaster". The book of Jonah shows "how thanks to grassroots mobilization, good leadership, and an effective prophet, a giant ship of state–including all the people and even the animals–can abandon its destructive course in time to avert a catastrophe. We can learn from Nineveh" (Rossing 2009, p. 142).

Astronomers have calculated that, at the very latest, life on earth will cease to exist when our Sun, after having spent most of its hydrogen through thermonuclear fusion, will turn into a Red Giant that will engulf the earth and extinguish all life on our planet. That is estimated to take place some four billion years from now. But life may come to an end much sooner: when another big meteor or asteroid hits the earth; or when an atomic war flares up and gets out of hand; or when enough insects have become extinct and fertilization of vital plants is no longer working; or when another virus emerges that is more infectious and fatal than the ones we have encountered so far. There are many reasons why life on earth could be erased. Life is fragile and delicate, individually and collectively.

The Covid 19 pandemic taught us that exponential growth of any kind, if not contained, eventually leads to the crossing of certain points at which vital systems are bound to collapse. We are, unfortunately, right in the midst of a very dangerous exponential growth curve: the population explosion of mankind. When exactly this curve will peak, we don't know yet. We only know that it cannot increase indefinitely. It is likely that the human population will hit various boundaries and tipping points that might eventually threaten life on earth. Due to its exponential growth rate, the human race has become a serious threat to the climate, to the global environment, to mankind itself, and to the whole living world. Climate change experts have for some time spoken of "tipping points" which, if crossed, may give rise to some unforeseen upheavals and disruptions of the earth's ecosystem that may alter and endanger, if not destroy, life on earth. As stewards of God's creation we have a responsibility to prevent this through concerted 'creation care'.

During the initial apex of the corona pandemic in Europe, when the "Fridays for Future" campaign (to save the planet) had come to a sudden standstill, the internationally-renowned climate researcher Hans Joachim Schellnhuber of the Potsdam Institute for Climate Impact Research (PIK) was quoted as saying: "He who thoughtlessly passes on the virus, jeopardizes the life of his grandparents; he who thoughtlessly releases carbon dioxide, jeopardizes the life of his grandchildren"[6] (DTS 2020).

References

Bangert K (2016) Muhammad. Eine historisch-kritische Studie zur Entstehung des Islams und seines Propheten. Springer, Wiesbaden

Bangert K (2015) Und sie dreht sich doch! 50 Antworten auf die Frage wie alles begann. Theiss/WBG, Darmstadt

Becker M, Öhler M (eds) (2006) Apokalyptik als Herausforderung neutestamentlicher Theologie (Wissenschaftliche Untersuchungen zum Neuen Testament). Mohr Siebeck, Tübingen

Bible. Versions cited: English Standard Version (ESV), New International Version (NIV), King James Version (KJV), New King James Version (NKJV)

Bischof N (1996/2004) Das Kraftfeld der Mythen. Signale aus der Zeit, in der wir die Welt erschaffen haben. Piper, München

Bloomquist KL (2009) God, creation and climate change, LWF studies. Lutheran University Press, Geneva

Bojanowski A (2010) Der Wassersturz am Bosporus. Die Theorie erhärtet sich, dass die biblische Sintflut auf eine reale Überschwemmung vor 7500 Jahren zurückgeht. In: Süddeutsche Zeitung, 10th May 2010; accessible online (as of 5 Oct 2020) under: https://www.sueddeutsche.de/wissen/vor-7500-jahren-der-wassersturz-am-bosporus-1.633396

Cooper A (1993) In praise of divine caprice: the significance of the book of Jonah. In: Davies PR, Clines DJA (eds) Among the prophets. Language, image and structure in the prophetic writings. JSOT, Sheffield, pp 144–163

Danz C, Schuessler W (2014) Die Macht des Mythos. Das Mythosverständnis Paul Tillichs im Kontext. de Gruyter, Berlin/Munich/Boston

Dimitrov P, Dimitrov D (2004) The Black Sea: the flood and the ancient myths. Varna, Slavena

DTS News Service, May 2020

Eliade M (1963) Myth and reality. Waveland Press/HarperRow, Prospect Hights, IL

Fair H (2018) Three stories of Noah: navigating religious climate change narratives in the Pacific Island region, geography and environment (Open Access) 5(2) (July–December 2018). Retrieved from https://rgs-ibg.onlinelibrary.wiley.com/doi/full/10.1002/geo2.68 (22 Sept 2020)

Fischer H (2019) Die Wunder Jesu. Wie die Geschichten heute zu verstehen sind. Philia/BoD, Bad Nauheim

Gertz, J. Chr. (2019) Hintere Propheten (Nebiim). In: Jan Christian Gertz (Hg) Grundinformation Altes Testament. Eine Einführung in Literatur, Religion und Geschichte des Alten Testaments,

[6]Dts News Service, May 2020.

in Zusammenarbeit mit Angelika Berlejung, Konrad Schmid und Markus Witte. Vandenhoeck & Ruprecht, Göttingen, 6. Aufl., pp 313–413
Grey E (2013) End of the world: the revelation prophecy. PeDante Press, Danbury, CT
Haarmann H (2003) Geschichte der Sintflut. Auf den Spuren der frühen Zivilisationen. C.H.Beck, München
IPCC (2015) Climate change 2014: synthesis report. the fifth assessment report of the intergovernmental panel on climate change [Core Writing Team, Pachauri RK, Meyer LA (eds)]. IPCC, Geneva, Switzerland
Isaak M (1996–2002) Flood stories from around the world. Retrieved from http://www.talkorigins.org/faqs/flood-myths.html#Greek (accessed Oct 2020)
Jonas H (1999) Gnosis. Die Botschaft des fremden Gottes. Verlag der Weltreligionen/Insel Verlag, Frankfurt/Leipzig
Kittel G (ed) (1964) Theological dictionary of the new testament, vol I. Eerdmans, Grand Rapids, MI
Kümmel WG (ed) (1966) Introduction to the new testament, founded by Paul Feine and Johannes Behm. Abingdon Press, Nashville/New York
Lindsey H (1970) The late great planet earth. Grand Rapids, Zondervan
Lindsey H (2015) Tonight on the Hal Lindsey Report; cf. https://www.hallindsey.com/hlr-2-27-2015/
Lücke F (1852) Versuch einer vollständigen Einleitung in die Offenbarung des Johannes oder: Allgemeine Untersuchungen über die apokalyptische Literatur überhaupt und die Apokalypse des Johannes insbesondere. Bonn, Eduard Weber
Luetz JM, Buxton G, Bangert K (2018) Christian theological, hermeneutical and eschatological perspectives on environmental sustainability and creation care—the role of Holistic education. In: Luetz JM, Dowden T, Norsworthy B (eds) Reimagining Christian education, cultivating transformative approaches. Springer, Singapore, pp 51–74
McFague S (1993) The body of god: an ecological theology. Fortress, Minneapolis
The New Bible Commentary (1970) Revised. Eerdmans Publ, Grand Rapids, MI
The Interpreter's Bible (1956) Vol. VI. Abingdon Press, New York/Nashville
O'Leary S (1994) Arguing the apocalypse. A theory of millennial rhetoric. Oxford University Press, Oxford/New York
Pitman W, Ryan W (1998) Noah's flood. The new scientific discoveries about the event that changed history. Simon & Schuster, New York
Pratt R L Jun (1993) Historical contingencies and Biblical predictions. An inaugural address presented to the faculty of reformed theological seminary, 23 Nov 1993; see under: https://thirdmill.org/magazine/article.asp/link/https:%5E%5Ethirdmill.org%5Earticles%5Eric_pratt%5ETH.Pratt.Historical_Contingencies.html/at/Historical%20Contingencies%20and%20Biblical%20Predictions (retrieved 7 Oct 2020)
Rauschenbusch W (1907) Christianity and the social crisis. Macmillan, New York
Rossing B (2009) God's lament for the earth: climate change, apocalypse and the urgent Kairos moment. In: Bloomquist K (ed) God, creation and climate change. Spiritual and ethical perspectives, LWF studies. Geneva, Lutheran University Press, pp 129–143
Roth M (2005) Israel und die Völker im Zwölfprophetenbuch. Eine Untersuchung zu den Büchern Joel, Jona, Micha und Nahum, Vandenhoeck & Ruprecht, Göttingen
Roy S, Mukhopadhyay M (2014) Geography behind myth—the tale of Noah's Deluge, ACB Publication: Kolkata; cf. also their paper "Truth behind Noah's Deluge—an implication of climate change adaption", presented at the XVIth World Water Congress, Cancun, Quintana Roo, Mexico, 29 May to 3 June 2017; see under: https://iwra.org/member/congress/resource/ABSID500_ABSID500_Saswati_RoyID_500Full_paper.pdf (as of 21 Sept 2020)
Rubow C, Bird C (2016) Eco-theological responses to climate change in Oceania. Worldviews: Glob Relig Cult Ecol 20:150–168. https://doi.org/10.1163/15685357-02002003
Rudiak-Gould P (2009) The fallen palm: climate change and culture change in the Marshall Islands. VDM Verlag, Saarbrücken

Schmid K (2019) Theologie des Alten Testaments. Mohr Siebeck, Tübingen
Schoppe CM, Schoppe SG (2004) Atlantis und die Sintflut. Die erste Hochkultur versank 5510 vor Christus im Schwarzen Meer. BoD, Norderstedt
Simpson J, Speake J (2009) (eds) The Oxford dictionary of proverbs, 5th edition. Oxford University Press, Oxford
Tillich P (1957/1975) Systematic theology, Vol. II. University of Chicago Press, Chicago
Tilly M (2012) Apokalyptik. A. Francke, Tübingen
Von Rad G (1978) Theologie des Alten Testaments, Vol. I. Chr. Kaiser Verlag, München
Vox L (2017, June 2) Why don't Christian conservatives worry about climate change. Washington Post. cf.: https://www.washingtonpost.com/posteverything/wp/2017/06/02/why-dont-christian-conservatives-worry-about-climate-change-god/
Walvoord JF, Walvoord JE (1974) Armageddon: oil and the Middle East. Zondervan, Grand Rapids
The Westminster Confession of Faith (1647) 3.1 as found in P. Schaff, The Creeds of Christendom (1877, reprinted; Baker Book House, Grand Rapids 1969)
Widengren G (1965) Die Religionen Irans (Die Religionen der Menschheit 14). Kohlhammer, Stuttgart
Zager W (1996) Gottesherrschaft und Endgericht in der Verkündigung Jesu. Walter de Gruyter, Berlin

Chapter 10
Standing in the Middle of the Ocean: Time for a Public Theology

Clive Pearson

Abstract The present is a *kairos* moment for the islands of 'the liquid continent': the prospects of mitigation, adaptation and resilience in response to the climate emergency must now wrestle with the implications of the Anthropocene. The most recent communiqués coming out of the Pacific Islands Forum Secretariat (2019) address multiple 'geostrategic' issues: there is no reference to the biblical and theological traditions which indwell the lived experience of island communities. For future policies to be well-grounded in communal life there is a pressing need for scientific, political and sociological discourses to be matched with the best contemporary theological thinking. The works of ecumenical bodies recognize that an appropriate theology must be set alongside the building of social capital and disaster relief. It is time to move beyond the ways in which biblical texts like the rainbow covenant with Noah and the plight of the innocent victim to be found in Job are employed. The reference to a *kairos* moment is an invitation to construct a public theology relevant to island experience and future challenges.

'What Are We to Preach?'

Some years ago I was asked to deliver a public lecture on theology, climate and the Anthropocene in the public hall on the island of Funafuti (Tuvalu). It was a daunting request: it is far from clear what the future holds for such islands in terms of mitigation and adaptation. The invitation was an opportunity, nevertheless, to speak on behalf of a public theology. That is a theology which is interdisciplinary and seeks to address matters of public concern. It is designed to participate in conversations and deeds that serve the nurture of a civil society and the common good. At the conclusion of the lecture one of the senior ministers (*faifeau*) asked me: "What are we now to preach?".

That response should come as no surprise. In a way that is often under-appreciated by climate and social scientists islands like those of Tuvalu and Kiribati are deeply

C. Pearson (✉)
Charles Sturt University, Parramatta, NSW, Australia
e-mail: cpearson@csu.edu.au

J. M. Luetz and P. D. Nunn (eds.), *Beyond Belief*, Climate Change Management,
https://doi.org/10.1007/978-3-030-67602-5_10

religious. Biblical ideas weave their way into customary knowledge. For a theologian like myself the task becomes one of seeking to discern what pattern of religious beliefs are dominant and being acted out. Is there an alternative that may assist those charged with the task of exploring possible mitigating and adaptive strategies? Is there a potential frame of religious belief which can sit alongside geopolitical strategies and do so in a way which coheres with the practices of faith?

The task of a public theology is to discern urgent issues that demand policy and debate. In this particular instance it is to be respectful to peoples most affected by the threats posed by a changing climate and rising sea levels and enable their voice: hence this work begins with the witness of an i-Kiribati—a former student of mine. His desire to find an appropriate theology for a time of climate change sits with that request to do with preaching. Both are set within the well-established pattern of a public theology which places that which is local inside a sympathetic global praxis. In this case the focus of attention falls on the role of Scripture which is of profound significance in island life. The dilemma is whether the excessive attention given to the Noah saga and the book of Job is becoming an obstacle to a more constructive religious voice that may emerge from a public theology.

'I Stand in the "Middle of the Ocean"'

It is fast becoming a staple of public theology to identify the climate crisis as a *kairos* moment for the 'global community' (Pearson 2010; Fretheim 2016; Conradie and Koster 2020; Deane-Drummond et al. 2017). This designation carries allusions to 'sign of the times', a crisis, an opportunity. For those who find themselves 'stand[ing] in the Middle of the Ocean' like Timoti Timon from Kiribati signs of that crisis are ever-present: loss of lands, villages relocated, the compromising of fresh water supplies, and the loss of food supply also due to the impact of saltwater on fruit trees and plant life (Timon 2018, 2019). It is a narrative that can be repeated in Tuvalu (Lusama 2004; Talia 2009; Curlew 2012) and any other low-lying island in the Pacific. In terms of timing *kairos* differs from linear, calendar, *chronos* time to which dates and years are assigned. For the Ecumenical Patriarch Bartholomew a *kairos* moment is a 'decisive time in our relationship with God's creation'. It is one in which 'choices that will affect our whole lives' must be made. 'Will we act in time to protect life on Earth … or [will] we fail to act'? (Fretheim 2016, p.132).

Timon is only one of the latest in a series of Pasifika theologians that includes Maina Talia (2009) and Tafue Lusama (2004), who are seeking to address diverse aspects of climate change. Like them Timon does not use the rhetoric and method of a public theology which is only slowly growing a foothold in island cultures. The benefit of this coming together of a public theology and a world Christianity—that is, non-western expressions of faith (Sanneh and McClymond (2016)—is that it frees the former from the constraints of its western origins: 'World Christianity has its own public face, voice and reason' (Barreto et al. 2017, p. xvii). It recognizes in a glocalized form that what issues arise, how they are addressed in the local idiom, and indeed how the public sphere—or its equivalent—is constructed is particular to culture. Within the setting of low-lying Tuvalu it allows Talia to draw upon indigenous customary knowledge—the *muna o te fale* [the wisdom of the house]—as well as the medium of the *fatele* (the traditional dancing song). Within the context of densely populated South Tarawa (capital of Kiribati) it allows Timon to make the case for an indigenous coconut theology of climate change[1] while making room for crisis-inspired songs that express a theology of resistance, lament, and "only God knows" [providence] (Pearson 2020).

That standing in the Middle of the Ocean is the first line of one such song: it conveys the ambiguity of seemingly having nowhere to stand because the word *marawa* represents 'the deep sea or deep void where feet cannot stand'. It suggests a lack of help and the tragic irony of having to live 'without a ground or strong foundation to stand'. It carries an intimation of migration and relocation which is then bound to a summons to resistance: *Koburake, koburake* 'arise'. For Timon this summons is a bid to make a stand through indigenous 'strength, skills and knowledge'. In terms of a *kairos* moment that affirmation of 'I stand' may nevertheless be construed as an opportunity for a public theology. It represents a claim to agency, a voice, a capacity for the other to hear, and perhaps holds out a hope that will find solidarity beyond the song's lament that there is 'no one to help me' (Timon and Kaunda 2019, pp. 4–5).

The Blue Pacific

Those charged with the task of overseeing regional security in the Blue Pacific are well aware of the deepening imperative to act with the sense of urgency that the word *kairos* implies. The Communiqué emanating from the 50th session of the Pacific Forum held on Tuvalu (13–16 August, 2019) described 'escalating climate change impacts': climate change is deemed to be 'the single greatest threat to the livelihoods, security and wellbeing' of peoples in the region. (Paragraph 14). The Communiqué built upon the Boe Declaration and the SAMOA Pathway and dealt with multiple geopolitical issues: its primary concern for climate change, though, was evident in its turn to the *Kainaki II Declaration for Urgent Climate Action Now* (Paragraph 15). The rhetoric permeating these documents is one of the need for resilience, security, health, and solidarity.

The strategic action recognized the 'need for urgent, immediate actions' while seeking a 'collective commitment' to realize a long-term vision—the *2050 Strategy for the Blue Pacific*. It was set within the framework of IPCC and UNFCC declarations, most notably, the Paris Agreement on carbon emissions, 2015. The Communiqué sought to be inclusive and agreed that the *Strategy* 'must ensure social, cultural,

[1] The term *coconut theology* was first employed by Sione 'Amanaki Havea'. It signifies the desire to construct a Christian theology making use of indigenous Pacific Island symbols, customs and ideas. The coconut is employed as the descriptive term as an alternative to the way in which Jesus made use of bread in his teaching. It thus represents the sustenance of life.

environmental and economic integrity, sovereignty and security'. The 'future of the Blue Pacific could not be left to chance'. The present required 'strong political leadership' for the sake of a sustainability informed by science. (Paragraph 9) According to the Prime Minister of Tuvalu, Enele Sopoaga, it might indeed be best for a United States of the Pacific to be set up under the umbrella of the Forum: such a move would allow Pacific Island countries (PIC) 'to amplify their concerns about climate change on the global stage' without being compromised by the political intervention and might of Australia (Tavola 2019). For these documents the affective words of urgency and seeing first-hand implications were the equivalents of now being a *kairos* moment.

Superwicked Problems and the Anthropocene

This emphasis on resilience and security is understandable. Whether it captures the full force of the *kairos* moment is a moot point on two counts. From a geopolitical perspective it is hard to imagine how inter-governmental and regional protocols—like the Communiqué—can manage what is, in effect, a superwicked problem (Lazarus 2010; Levin et al. 2012; Pearson 2019). Such problems are in urgent need of resolution but are extraordinarily complicated across multiple disciplines. The tendency is often to privilege immediate vested interests and thus delay the longer-term good. The 'redistributive implications are enormous'. (Lazarus 2010, pp. 10750–10751) The urgency i n which the Communiqué and the Kainaki Declaration are cast, furthermore, makes no reference whatsoever to the present moment being one of transition from the Holocene epoch to the Anthropocene. This paradigmatic change has been taken up by Earth Systems scientists like Will Steffen who find this geological term helpful for the way in which planet's trajectory is towards a 'hothouse Earth' for which there is no analogue (Steffen et al. 2018). Writing at the close of 2019 Timothy Lenton, Steffen et al. inhabit a different theoretical space from the one expressed less than a year before by Forum Secretariat: the grim prospect is one of nine out of fifteen tipping points in the Earth's System having now been exceeded (Lenton et al. 2019). In an interview reported in July 2020 Steffen warned that 'eternal growth is not possible': the most likely outcome is the collapse of civilization (Moses 2020). Mitigation and adaptation are both at risk.

There is a world of epistemic difference between Timon's coconut theology of climate change and this talk of superwicked problems and the Anthropocene. The respective discursive fields could hardly be more removed from one another; the same is true for the songs of lament, resistance and providence with the Communiqué and the Kainaki Declaration.

The hermeneutical problem that arises as a consequence becomes a variation on a theme already addressed separately by Hannah Fair and Wolfgang Kempf. Each was seeking to delineate the way in which islanders' use of songs and biblical texts enable a form of agency in the face of confusion and deepening crisis. The common practice has been to dismiss such customs and how isolated biblical texts are used as a naïve form of climate denial 'rooted in Christian teaching'. The recourse to such is seen as evidence of a deficit of scientific knowledge. It can seem like a 'fundamentalist extravagance' (Kempf 2017, p. 23). For her part Fair takes issue with 'the reductive approach common to academic accounts of religious responses to climate change in Oceania'. Religion becomes a 'barrier'; the church is slow to become 'a climate change actor' and is more content with its internal maintenance and the provision of worship and pastoral care (Fair 2018, pp. 4–6). In order to meet this criticism Kempf insists upon there being many different types of 'discursive constructions' surrounding climate change and sea-level rise (p. 30). Fair effectively agrees: she duly argues the case for storytelling and the possibility of what she described as a 'double knowledge' approach to beliefs where science and religion can work together (p. 7).

The Problem with Noah

The particular biblical text most often cited has to do with the covenant God made with Noah at Gen 9:11: "never again shall all flesh be cut off by the waters of a flood, and never again shall there be a flood to destroy the earth". The promise embedded within this covenant presumes earlier movements in the Noah cycle (Gen: 6.9-9-17), most notably the narrative of the flood itself. For Bible-conscious peoples there is much to commend a literal reading of this covenant. The hermeneutical distance in terms of time and place between its original setting and the islands of Kiribati and Tuvalu can be ignored. There is no need to delve into its Yahwist and Priestly sources, how those sources were woven together and whatever level of indebtedness there might be to other ancient Mesopotamian epics (de la Torre 2011, p. 112; Habel 2011, pp. 90–102). Nor is there any need to become familiar with how the whole Noah cycle has been interpreted in the light of ecojustice principles for the sake of a biblical ecotheology (Habel 2011, pp. 90–114). There is no need to place this covenant established through a bow in the sky into canonical discussion with others to do with circumcision, the sabbath and the new covenant in Christ.

Read at face value the Noah covenant seems to establish grounds for hope and some solace or comfort. This reading offers a more sympathetic theological response than the default tendency to regard natural disasters as a type of righteous punishment (Fair 2018, p. 5). That is necessary because without the covenant the flood stands as 'the archetype of human catastrophe'. James Okoye reckons that it represents God's absolute "no" to a corrupt and violent world' (Okoye 2018, p. 96). It presumes the indictment that follows. It is not hard to see how the text can then easily become either misused—or abused. The flood saga is ripe for preying upon vulnerabilities.

What unfolds in the Hebrew text is rather different. The Hiphil stem for the Hebrew word 'to establish' (Gen: 9.17) resonates with Timon's standing in the middle of the ocean. It means 'to make stand, to erect'. It is designed to convey a sense of the covenant being fulfilled, carried out and kept (Hamilton 1990, p. 316). The literal reader is able to embrace the first covenant to be made with 'all flesh'—with the whole of humankind and the rest of creaturely existence—and not just one with the people of Israel (de la Torre 2011, p. 124).

This need to create some distance away from the possibility of punitive theories of retribution lies behind Lusama's appeal to the book of Job. The problem is now one of the perceived absence of God in the face of innocent suffering: 'Why do the poor and the marginalized suffer undeservingly while the rich who gain their wealth through wicked ways of manipulation and exploitation, enjoy life abundantly?' (Lusama 2004, pp. 88–94). This reading is consistent with the 'Otin Taai Declaration of 2004. The Pacific Council of Churches rejected the idea that the covenant with Noah provides a divine 'guarantee' that God 'would not flood the earth again': high tides and surges 'are not an act of God'. They are 'a result of human economic and consumer activities that pollute the atmosphere and lead to climate change' (World Council of Churches 2004).

It is not then difficult to find examples of claims being made that this promise of no more floods leads to passivity, 'an avoidance of praxis' and a particular type of climate denial (Kempf 2017, pp. 28–29). In a report featuring Kiribati published in the same year as the Communiqué Richard Johnson noted that '[f]or a deeply religious population, this story is as good as science' (Johnson 2019). To understand the Noah covenant in this way does not mean islanders who hold this position are unaware of rising sea-levels and the risks associated with water and food supplies being compromised through salinization. It does not lead to a literal denial of climate change as a matter of course. In the three-fold pattern discerned by the sociologist, Stanley Cohen, this form of denial may be classified as interpretative (Cohen 2001, pp. 7–9). It accepts that there is climate change but it is understood through the lens of the Noah saga and there is little that can be done to mitigate the risk.

The signs of climate change are evident but viewed through a counter-narrative. That alternative may assume the Noah covenant but be expressed in another form—as it was in the case with President Taneti Mamau who insisted that 'Kiribati will [not] sink like the Titanic ship'. That will not happen because '[o]ur country, our beautiful lands, are created by the hands of God' (Doane 2017). The contrary interpretative rather than literal form of denial may be then supported by regular reminders of the Noah story like the appearances of rainbows and the occasional scientific claim suggesting that either Tuvalu or Kiribati is growing, rather than shrinking (Birbo and Woodroffe 2013; Kench et al. 2018). Strictly speaking, the bow in the text is not necessarily to be understood as a rainbow: it may indeed be a warrior's bow being laid down as a sign of peace (Okoye 2018, pp. 112–113; Hamilton 1990, pp. 317–318; de la Torre 2011, p. 124).

These alternative ways of coming at the text demonstrate the importance of culture and life-setting in the reception and appropriation of Scripture. Writing with reference to the Noah narrative Talia argues that the way in which this story has been read is through the legacy of western missionary endeavour. It carries with it as such 'certain ideologies' that lead to the silencing of indigenous voices. Those 'patterns of belief' limit the need to 'engage in dialogue' and the capacity for the text to 'speak to us' in order to 'locate the will of God' (Talia 2009, p. 70).

The merit of Kempf's approach lies in the manner in which he situates the Noah story within its life-setting in Kiribati: he is not seeking to deflect the critique of media headlines or the predictions of western science. In a way he is anticipating the kind of thinking which is to be found in Timon's songs of resistance, lament and providence. What Kempf is mediating is the confusion of the i-Kiribati on first becoming aware of the 'scenarios of inundation and disappearance' immersed in the alien language of climate science and its subsequent revision. Kempf anchors his enquiry through the 'anticipatory power' of the words of 'locally composed songs'. Rather than being left standing in the middle of ocean with Timon Kempf selects songs of the Bitibiti singing group based on South Tarawa (p. 35). One in particular, *Taekan Rabakau*, was designed to address the scientific theory that 'unsettled us deeply' being handed on. Where Kempf differs from Timon—and this difference is made possible through his choice of song—is how he ties the exegesis of the lyrics back into the problematic Noah text. In his coconut theology of climate change the only biblical references Timon made were to Psalm 137 and John 10:10.

At first the words of *Taekan Rabakau* drew upon the Noah story as a counternarrative: it sought to advance the claim that the 'Christian God is above such things as [Western] science and the statements of its representatives' (pp. 36–37). Over the subsequent years with further knowledge the song is less of a song of denial of climate change per se: it becomes rather a call to 'stick to our faith and believe that something is alright' (p. 37). God would not let the people down (p. 38).

The standard critique of how the Noah story can inform a contemporary audience is inclined to ignore the nuanced movements within the narrative. Kempf draws out the potential for such through his second song, *Ai kamira kanoan te bong*. The emphasis moves away in time from the rains falling, the waters rising, bows in the sky and promissory covenants. In place of passivity is a theme of preparation. Noah was warned of the imminent flood and given time to make ready. The cultural backstory for the text has become one of climate change rather than the need for an aetiological myth. Now the warning with regards the incursion of the sea comes via scientific prediction (p. 41). Through interpreting Noah by way of these songs Kempf concludes that a 'local way out of the valley of negation' has been found (p. 43).

This prophetic way in which to interpret the Noah story represents the second of three ways in which Fair concluded the narrative could be read. The first had been which took the covenant as a sign of continuity of the world as it is known (Fair, p.9). In this second option Noah becomes an 'icon of preparation'. It is possible to set him alongside the patriarch Joseph whose capacity to interpret dreams enabled him the prepare Egypt and his own family for famines (pp. 9–10). Fair's theory of multiple perspectives and accent on storytelling is allowing for much more variegated reading of Noah to emerge than is commonly believed to be the case in cross-disciplinary enquiries. The full advantage of placing the Noah narrative inside its life-setting in an indigenous culture is then evident in Fair's most radical option. She assigns responsibility for this reading to an 'Ezekiel' (who actually turns out to be Talia) who wished the reception of biblical stories to be released from the 'missionary enterprise'. Talia is seeking a far more liberative and subversive reading. Now the ark itself becomes a vehicle for the rich and the powerful. This becomes a metaphor for the geopolitics of climate change: the animals on the outside are the poor and the vulnerable (p. 11).

The benefit of the analysis made by Fair and Kempf is that it placed the reading of the Noah narrative within the 'lived religion'—and thus the experience and self-understanding of a Bible-conscious people. It recognized that there are multiple ways in which this text can function in the response to sea-level rise and the climate emergency. The presenting issue is not a simple form of climate denial or a pitching of religious claims over and against those of a western science. Kempf was also able to set the way in which the Noah cycle is put to use within the performative practice of song and thus indigenous patterns of customary knowledge. The peoples of Tuvalu and Kiribati are finding voice, laying claim to identity and offering an alternative way of knowing and agency.

The Bible and Disasters

The merits of this type of discourse should not be ignored. It is arguably the case, that this re-reading of the intersection of the Noah story and climate change does not go far enough, however. Writing in their report on resilience a working group led by the Pasifika theologian Seforosa Carroll argued strongly for the need for concerted theological attention to be given to diverse forms of disaster. The purpose of this focus was not to construct a counternarrative as *per* Kempf, at least not in the form that he mapped. It was rather driven by the role the Bible plays in a lived way in the life of the community. It was anticipated that 'changes to the environment will inevitably have an impact on the spiritual and cultural lives of communities' (Carroll 2019, p. 4). It was recognized that the church which is the institution through which the Bible becomes an actant is 'uniquely situated in their communities' to help prepare for what is to come.

The 'current ecological crisis' is seen to be a 'reflection of our broken relationship with God each other and the creation'. There had been a 'failure to be "good custodians of the earth"; there had been a loss of sight as to how creation is "God's abundant gift". Through anthropogenic climate change the 'earth system is at risk'. There was an open acknowledgement of the findings of 'scientists and climatologists' (p. 4).

Not surprisingly that framing of the task at hand led to a series of questions that would normally be found within a systematics treatment of the problem of theodicy and the plausibility of divine providence. Lying behind the particular questions identified was an overarching fundamental concern having to do with the 'nature and character of God'. Writing for a religious audience the focus of attention fell upon 'what then can disasters teach us about God, our relationship with God within creation and our relationships with our fellow human beings?' (p. 3). The range of implicit theological themes covered the doctrinal territory of creation, 'brokenness' or sin.

What was most noticeable about the line of approach adopted by the Carroll report was the ease with which the intersecting matters of resilience and theology turned to 'The Bible and Disasters'. The underlying assumption was the need to wrestle with the biblical witness while being well aware of the need for a contemporary hermeneutic. For the present purpose two matters stand out. The first was the recognition of the call for a relevant theology and the employment of Biblical references in doing so. There was an obvious need to counter notions of divine retribution and the occasional acts of transference in assigning responsibility for such. The second was the sheer diversity of texts actually cited. The focus was not confined to Noah or to Job. In a manner which frequently has not happened in the response of those inhabiting low-lying islands there were references to the synoptic gospels (especially Luke), John, and Paul's epistles as well as 1 Peter. There was an evident willingness to draw upon the New Testament as well as the Hebrew Bible/Old Testament. With regards the latter it was the book of Psalms that furnished examples of lament and protest. The impact of climate change was set alongside Paul's understanding of a 'groaning creation' (Romans 8).

From the vantage point of a Christian ethics the biblical texts were here being used in a way other than the prescriptive mode in which the Noah cycle is often used or dismissed. The intention—whether realized or not—was to tap into what has been described as the instinct of the faithful. The rhetoric was designed to constitute a 'we' and use the biblical texts in a way which adopted an amalgam of what has been described as illuminative and formative types. Whereas the prescriptive mode is likened to evoking a mirror response the illuminative serves as a window. There is no direct correspondence between the text and the current issue under scrutiny. The text releases ethical directives and principles, an orientation, through which a response to the climate emergency, in this instance, may be viewed. The frequent exhortatory tone—itself a standard feature of the Greek epistolary tradition in which the apostle Paul stands—coupled with default-like use of the first person plural pronoun—we—ensured that the Carroll report had effectively located the call for resilience within a formative use of Scripture. The citing of Scripture is now not primarily for the explicit resolution of a problem. Nor is it designed to be the equivalent of a Bible-based apologetic which seeks to explain and maybe justify the ways of God in uncertain times.

The formative use of the Bible in a theological ethics fastens attention upon personal and communal character. It involves consciousness-raising, a moral predisposition, and the prospect of appropriate action. Here the Carroll report referred specifically to the epistle of James and, more tellingly, to the close of the parable of the Good Samaritan where Jesus commands those with ears to hear to "go and do likewise". It means, in the first place, that the fundamental analogy between the biblical text and the current context lies not in some ethical situation or event but in the believing community itself. The distinctive social reality of the Christian community is the primary form of continuity with the biblical world.

The selection of texts was woven in and through the unfolding narrative of resilience. It was coloured by a desire to make a faithful response, to be pastoral, and thus present the opportunity for faithful action (p. 4). There was a tacit acknowledgement then of how the Bible functions in the Christian life—and a Christian community—not by way of isolated texts which may be invoked out of confusion. The one time the Noah cycle was cited had to do with the calling to preparedness (p. 7).

The Public Bible

The emphasis the Carroll report placed on 'opportunity' and 'preparedness' is a point of entry into the discipline of a public theology. The term has not been widely used throughout the Blue Pacific. It does not feature in Carroll's work on resilience: the idea of opportunity though conveys an allusion to a moment of *kairos* and crisis. These two words come together in the case the Norwegian theologian, Kjetil Fretheim, has made for a kairotic public theology in response to the global climate crisis (Fretheim 2016, pp. 132–140).

The purpose of a public theology is designed to address issues that impinge upon the common good and the flourishing of all. It recognizes that the right to be heard in the diverse spheres of the public domain needs to be earned rather than assumed. Its tone is persuasive rather than dogmatic as it seeks to address matters of public affair in an interdisciplinary mode. It must of necessity sift the theological heritage of Christian beliefs, ideas and symbols in order to fulfil a bilingual vocation. There is a responsibility here to draw upon that inheritance for the sake of representing a position in the public sphere; there is also an imperative to talk back into the life of the professing Christian community with insights and expertise that others in the public domain have contributed. Being bilingual means being able to address an issue via an explicit Christian discourse as well as in a universally accessible language of reason.

In terms of a response to the climate emergency—and, by way of extension, to the Anthropocene—the role of a public theology is limited. It is not one of the sciences that feature in the deliberations of the IPCC; it is absent from work of the Forum and its vision for the Blue Pacific. And yet the way in which the Noah story has functioned in the region suggests that it may well have a role to play primarily in the life of those who dwell in the most vulnerable locations. It presumes that Scripture and belief have a role to play. The Blue Pacific is a regional context in which the Bible has a public as well as a personal and ecclesiastical role. It possesses a function which can speak into communal life as well as mediating spiritual and ethical perspectives, rituals and values into public life.

That is the line of argument pursued by Paul Hanson. The complexity of the task is not ignored. The Bible is ambiguous: for that reason Hanson realizes the critical importance of a 'trustworthy hermeneutics that provides safeguards against arbitrariness and guidelines for appropriate application' (Hanson 2017, p. 25). There is an obvious need to come to terms with what the Bible is and what it is not. Hanson duly makes a distinction between what he calls the warp and weft of the Bible.

Under the first sub-heading comes the story. In the context of the Blue Pacific the Noah narrative is an individual story that 'pulsates with meaning'. The particular movements within the storyline must be respected but not in a way that leads to 'the atomization of Scripture into a disjointed anthology' (p. 26). This warp ought to situate the particular within a vast and living drama able to evoke a lifelong engagement. These stories represent 'episodes arising out of the experiences of a people struggling to shape a common life in relation to its God'. They are instalments into an 'identity-shaping whole' (p. 27). The weft constitutes the theocratic principle that further binds the story together into canonical meaning. Hanson privileges the idea of a covenant (p. 28). For its public role this model of warp and weft—or something like it—prepares a community to draw upon the resources they have inherited: they can set about the task of

> translating ideals and strategies from the comfort zone of [their] own communal traditions and practices into a language comprehensible to other communities populating a diverse universe. (pp. 35–36)

Hanson had been writing in response to the history of the Bible in the politics of the United States. His warp and weft theory is able to hold in tension how the Bible is not unilinear in its ethical direction: it possesses a canonical quality of rehearsing a plot. For J. Richard Middleton that plot extends 'the full sweep' from creation to the eschaton. It holds out a vision of holistic salvation, cosmic renewal and the redemption of all things. At one level Middleton's thesis may seem far removed from rising sea-levels and the difficulty of now growing *pulaka*. There is only a passing reference to Noah and Job (Middleton 2014, pp. 47, 552, 54, 61, 92, 93, 99, 117, 133). What is even more surprising is his failure to consider the current threats to the planet—the site of his renewal—in the appendix 'Whatever Happened to the New Earth?' Where his writing is helpful is the way in which he moves along a different trajectory from Hanson: in actual practice he weaves together many texts into a plot that generates an internal canonical discussion and does not leave individual biblical stories stranded. That has been one of the problems of the Noah story. As a consequence the text has been open to the criticism of climate denial: the inhabitants of low-lying islands were left far too dependent upon a highly restricted range of biblical narratives. It is time for redress.

The Present *Kairos*

The present is a *kairos* moment in which to make a stand for a broader based biblical response to the climate emergency, the trajectory of a hothouse Earth and the Anthropocene. Bruce Birch and Larry Rasmussen et al. have called for a 'new conversation' on the Bible, ethics, and the 'planetary changes [that] are epochal and probably deeper and more extended in space and time than we realize' (Birch et al. 2018, p. xiii; Rasmussen 2013). These authors assume 'the inadequacy of literalism', the need for an overarching hermeneutic, the role of the canon, the importance of formative stories for communities and the imperative of engaging with the moral imagination. They are deeply mindful of the eclipse of the Euro-Western Christian temperament or mindset.

The Carroll report demonstrated the potential for a Pasifika theology of resilience in the face of disasters, especially those associated with a changing climate. It captured much of the intent of a public theology. The bilingual nature of this emerging discipline ought to be nurtured for several reasons. For the sake of a sustainable and resilient vision—if that is indeed possible given Steffen's prognosis for the planet in the Anthropocene—then it is in the interest of the Forum's Secretariat to encourage an illuminative and formative biblical discourse that is in a constructive tension with the rhetoric of climate scientists and the preferred geopolitical strategy. It places a moral discourse of the heart alongside those of cultural identity and the litany of calls for nations, partners and industries to meet the agreements made. It strengthens the lines of adherence for ethnicities for whom a western climate science remains confusing (Pearson 2019). The other side to this coin is for the churches to take to heart Hanson's model of the warp and weft of the biblical message and look beyond the self-interest of the institution and engage in Birch and Rasmussen's new conversation.

The Bible acts in personal and village life. It is multifaceted in its outworking. For that reason the present is a *kairos* moment: it is time to weave a mat of many scriptural texts that speak into a wide range of life-situations arising out of the climate crisis. There is a wealth of biblical material that can address themes such as the care of creation (Colossians 1:15–23), migration (Psalm 137), being an alien in a strange land (1 Peter 2:11–12), dealing with world powers (Ephesians 6:12) and coming face to face with how to live in the shadow of apocalyptic endings (Matthew 24:36–44; 25:1–13; Luke 12:40). These biblical references are only the tip of what is possible. There is an exhaustive amount of biblical and theological scholarship on these diverse themes. This work can sit alongside Lusama's use of Job and Kempf and Fair's understanding of Noah as a counternarrative. These additional readings may well be able to stimulate more songs and *fatele*.

'Not Without My Neigbour'—Madrid, 2019

The practice of a public theology is to find organizing texts—that is, key texts that shape a theological response and around which other relevant biblical texts can assemble. For the kind of kairotic public theology Fretheim (2016) advocates Noah is not the right text. In his doctoral research Talia—who happens to be the current co-chair of the Global Indigenous Forum on Climate Change—has recognized the affective power of the parable of the Good Samaritan and its talk of neighbours—*tuakoi*. It offers so much. The very idea of a neighbour resonates with customary indigenous knowledge: the biblical notion presumes the inclusion of the stranger—as indeed does the *Qur'an*. Richard Carter and Samuel Wells have used the leading question of the parable, 'who is my neighbour?', to examine multiple political issues (Carter and Wells 2018).

In terms of mitigation and adaptation the impact of this biblical work will always be indirect. It is not so obviously practical in terms of an agenda of specific things to be done. Its purpose is to animate churches in such a way that fresh connections can be made between Scripture, a people's way of life and the daunting threats posed by a changing climate, let alone the Anthropocene. The significance of a shift in a biblical trajectory can be seen in the high level press conference Talia organized at the COP 25 Conference in Madrid. For the first time in a Tuvaluan presentation on the international stage the rhetoric of the neighbour was invoked: the parable of the Good Samaritan lay in the background. The theme was 'Not without my neighbour' conceived in a planetary manner. The panellists were a former President and Prime Minister (Tong and Sopoaga) as well as the Head of Greenpeace Pacific and the General Secretary of the Pacific Conference of Churches. Those speaking have been among the most ardent advocates for policies of mitigation and adaptation. In this example of a public theology they found themselves engaged in a media conference where they drew upon the theme of the neighbour to inject an ethical imperative based upon an appeal to those who live 'near by'. They were provocative: they were on the front foot in a way in which Noah and Job cannot do.

References

Barreto R, Cavalcante R, da Rosa WP (eds) (2017) World Christianity as public religion. Fortress Press, Minneapolis

Birbo N, Woodroffe CD (2013) Historical area and shoreline change of reef islands around Tarawa Atoll, Kiribati. Sustainability Science 8: 345–362. https://doi.org/10.1007/s11625-013-0210-z

Birch BC, Lapsley JE, Moe-Lobeda C, Rasmussen LL (2018) Bible and ethics in the Christian life: a new conversation. Fortress Press, Minneapolis

Carroll S (ed) (2019) A theology of disaster resilience in a changing climate. UnitingWorld, Sydney

Carter R, Wells S (2018) Who is my neighbour?. The Global and Personal Challenge, SPCK, London

Cohen S (2001) States of Denial knowing about atrocities and suffering. Polity Press, Cambridge

Conradie EM, Koster HP (2020) T&T Clark handbook of Christian theology and climate change. T & T Clark, London and New York

Curlew LK (2012) The cultural impacts of climate change: sense of place and sense of community in tuvalu. A country threatened by sea level rise. University of Hawai'i Press, Honolulu

Deane-Drummond C, Bergmann S, Vogt M (2017) Religion in the anthropocene. Cascade Books, Eugene

de la Torre M (2011) Genesis. Westminster John Knox Press, Louisville

Doane S (2017) As climate change threatens islands, Kiribati's President plan development. CBS News 20 November, https://www.cbsnews.com/news/climate-change-kiribati-president-taneti-maamau/. Accessed 15 June 2020

Fair, H (2018) Three stories of Noah: navigating religious climate change narratives in the Pacific Island region. Geo: Geography and Environment e00068, 1–15. https://doi.org/10.1002/geo2.68

Fretheim K (2016) Interruption and imagination. public theology in times of crisis. Wipf & Stock, Eugene

Habel N (2011) The birth, the curse and the greening of the Earth: an ecological reading of Genesis 1–11. Sheffield Phoenix Press, Sheffield

Hamilton VP (1990) The book of Genesis: chapters 1–17. Eerdmans, Grand Rapids

Hanson P (2017) The Bible and public theology. In: Kim S, Day K (eds) A companion to public theology. Brill, Leiden

Johnson K (2019) Climate change could drown Kiribati, but the nation looks for Noah's Ark, ABC Religion and Ethics, https://www.abc.net.au/news/2019-07-21/kiribati-hopefull-in-face-of-climate-change-rising-sea-levels/11228658. Accessed 15 June 2020

Kempf W (2012) The rainbow's power. climate change sea level rise and reconfigurations of the Noah story in Oceania. Pacific voices. Local governments and climate change conference papers. University of the South Pacific, Suva. 68–84

Kempf W (2017) Climate change, Christian religion and songs: revisiting the Noah story in the Central Pacific. In: Dürr E, Pascht A (eds) Environmental Transformations and Cultural Responses. Ontologies, discourses and practices in Oceania. Palgrave Macmillan, New York, 19–48

Kench PS, Ford MF, Owen SD (2018) Patterns of island change and persistence offer alternate adaptation pathways for atoll nations. Nature Communications 9:605. https://doi.org/10.1038/s41467-018-02954-1

Lazarus RJ (2010) Super wicked problems and climate change: restraining the present to liberate the future. The Environ Law Reporter 40(8):10749–10756

Lenton TM, Rockström W, Gaffney O, Rahmstorf S, Richardson K, Steffen W, Schellnhuber HJ (2019) climatetipping points—too risky to bet against. Nature 575(7784):592–595. https://doi.org/10.1038/d41586-019-03595-0

Levin K, Cashore B, Bernstein S, Auld G (2012) Overcoming the tragedy of super wicked problems: constraining our future selves to ameliorate global climate change. Policy Sci 45:123–152. https://doi.org/10.1007/s11077-012-9151-0

Lusama T (2004) Punishment of the innocent; the problem of global warming with special reference to Tuvalu, Master of Arts and Religion thesis, Tainan Theological College and Seminary

Middleton JR (2014) A new heaven and a new earth: reclaiming a biblical eschatology. Baker Academic, Grand Rapids

Moses A (2020) 'Collapse of civilisation is the most likely outcome': top scientists claim. Voice of Action, https://voiceofaction.org/collapse-of-civilisation-is-the-most-likely-outcome-top-climate-scientists/. Accessed 8 July 2020

Okoye JC (2018) Genesis 1–11: a narrative theological commentary. Cascade Books, Eugene

Pacific Forum Secretariat (2019) Fiftieth Pacific Islands Forum Communiqué https://www.forumsec.org/wp-content/uploads/2019/08/50th-Pacific-Islands-Forum-Communique.pdf?fbclid=IwAR1WZvpF0MASWkCRavx6DkreOTlLWKb3t9nQHNjhDVevXIiVrTZdmW_yzUc. Accessed 1 June 2020

Pearson C (2010) The purpose and practice of a public theology in a time of climate change. Int J Public Theol 4(3): 356–372

Pearson C (2019) Acting justly in the Anthropocene. In: Pearson C (ed) Enacting a public theology. SUNMedia, Stellenbosch, 87–98

Pearson C (2020) God's continued providence. In: Conradie EM, Koster HP (eds) T&T Clark handbook of Christian theology and climate change. T & T Clark, London & New York, 395–405

Rasmussen L (2013) Earth-honoring faith. Religion ethics in a new key. Oxford University Press, Oxford and New York

Sanneh L, McClymond MJ (eds) (2016) The Wiley Blackwell companion to World Christianity. John Wiley Sons, Malden & Oxford

Steffen W, Rockström W, Richardson K, Lenton TM, Folke C, Liverman D, Summerhayes CP, Barnosky AD, Comell SE, Crucifix M, Donges JF, Fetzer I, Lade SJ, Scheffer M, Winkelmann R, Schellnhuber HJ (2018) Trajectories of the Earth system in the anthropocene. PNAS 115(33):8252–8259. https://doi.org/10.1073/pnas.1810141115

Talia M (2009) Towards *fatele* theology. A contextual theology response in addressing threats of global warming in Tuvalu. Master of Theology thesis, Tainan Theological College and Seminary

Tavola K (2019) 2050 strategy for the Blue Pacific continent. A sea-change? https://www.islandsbusiness.com/component/k2/item/2657-2050-strategy-for-the-blue-pacific-continent-a-sea-change.html. Accessed 1 June 2020

Timon T (2018) Towards an emerging "coconut tree missiological imagination". An enquiry into climate change and its relevance for ministerial formation at Tangintebu Theological College. Unpublished PhD thesis. University of KwaZulu-Natal, Pietermaritzburg

Timon T, Kaunda C (2019) "I stand in the middle of the ocean": the emerging coconut theology of climate change in Kiribati. Pharos J Theol 100, 1–13

World Council of Churches (2004) The 'Otin Taai Declaration. https://www.oikoumene.org/en/resources/documents/wcc-programmes/justice-diakonia-and-responsibility-for-creation/climate-change-water/otin-tai-declaration. Accessed 15 June 2020

Part IV
Weaving Worldviews: Science and Spirituality Together

Chapter 11
Paddling on Both Sides of the Canoe: Toward a Consilience of Science and Spirituality in Climate Change Response

Rowan Gard

Abstract Oceania is especially vulnerable to the effects of climate change. Policymakers and non-governmental relief organisations (NGOs) understandably focus on science to inform and answer questions of how best communities should address the grand challenge of climate change, yet scientific knowledge is just one component in a range of valuable risk-management strategies. Many Pacific Islanders have deep held spiritual beliefs that exist in a dynamic social milieu that in turn influences the ways in which individuals and communities respond to climate events and engage with aide relief organisations. This chapter explores aspects of faith-engaged adaptation practices in the Pacific Islands around varied responses to extreme weather (e.g. Tropical Cyclone Winston in Fiji) and long-term climate change. To achieve true embedded climate-change adaptation and long-term sustainable development, a community-focused bottom-up approach is required by policymakers and NGOs, with special consideration given to the significance of faith and faith-based organisations (FBOs) in Oceanian communities. Specifically, an integrated discourse and risk-management approach that combines science and spirituality, much like a *vaka* (Vaka or waka means "boat" or "canoe" in most Polynesian languages) or voyaging canoe that must be paddled on both sides, is vital in the negotiation and management of climate change in the Pacific.

Introduction

Scientific efforts to facilitate the reduction of human emissions of greenhouse gases—carbon dioxide (CO_2), nitrous oxide (NO), methane (CH_4), and others—have had mixed success on a wider scale, though some take a more pessimistic perspective stating that "science has failed" in addressing the grand challenge of climate change and initiating change in response to the crises (Finucane 2009: 1). For example, some researchers have called for stratospheric aerosol injection, a process by which

R. Gard (✉)
University of St. Andrews, St. Andrews, Scotland
e-mail: arg21@st-andrews.ac.uk

J. M. Luetz and P. D. Nunn (eds.), *Beyond Belief*, Climate Change Management,
https://doi.org/10.1007/978-3-030-67602-5_11

particles would be sprayed into the atmosphere in an attempt to block solar radiation from reaching the Earth; thus, theoretically cooling the planet (Krishnamohan et al. 2020). Troubled by this suggestion? You are not alone. As a singular resource of mediation for communities, nations and regions in response to the climate crises, this type of technocratic scientific approach is tremendously ambitions yet woefully ineffective, not to mention potentially dangerous. Unsurprisingly these top-down tactics are prone to failure as they tend to disregard the dynamic and multifaceted interplay of cultural, psychological, spiritual, political and economic elements—or the ethnoscapes, ideoscapes and financescapes as they have also been termed (Appadurai 1990)—that impact response to climate change.

A risk-management approach and a wider academic discourse that integrates both science with an appreciation of scientific understanding in the communities where response is most needed to climate change is critical. This is especially true in Oceania where Pacific Islanders and their communities are suffering as a result of the climate crises (Kumar et al. 2020; McNamara et al. 2020; Scott-Parker and Kumar 2018). In considering this troubling reality of in Oceania there is much that can be learned from human response to risk and the role of faith, faith-based organisations and the social dynamics associated with climate resilience. In the following sections a consideration of climate change and extreme weather in Oceania, is exemplified in a case study of Tropical Cyclone Winston in Fiji, with an emphasis placed on adaptive risk-management responses and tools that embrace the role of FBOs in communities. The remained of the chapter reflects on intuitive worldviews and the meaning of life from a Melanesian Christian perspective. While the concluding sections culminate in understanding risk as a social construct, community context and the critical importance of tools for engagement that offer trustworthy climate change information and resources in an FBO setting to communities.

The Realities of Climate Change—Rising Sea Levels and Intensified Storms

Given the current levels of greenhouse gases in the atmosphere, the end of the twenty-first Century is likely to experience a "1 meter sea level rise— [it] seems unavoidable… although there may be an increase of 2 meters" (Carr et al. 2014: 37). For an island that is only a few meters above sea level at high tide this is an exceedingly troubling projection. Beyond this, storms can cause damaging "king tides" or exceedingly high tides, though high tides and flooding are not the only means by which climate change is damaging and endangering life in Oceania.

Small-islands are considered among the most vulnerable to climate change because of their limited size, proneness to natural hazards, physical isolation, low adaptive capacity, and high adaptation costs relative to gross domestic product. This was highlighted in the Maldives following the Boxing Day Tsunami in 2004, "which in a matter of hours inflicted macro-economic damage on the order of 62% of the archipelago's GDP" (Luetz 2017: 45). When disaster strikes, a domino effect can occur, causing one vulnerable sector to influence another. The effects of less dramatic climatic events may accumulate over time, also setting in motion a chain of problems. For example, fresh water is a key resource yet when supply is affected by climatic events, food security and public health are threatened such as when an island's freshwater lens becomes inundated with sea water thus becoming brackish and undrinkable.

Under climate change scenarios, tropical cyclones are predicted to have greater intensity, and droughts and floods may occur more frequently (Hashim and Hashim 2016; Shope et al. 2016; Sugi et al. 2017). The droughts lead to shortages of drinking water and impair food supplies; the floods damage infrastructure, contaminate freshwater and agricultural soils with saltwater, and may compel evacuation. Both droughts and floods can damage local ecosytems and reduce biodiversity. Of further concern is the absorption of carbon dioxide by the oceans leading to ocean acidification, which harms marine organisms with shells, including coral reefs that are critical physical barriers for atoll islands serving as protection against storm surges and high waves (Gerrard and Wannier 2014: 4).

Globally more than 600 million people, or 8.6% of the world's population, live in coastal zones (i.e. generally considered to be zero to ten metres elevation), which comprise only 2% of the world's land area. Of these 600 million people more than half, ~360 million, live in urban centres. This amounts to 13% of the world's urban population living in the highest-risk zones of sea level rise (McGranahan et al. 2007: 17–20). Thus, Pacific Island nations, most especially the low-lying atolls, and their urban centres are exceptionally vulnerable to the adverse effects of human-induced climate change and extreme weather events such as Tropical Cyclone Winston in 2016.

The +30,000 islands of the Pacific are spread across five time zones in both the Northern and Southern Hemisphere (Andrew et al. 2019). Approximately 12.4 million people live in the 22 Pacific Island Countries and Territories (PICTs) of Melanesia, Polynesia, and Micronesia (United Nations 2019). This figure does not include a further 4.9 million people residing in Aotearoa New Zealand and another 1.4 million residing in Hawai'i (United Nations 2019). Troublingly the populations and communities o f he Pacific impacted by weather-related disasters, such as category five cyclones (e.g. Tropical Cyclones Pam in 2015 & Winston in 2016), have increased dramatically in the last several decades. Data and climatic projections from the Intergovernmental Panel on Climate Change (IPCC) point to increasingly damaged environments, such as coral reefs, overall loss of terrestrial and marine life and ultimately decreasing biodiversity (IPCC 2018).

Juxtaposed against the worsening environmental conditions, the dense urban and economic centres of the Pacific are located in coastal zones and are also vulnerable to rising sea levels and other extreme events. Beyond tropical cyclones, these include floods, tidal surges, droughts and landslides. Though sea level rise is commonly misunderstood as being a uniform process, it is a far more complex process with various factors, such as the formation of the Earth's crust, ocean temperature and salinity reflecting regional variability, as well as the gravitational and rotational changes driven by the ice-ocean mass exchange that all influence and exacerbate sea level in regional areas (Farrell and Clark 1976; Mitrovica et al. 2001: 1026–1029; Mitrovica et al. 2009: 753). Ultimately, this results in uneven rises and falls in overall sea level rise, with some areas experiencing an amplified rise, especially in the equatorial regions of the Pacific (Milne et al. 2009). While these processes have occurred for millennia, humans have affected sea level rise via direct means of manipulation of ground and surface water flows and indirectly through greenhouse gas emissions that increase global temperatures (NOAA 2019). These warmer temperatures impact the mass and volume of seawater through increased terrestrial ice melt and warmed ocean temperatures, both of which compound rising global sea levels.

The Consequences of Climate Change—Fiji as an Exemplar

In early February 2016, in the two weeks preceding the Category Five Tropical Cyclone Winston, thousands of dead fish washed up on the Coral Coast of Fiji, which stretches from Suva to Sigatoka. According to a local marine ecologist, the "fish kills" were a result of the unusually high ocean temperatures of ~35º Celsius in the shallower reef areas along the Coral Coast (Gard 2016, np). The Ministry of Fisheries and Forests issued a warning "advis[ing] people living in villages along the Korolevu to Namada coastline to refrain from eating any dead fish along their coasts as they may be harmful to human health." This warning was issued following some villagers falling ill and then being hospitalised after eating the fish (Gard 2016, np). The Fijian Government instructed citizens to gather the dead fish into piles and burn them, then bury or dispose of the remains. In visiting the beach in the days after the fish kills, the stench of dead fish remained and I had a feeling of having passed into a post-apocalyptic world. As someone who has spent most of their life near the Pacific Ocean, this experience was one of the most surreal and deeply disturbing of not just my fieldwork to date, but my life.

Coinciding with the heat wave and fish kills was a coral bleaching event where the thermal stress of the high oceanic temperatures caused the coral to lose its symbiotic algae that live inside its calcium carbonate skeleton. Consequently, the coral turned pale, leaving ghostly white sections of reef. Local experts cited the "calm seas and clear skies [as causes for the]… seawater temps on the nearshore reefs to have heated up to 35–36 C + during low tides, causing corals in these nearshore lagoon areas to become severely bleached over the last week and many have already died. Corals in deeper areas are now also showing signs of heat stress as mean seawater temperature starts to exceed 30C" (Gard 2016, np). It was a troubling time and as I spoke with friends, colleagues at the University of the South Pacific (USP) and locals at the beach regarding the fish kills and bleaching event, everyone agreed that they had never seen anything like it before.

The Big Monster - Tropical Cyclone Winston as Case Study

While conducting fieldwork in Fiji in 2016 I was a survivor of Tropical Cyclone Winston and in the aftermath a relief worker with Fiji Red Cross (Gard and Veitayaki 2017). On Saturday the 20th of February Winston made landfall in Suva and cut a path of destruction across the Fijian Islands. I recall sitting with my neighbours in their living room illuminated with only a few candles and wondering if the roof would peel away in the incredible winds. Winston generated gusts of wind recorded at 200 MPH and average wind speeds of 143 MPH (Westcott 2016). The tall palm trees in the backyard crashed to the ground, as Winston was a Category 5 superstorm of unprecedented strength and destruction. Forty-four people were killed, buildings were stripped to their foundations and towns were reduced to rubble, leaving many in abject poverty.

For months after Winston, families struggled to procure basic necessities, such as food and clean drinking water for their children (Bolatagici 2016). My own neighbourhood of Suva Point was without electricity for one week after Winston, though this was nothing when compared to rural communities that faced far worse difficulties, some going months without electricity and running water (Round 2016). In the days and weeks following Winston, I volunteered with Fiji Red Cross assisting with damage and need based assessments and emergency aid supply deployment. I also joined a dozen colleagues from the University of the South Pacific (USP) in gathering and purchasing relief supplies for the fishing community on Qoma Island located to the northeast of Suva (Fig. 11.1).

Qoma was especially heavily hit and after gathering supplies we made the two-hour trip to the island via car and then boat. After we had unloaded the relief supplies from the boats, which ranged from freshly baked bread to plastic sheeting, we were formally greeted by the village elders on the concrete foundational slab that had once served as a community gathering space and village hall. Following our formal greeting, I wandered through the community, speaking with villagers and listening to their Winston, or "big monster" as they called it, survival stories. It was clear that the devastation was near-total on Qoma. Many homes and buildings had been stripped to the concrete slabs, while other buildings, such as the Methodist Church, the largest building on the island, had a roof that had collapsed and was unusable (Gard and Veitayaki 2017: 160–161) (Fig. 11.2).

> As I surveyed the damage and spoke with villagers, I was especially touched by the story of one family who had sheltered together in the bathroom, which had partial concrete walls and thus was stronger than the rest of their home. As I spoke with the middle-aged father he shared with me that 'God kept us safe; he is always with us' and it was the love of his family that had enabled him to face the future (Gard 2016, np). Drawing upon faith and family, he was already in the process of rebuilding his family's former home. (Gard and Veitayaki 2017: 160–161)

After we had spent the afternoon on Qoma, my USP colleagues and I returned to the boats for the short journey back to our vehicles on Viti Levu. On the drive back to Suva, we discussed ways in which Fijian communities could be made more resilient

Fig. 11.1 Suva Point experiencing an unusually high tide following Tropical Cyclone Winston, February 2016 (*Source* Image courtesy of the author)

for future extreme weather events and long-term climate change. In previous fieldwork on Qoma, one USP colleague had encouraged the villagers to plant mangroves as a means of stabilising their shoreline. This had proven advantageous, as much of their shoreline was in a superior condition compared to that of their neighbours.

Regrettably the villagers' commercial seaweed and terrestrial gardens had been destroyed. They asked a colleague for assistance in securing seaweed starters to replant their aqua gardens, as well as seeds for their terrestrial gardens (Gard and Veitayaki 2017). The desire to rebuild and replant is a powerful one for those who live through the trauma of an extreme weather event, and the need to positively plan for the future is vital.

> It occurred to me how markedly different sharing seeds and plant starts with a community was when compared to more typical Western aid deployment. This style of aid often provides multiple rounds of canned Western food items for immediate consumption. Though immediate needs are met, the future is not considered in the same way, nor does it offer hope the same way that planting seeds in a garden encourages the gardener to imagine a future in which a bountiful harvest is possible. (Gard and Veitayaki 2017: 160)

Fig. 11.2 A Bible is open on the arm of a sofa while a family rebuilds their home on Qoma Island, Fiji. February 2016 (*Source* Image courtesy of the author)

Following Winston there were issues around the deployment of relief assets. When speaking with friends, colleagues and community members working in relief deployment, there were rumours of Fijian military personnel taking and selling supplies intended for Winston victims. It was also known that representatives of the Fijian government had distributed faulty information associated with the national declaration of emergency and aid request. More specifically, some of the bank account numbers for donations had been altered so that funds were being syphoned into third-party accounts. Some of these corruption charges were briefly acknowledged by the Fiji Independent Commission Against Corruption (FICAC) and the National Disaster Management Office (NDMO) in a press release in March (Fiji Independent Commission Against Corruption (FICAC) 2016). Little wonder given such incidences that the people I observed seeking assistance beyond their own families and trusted community members were more comfortable relying on their local churches, parishes and faith-based organisations. (Gard and Veitayaki 2017: 161)

Faith as a Floatation Device

In the twentieth Century religion and the role of faith-based organisations (FBOs) have often been ignored by those involved with humanitarian aid and development. One scholar went so far as to call it "a development taboo" after completing research that explored the paucity of studies on the roles of religious organisations in development (Alan ver Beek 2000: 31). In the twenty-first Century the academic stigma associated with FBOs has lessened, and now the academic community and the wider international aid community have begun to explore in greater depth the role FBOs play in humanitarian aid (Peterson 2010).

Though FBOs have a long history of humanitarian works, the last two decades have seen a proliferation in religious charities, NGOs and other faith-based associations. Importantly, the World Bank has estimated that as high as 50% of all health and educational services globally are provided by FBOs (James 2009: 7). The World Bank's study entitled *Voices of the Poor* concluded that many people in lower socioeconomic conditions demonstrated more confidence in religious organisations than in secular, governmental or military governing bodies (Narayan et al. 2000). These finds echo my own fieldwork experiences whereby community members were primarily interested in turning to their families, both those that lived locally and abroad, and local faith-based organisations for assistance in the aftermath of Winston (Gard 2016, np).

Estimating the total number of NGOs operating in the world is difficult and data are incomplete, yet "according to the United Nations Development Program, there are approximately 40,000 non-governmental organizations in the world in addition to the community-based organizations which number in the hundreds of thousands" (Leverty 2008: 1). More than 5,000 NGOs have been granted consultative status with the Economic and Social Council (ECOSOC) in the United Nations (UN) (United Nations 2020). Embedded within Article 71, Chap. 10 of the UN's Charter, authority is given to "make suitable arrangements for consultation with non-governmental organizations which are concerned with matters within its competence" (United Nations Charter 1945). This is advantageous as many FBOs often have deeper histories than other NGOs, which can in turn strengthen their connections with local communities.

In catastrophes, FBOs are able to build on established foundations of trust and expedite community support:

> I observed this first-hand while in Fiji while engaging with the congregation of Saint Luke's Anglican Church and the aid they provided to the wider Fijian community following Winston. Church members and staff collected supplies and donations – both local and international – and redistributed the goods and money to those in rural communities most heavily impacted. Although past relationships and pre-established sympathies may impact a FBOs abilities to engage with the wider community in an unbiased and unpolitical manner, especially given the competitive dynamics of religious organisations in some Pacific communities, an increased integration of FBOs at all levels of extreme weather and climate change response can nonetheless prove advantageous, as they have much to offer the communities in which they are embedded, as well as the wider humanitarian community. (Gard and Veitayaki 2017: 162)

Understanding the Oneness of Life and Intuitive Worldviews

There are many commonalities in the distinct cosmologies of the Pacific, most striking perhaps is that all cosmologies embrace the sacred elements of both the spiritual and natural world. From the earliest Austronesian explorers voyaging increasing distances into the great Pacific Ocean, Pacific Island peoples have amassed a wealth of environmental knowledge. This environmental knowledge when taken in conjunction with cultural and spiritual beliefs, apprises their mental models of how the natural world functions.

Consider the Melanesian concept of creation, as expressed in *Making Sense of the Oneness of Life: a Melanesian Christian View on Creation,* "in terms of people's relationship to the earth and the environment, with all their contents, visible and invisible. It is an affirmation of life's unity – the interrelatedness, and the interdependence, between human life and nature" (Ole 1990: 33). From this perspective all things are connected and dependent on each other, and greater still, "each is made possible by the other. The interaction, that takes place between nature and people, provides a meaning of life" (Ole 1990: 33).

This is further exemplified in the Papuan mythology of the origin of the coconut tree, which is essential to Papuan agriculture and spirituality. In the distant Papuan past, there was once a village near the sea where many fishermen lived, and in their midst was one exceptional fisherman who would leave the village and fish alone, always returning with more than he could eat. The other villagers grew curious as to why this man was so exceptional at catching fish. After much discussion, it was decided that a young boy would follow him to see how exactly he achieved such success. This is what he saw:

> The man put down his fishing basket, and then took hold of his head with his two hands. He pulled and pulled, till his head came off. He placed his head on the sand, and he walked into the sea, till the water came up to the middle of his body. After he had stopped in the water, the man bent down till his throat was level with it. Then the fish came swimming in great numbers, and swam down the man's throat. When he had enough, he walked slowly back to the shore. He shook the fish out, and put his head on again. The boy, who witnessed all this, was afraid, and he ran home quickly, and told others about it. The next day, all the men went quickly to the seashore, to see what happened. They saw the man take off his head, and get into the water. Then, one of the men ran out from the trees, and, taking the head, he threw it into the bush. After a short while, the owner came out of the water, and began feeling for his head. When he could not find his head, he ran back into the sea, and changed into a big human fish. He swam out of sight.
>
> A few days later, the boy, who first saw the man, started to think about the head. He went into the bush, to try to find it. When he came to the place, where it had been thrown, he found a palm tree growing. Nobody knew what the tree was. It had nuts on it. The men, who saw this, then were afraid to eat them. But one of the women took a nut, and ate the inner part. When the others saw that she was not harmed, they ate some as well. Melanesians say, if you look at the nut, when the husk has been taken away, you can see the face of a man. I can tell you, it is the face of the man, whose head became the first coconut. (Ole 1990: 34–35)

From this creation myth, a perspective on the wholeness of life is given narrative form and reflects a perspective that embraces both nature and humanity, as the fisherman in the myth literally opens himself up to nature in a conflation of interior and exterior, man and animal, terrestrial and aquatic. Through human–environment interdependence and interaction, all life on the planet is integrated. This is also what Bernard Narokobi speaks of in *What is Religious Experience for a Melanesian*? when he states one's "encounter with the Spirit, the law, economics, politics, and life's own total whole… [is the] Melanesian's vision of cosmos, and its relationship with it" (Narokobi 1977: 8). From this perspective life is understood to be all encompassing, transcendental, sacred. In this spiritual world, "[l]ife is consecrated towards upholding that religious order… within the world of spirits" (Ole 1990: 36). It follows, then, that life emerges from a divine source and thus it is only right that human endeavours be centred on a spiritual path embedded within a religious framework. One might imagine that this path is circular and that life both emerges from the divine and submerges within the divine over the course of an individual's lifetime.

From this Melanesian spiritual and religious framework it is life itself, and through life, that all things are connected; such that to separate a human life from its surrounding environment results in a false dichotomy. Consider that for many Melanesians, "[i]f we lose our land, for example, we lose our place in creation. If we find ourselves alienated, or exiled, from our land, we find ourselves at odds with ourselves and all creation" (Ole 1990: 36). This is a view of the natural world that contrasts with the Western scientific view of the natural world that can be rationalized and comprehended without reference to spiritual forces that cannot be observed directly.

Drawing upon my own experiences of living and working in the Pacific and my recent fieldwork in Hawai'i, Fiji, Vanuatu and Aotearoa New Zealand (Gard 2015–2017, np), I found the spiritual reality to be far more complex than one might initially think. Though the trope of rupture is embedded in Christianity in such elements as conversion or the second coming of Christ, rupture is too simple a way of describing the complex synthesis of spiritual beliefs that arise in many Pacific populations following contact with Christian missionaries from the seventeenth Century onwards. Scholars describe this complex synthesis as a mixing of various Christian beliefs and aspects of traditional Oceanian animistic beliefs. For example, this complex synthesis can take the form of continued belief in indigenous deities and spiritual entities (Meyer 1999), such as in Hawai'i where many who consider themselves Christian also believe in *Pele*, the goddess of fire, volcanoes and the creator of the Hawaiian Islands. In fact, when I was preparing to leave Hawai'i several colleagues and friends encouraged me to make an offering to "Madame *Pele*" or *Ka wahine 'ai honua* ("The earth-eating woman") to make sure I was "right" with her before my departure from the islands, as this was deemed sound advice and would ensure safe travels. My colleagues and friends did not find this suggestion, nor belief in *Pele*, to be in conflict with their Christian faith or religious practices, as "*Tutu Pele*" ("Grandma Pele") is a part of everyday life and coexists within the same worldview as the Divine Godhead or Trinity in Christianity.

On Integrating Diverse Worldviews

Although unique in many ways, the Pacific region provides an exemplar of escalating vulnerabilities faced around the world as a result of climate variability and change. Smaller increases in global mean temperature are now projected to lead to significant negative consequences for most regions. Examining Pacific Islanders' responses to existing climate challenges can provide clues about how a host of psychological, social, and cultural factors may influence adaptation in other regions.

The merging of Christian ideologies and indigenous lifeways can often cause agonising ethical and social conflicts for recent converts, though this is not always the case. Often a false dichotomy between the arrival o f a 'Christian" way o f lfe and previous lifeways and social forms is drawn, and though conversion can be a break with the past, it does not automatically and completely replace previous beliefs and lifeways in converted people. I would argue that the peoples of the Pacific have been selective, skilled and creative in the process of appropriating Christian scriptures and themes into Oceanian worldviews that serve their own ends in a spiritual context, as well as in political and economic contexts.

Through fostering consilience of science with traditional environmental knowledge—much like if one is to imagine a double-hulled canoe—a higher potential for long-term climate change adaptation is attained. Further, if one considers the rich tradition of deep sea voyaging and navigation in Oceania, we realize that Pacific Island peoples have been successfully navigating climatic variability for centuries, such as with the El Niño–Southern Oscillation. One contemporary example can be found in the Polynesian Voyaging Society and the *Mālama Honua* ("to care for our Earth") Worldwide Voyage, where the traditional voyaging *vaka Hōkūle'a* ("Star of Gladness") without the aid of any modern instrumentation circumnavigated the globe from 2013 through 2017 (Polynesian Voyaging Society 2020). Further examples can be found in studies with Samoan elders which have documented a high awareness of weather and climate knowledge, and the elders' ability to forecast extreme weather events (Lefale 2003). Still another example is the traditional Hawaiian practice of the *ahupua'a*[1] ("a large traditional climatic, geologic and socioeconomic subdivision of island land"), as a highly integrated watershed management system (Chirico 2015).

That is not to say that integrating traditional knowledge with scientific and technological knowledge is easily achieved. It can be challenging. For instance, difficulty can arise in finding scientific terminologies and descriptors for indigenous knowledge and words. One simple example in which I achieved parity while working at the Bishop Museum—The Hawai'i Museum of Cultural and Natural History—was through creating activities and educational materials for school children and the general public that named and described biological data and the natural world through a trilingual approach, where the local or common name, scientific or Latin name, and

[1] If you are not familiar with ahupuaas it may prove useful to imagine a high island as a roughly circular "pie" where each slice of the pie represents an ahupua'a or integrated ecosystem that is holistically viewed from the centre uplands or highlands of the island to the coastal area terminating in the shore at the outer edge or "crust" of the pie.

Hawaiian name were all learned and used.[2] Through encouraging the local community regardless of age to engage in environmental education and activities that are meaningful and useful to them, as well as placing their local and indigenous knowledge on par with scientific knowledge is of benefit to all involved. I have also found this approach to be a powerful tool to tackle various stigmas associated with who can "do science."

It is important to acknowledge that applicable climate and environmental knowledge does not only come from science, rather important insights and information pertaining to climate risks can come from diverse perspectives. Pacific Islanders exhibit diverse cultural and spiritual beliefs in relation to the natural world and their connection with it. This in turn informs mental models that can shape responses to climate events and risks. Through working to understand and address dissimilar perspectives on climate change governmental policymakers, NGOs and FBOs are better equipped to utilize tools best placed to advance and support community action. Though challenging, this is a most worthy challenge, as it is through integrating diverse belief systems and mental models that we find a collective way forward and assure survival of the species on an increasingly polarized planet.

Risk as a Social Construct

In the Pacific, climate impacts are compounded by socioeconomic stressors. For example, in the aftermath of Tropical Cyclone Winston, there was rapid and unplanned movement of rural villagers and outer-island residents (i.e. primarily women and children) to urban centres and an out migration of those working to physically rebuild villages (i.e. primarily men) to the remote areas of the Fijian Islands. Further destabilizing factors in the PICTs include rising urban poverty and political instability all of which impact a range of social services such as medical care, education, and natural disaster risk management. When thinking of climate change risk factors it is important to move beyond the environmental aspects and consider the often more nuanced socially constructed aspects of risk (Power et al. 2017; Walshe et al. 2018).

Addressing issues of distrust in risk information is critical for policymakers, NGOs and FBOs to tackle (Scott-Parker et al. 2017). Recall the example from Fiji following Tropical Cyclone Winston where fellow relief workers shared concerns regarding representatives of the Fijian government distributing faulty information associated with the national declaration of emergency and aid request and that some of the bank account numbers for donations had been altered so that funds were being syphoned into third-party accounts. More broadly, research has found that distrust appears higher among people in economically and socially disadvantaged groups

[2] It is important to note that social science approaches, such as ethnographic interviewing of community members, can yield rich scientific data related to flora and fauna when correlated with pre-existing scientific data and knowledge.

(Narayan et al. 2000). These findings may reflect alienation and fear associated with removal from the processes associated with implementing meaningful governmental and policy change, as well feelings of powerlessness to challenge authoritative entities and structures.

One means of tackling issues of distrust and equity in relation to climate change risks can be through wide ranging engagement focusing on community education programmes and FBO partnership building (Mackay et al. 2018; McIver et al. 2017). Through dialogue with a wider range of participants one can begin to understand the dynamic social context in which societal risks develops. Events such as workshops that provide a mix of informative formal presentations and participant discussion, as well as coffee breaks and meals, are especially meaningful tools in Oceania, as food sharing and hospitality are culturally significant.

Through documenting and understanding the specific risk factors impacting communities on the frontlines of climate change, communities are better placed to engage in critical decisions and adapt to localized climate impacts. Through an integrated and collaborative approach that draws upon the connections already in place by partners, such as FBOs, trustworthy climate information can be delivered to communities that strengthen and increase resilience in Oceania.

Community Context & Tools for Engagement

It is critical for policymakers, NGOs and FBOs to select tools which will prove most effective for the specific communities in which they are engaging. Local context will dictate which risk-management tools will provide the most effective support in a given locale. For instance, both formal and informal public education programmes can prove effective in creating lasting partnerships that merge diverse belief systems with support for local resilience actions, such as in the informal approach of my Fijian colleague encouraging the community on Qoma to plant mangroves for increased shoreline stability. Public education programmes provide important opportunities to improve communities' general understanding of climatic processes and specific risk adaptation strategies in ways that are meaningful to them. An excellent example of this can be found in the community focused sustainable development programme on Gau Island, Fiji.

The "Lomani Gau ('guardians of the island') project, as it is called, does not attempt a clumsy separation of 'climate change' issues from 'environmental' issues more generally or from 'economic' issues; all are integrated" (Veitayaki and Holland 2018: 1). Recall that this approach is reminiscent of Bernard Narokobi's thoughts on the oneness of life from a Melanesian spiritual perspective (Narokobi 1977: 8). Through the work of the Lomani Gau project Pacific Islanders demonstrate that communities do "respond proactively and positively to climate change forecasts, but only if communicated properly: the forecasts must be made to speak to existing local concerns and to acknowledge the many other drivers of change. According to

Veitayaki and Holland, the English phrase 'climate change' or a vernacular translation should not even be used in the initial phases of the project; it should be introduced later, once the community has already shown interest in drafting a community resource management plan" (Veitayaki and Holland 2018: 1). More broadly, climate change narratives tend to use insider-outsider perspectives and conceptual language, such as using terms like "climate refugees" that Pacific Islanders may not find relatable or even understandable (Luetz and Merson 2019). According to Luetz and Merson "inclusivity in coining conceptualizations has already made advances in discourses about disability, and there is the hope that "inclusion" may be similarly normalized in the climate migration domain" (2019: 6). As previously noted, the "'nothing about us, without us' (Charlton 2000) cry within the disability discourse, calling for representation in a bureaucratic system of oppression and disempowerment, is hauntingly relevant" in the climate change discourse (Luetz et al. 2019: 120).

Creating an effective risk education program requires a thorough understanding of alternative mental models of climate risk and social context. Some communities and lay individuals will have sophisticated mental models and understanding of risk, but often mental models may be limited and include elements of misinformation or misperceptions which may in turn encumber effective decisions and responsive actions. Thus, through weaving local and indigenous knowledge from a communities collective experiences, with educational and decision-support systems, increased uptake and best results are probable. For instance, the traditional example of *tabu, tapu*[3] or *kapu* and limiting the usage of fish ponds until a population has returned to a sustainable population in Hawai'i serves as inspiration in current waterways management (Chirico and Farley 2015).

It is also worth noting that engagement tools which are dependent on electricity and internet access, such as websites and email, may not be available to communities, especially those in rural and remote areas. For instance, in the aftermath of Tropical Cyclone Winston I found verbal communication via mobile phone and in person (often at the same time daily or weekly), to be most effective, as most were without reliable access to electricity and internet.

Moving forward policymakers will benefit from community partnerships that encourage shared, informed decision making such as in the *Lomani Gau* project. There is tremendous value in collaborative, place-based approaches to identifying and tackling communal needs and associated risks. Through working with scientists, cultural and religious leaders, and community groups meaningful, collaborative and sustained bottom-up response to the climate crisis can be achieved.

[3]Tabu, tapu or kapu is a traditional concept prevalent throughout Polynesia that denotes an area or object is sacred, with "spiritual restriction" or "implied prohibition." It may also denote prohibitions or rules associated with being "out of bounds" and is the origin for the Western concept of taboo.

Conclusion—Towards Consilience

Oceania is an exemplar of the intensifying impacts of a changing climate on frontline communities around the planet, especially those in coastal zones. Pacific communities hold a range of beliefs regarding the natural world and climatic processes, as well as their connection to both. Behavior studies indicate that people make choices, both rational and irrational, based on a range of data that ultimately resonates and with their previous experiences and intuitive worldviews. Through engaging with communities in ways that are meaningful to them, such as drafting a community resource management plan (Veitayaki and Holland 2018), and improving climate literacy and decision processes support through educational programmes in FBO settings, more climate-resilient communities can be established.

Consequently, in conjunction with the international top-down negotiation processes of the Paris Agreement and subsequent UN Conference of the Parties (COP)s (Gard 2018), a bottom-up approach that is informed by spiritual and religious perspectives is critically needed to build credible institutions, policies and community support measures. Science and technical knowledge is a significant aspect of what decision makers need to know, but the greatest potential for a successful response to the climate crisis will be achieved through an integrated approach to risk management, an approach that assists communities in integrating faith informed perspectives, ascertaining priorities, promoting actionable outputs, and facilitating successful adaptation to the rapidly changing world, such as in the example of Tropical Cyclone Winston in Fiji.

A dichotomy between that which is considered scientific and that which is deemed spiritual or religious is a determent and false in an Oceanian context. As Fair notes, the "rejection of religious perspectives appears to emerge from both a misunderstanding (and secular rejection) of religious thought and a desire to enforce the boundaries between the religious and the scientific" (Fair 2018: 5). There is at times a latent, and at others blatant, perception that present climate change adaptations must "be rectified by an increase in scientific information and a rejection of religious knowledge" (Fair 2018: 5). In response to this point, consider once again the Melanesian concept of creation, as expressed in *Making Sense of the Oneness of Life: a Melanesian Christian View on Creation*, "in terms of people's relationship to the earth and the environment… It is an affirmation of life's unity – the interrelatedness, and the interdependence, between human life and nature" (Ole 1990: 33). From this perspective all things are connected and dependent on each other, and greater still, "each is made possible by the other… [and] provides a meaning of life" (Ole 1990: 33). This holistic perspective further underscores the confusion and damage that can be wrought when climate change discourse and response force a false dichotomy between what in a Western context deemed to be exclusively scientific and religious knowledge.

As it been noted previously, the most influential and trusted communication channels are often those that involve FBOs and religious leaders, such as I noted in the response to Tropical Cyclone Winston (Gard and Veitayaki 2017; Nunn 2017). Religious leaders can and do profoundly influence community engagement and response to climate change adaptation initiatives. Thus, I maintain the greatest chance for a sustainable and successful climate change response lies in rowing on both sides of our canoe—which is to say that response efforts that are both scientifically rigorous and spiritually informed, while simultaneously being locally integrated and culturally meaningful (i.e. bottom-up), as well as being a part of a wider international (i.e. top-down) strategy are a critical area for future cultivation.

Opportunities for Future Research

There is considerable opportunity for future research in this area. As Hagevi notes, "[in] the future, it is essential to conduct more comparative research regarding the elusive relationship between religiosity and environmental opinion" (2014: 105). As part of this elusive relationship between religiosity and environmental opinion, there are two opportunities that immediately come to mind, though in truth there are numerous opportunities in this area. The first being to further explore the success of the Polynesian Voyaging Society, most especially the Mālama Honua ("to care for our Earth") Worldwide Voyage, where the traditional voyaging vaka Hōkūle'a ("Star of Gladness") without the aid of any modern instrumentation circumnavigated the globe from 2013 through 2017. In many ways, the Polynesian Voyaging Society has successfully integrated scientific principles, spiritual and religious beliefs and long-held traditional Hawaiian knowledge in a synthesis of climate change leadership that has inspired communities around the world to now take action in their own right. Whilst the second and more broad opportunity, explores the overlaps and interface of religious stakeholders and environmental campaigners in Oceania, such as with 350.org and the Pacific Climate Warriors, the Pacific Conference of Churches and World Conference of Churches.

References

Alan ver Beek K (2000) Spirituality: a development taboo, Dev Pract 10(1): 31–43

Andrew NL, Bright P, de la Rua L, Teoh SJ, Vickers M (2019) Coastal proximity of populations in 22 Pacific Island countries and territories. PLoS ONE 14(9):e0223249

Appadurai A (1990) Disjuncture and difference in the global cultural economy. Theory, Culture & Society. 7:295–310

Bolatagici L (2016) Food supply runs short. The Fiji Times, 21 May 2016. Online. Accessed 15 December 2016

Carr ME, Rubenstein M, Graff A, Villarreal D (2014) Sea level rise in a changing climate: what do we know? In: Gerrard MB, Wannier G (eds) Threatened island nations: legal implications of rising seas and a changing climate. Cambridge University Press, pp 15–54

Charlton J (2000) Nothing about us without us. Disability oppression and empowerment. University of California Press, CA

Chirico J (ed) (2015) Thinking like an island: navigating a sustainable future in Hawai'i. University of Hawai'i Press
Fair H (2018) Three stories of Noah: navigating religious climate change narratives in the Pacific Island region. Geo Geogr Environ 5(2)
Farrell WE, Clark JA (1976) On Postglacial sea level. geophysics journal international on behalf of the royal astronomical society. Oxford University Press. 46: 647–667
Finucane M (2009) Why science alone won't solve the climate crisis: managing climate risks in the Pacific. *Asia Pacific Issues*. Analysis from the East-West Center. No 89. August 2009
Gard R (2015) np. Unpublished field notes
Gard R (2016) np. Unpublished field notes
Gard R (2017) np. Unpublished field notes
Gard R (2018) Welcome to the age of sustainable development? In: Anthropol Today 34(1): 18–19
Gard R, Veitayaki J (2017) In the wake of Winston—climate change, mobility and resiliency in Fiji. In: Brebbia C (ed) Disaster Management. Wessex Institute Press, Southampton, UK, pp 57–68; and Int J Safety& Secur Eng 7(2): 157–168
Gerrard MB, Wannier GE (2014) Threatened island nations. Legal implications of rising seas and a changing climate. Cambridge University Press, Cambridge
Hagevi M (2014) Religion and the environmental opinion in 22 countries: a comparative study. Int Rev Sociol 24(1):1–19
Hashim JH, Hashim Z (2016) Climate change, extreme weather events, and human health implications in the Asia Pacific Region. Asia Pac J Public Health 28:8S–14S
James R (2009) What is distinctive about FBOs? The International NGO Training and Research Centre (INTRAC). Praxis Paper 22:7
Krishnamohan KS, Bala G, Cao L, Duan L, Caldeira K (2020) The climatic effects of hygroscopic growth of sulfate aerosols in the stratosphere. Earths Future 8(2)
Kumar L, Jayasinghe S, Gopalakrishnan T, Nunn PD (2020) Climate change and the Pacific Islands. In: Kumar L (ed) Climate change and impacts in the Pacific. Springer, Berlin, pp 1–32
Lefale P (2003) A samoan perspective on the role of traditional knowledge. In: Helander E, Mustonen T (eds) Adapting our understanding of climate change in snowscapes, dreamscapes: snowchange book on community voices of change. Tampere Polytechnic Publications 12, pp 495–515
Luetz J (2017) Climate change and migration in the Maldives: some lessons for policy makers. In: Leal Filho W (eds.) Climate change adaptation in Pacific countries, Springer Nature, Switzerland, pp 35–69
Luetz J, Bergsma C, Hills K (2019) The poor just might be the educators we need for global sustainability—a manifesto for consulting the unconsulted. In: Leal Filho W, Consorte McCrea A (eds) Sustainability and the humanities. Springer Nature, Cham, Switzerland, pp 115–140
Luetz J, Merson J (2019) Climate change and human migration as adaptation: conceptual and practical challenges and opportunities. In: Leal Filho W, Marisa Azul A, Brandli L, Gökcin Özuyar, Wall T (eds) Climate action. Encyclopedia of the UN sustainable development goals, vol SDG 3—Climate action. Earth and Environmental Science. Springer Nature, Cham, Switzerland, pp 1–13
Leverty S (2008) NGOs, the UN and APA. American Psychological Association, Office of International Affairs. Online. Accessed 12 May 2020
Mackay S, Brown R, Gonelevu M, Pelesikoti N, Kocovanua T, Iaken R, … Mackey B (2018) Overcoming barriers to climate change information management in Small Island Developing States: lessons from Pacific SIDS. Climate Policy, 1–14
McGranahan G, Balk D, Anderson B (2007) The rising tide: assessing the risks of climate change and human settlements in low elevation coastal zones. Environ Urban 19(1): 17–37. Online. Accessed 20 April 2020
McIver L, Bowen K, Hanna E, Iddings S (2017) A 'Healthy Islands' framework for climate change in the Pacific. Health Promot Int 32(3):549–557

McNamara KE, Clissold R, Westoby R, Piggott-McKellar A, Kumar R, Clarke T, Nunn PD (2020) An assessment of community-based adaptation initiatives in the Pacific Islands. Nat Clim Chang 10:628–639
Meyer B (1999) Commodities and the power of prayer: pentecostalist attitudes towards consumption in contemporary ghana. In: Meyer B, Geschiere P (eds.), Globalization and identity: dialectics of flow and closure. Blackwell, Oxford, UK
Milne GA, Gehrels WR, Hughes CW, Tamisiea ME (2009) Identifying the causes of sea level change. Nature—Geoscience 2: 471–472
Mitrovica J, Tamisiea M, Davis J, Milne G (2001) Recent mass balance of polar ice sheets inferred from patterns of global sea-level change. Nature 409:1026–1029
Mitrovica J, Gomez N, Clark P (2009) The sea-level fingerprint of West Antarctic collapse. Science 323:753
Narayan D, Chambers R, Shan MK, Petesch P (2000) Voices of the poor. World Bank
Narokobi B (1977) What is religious experience for a melanesian? In: Point Series 1. The Melanesian Institute, Goroka, pp 7–12
NOAA (National Oceanic and Atmospheric Administration) (2019) US Department of Commerce. Global ocean heat and salt content. Online. Accessed 28 May 2020
Nunn P (2017) Sidelining god: why secular climate projects in the Pacific Islands are failing. The Conversation, 17 May
Ole RT (1990) Making sense of the oneness of life: a melanesian Christian view on creation. Melanes J Theol 6–2; 33–41. Online. Accessed 10 May 2020
Peterson M (2010) International religious NGOs at the UN: a study of a group of religious organizations. J Humltarian Assistance. Online. Accessed 10 May 2020
Polynesian Voyaging Society (2020) The Mālama Honua Worldwide Voyage. Online. Accessed 15 May 2020
Power SB, Delage FPD, Chung CTY, Ye H, Murphy BF (2017) Humans have already increased the risk of major disruptions to Pacific rainfall. Nat Commun 8:14368
Round S (2016) Cry for more help from remote Fijian village. Radio New Zealand, 12 April 2016. Online. Accessed 27 May 2020
Scott-Parker B, Kumar R (2018) Fijian adolescents' understanding and evaluation of climate change: implications for enabling effective future adaptation. Asia Pacific Viewpoint 59(1):47–59
Scott-Parker B, Nunn PD, Mulgrew K, Hine D, Marks A, Mahar D, Tiko L (2017) Pacific Islanders' understanding of climate change: where do they source information and to what extent do they trust it? Reg Environ Change 17(4):1005–1015
Shope JB, Storlazzi CD, Erikson LH, Hegermiller CA (2016) Changes to extreme wave climates of islands within the Western Tropical Pacific throughout the 21st century under RCP 4.5 and RCP 8.5, with implications for island vulnerability and sustainability. Global Planet Change 141:25–38
Sugi M, Murakami H, Yoshida K (2017) Projection of future changes in the frequency of intense tropical cyclones. Clim Dyn 49(1–2):619–632
The Intergovernmental Panel on Climate Change (IPCC) (2018) Global Warming by 1.5°C: An IPCC Special Report on the impacts of global warming of 1.5°C above pre-industrial levels and related global greenhouse gas emission pathways, in the context of strengthening the global response to the threat of climate change, sustainable development, and efforts to eradicate poverty. Edited by Masson-Delmotte, V. et al. Switzerland
United Nations (2019) Elaboration of data by UN, department of economic and social affairs, population division. World population prospects: The 2019 revision. Online. Accessed 25 May 2020
United Nations (2020) Open call for NGOs to apply for ECOSOC consultative status. UN Department of Economic and Social Affairs News. 12 May 2020. Online. Accessed 25 May 2020
United Nations Charter (1945) Online Archives. Accessed 12 May 2020

Veitayaki J, Holland E (2018) 9 Lessons from Lomani Gau Project, Fiji: a local community's response to climate change. In: Pacific climate cultures living climate change in Oceania. De Gruyter, pp 121–136

Walshe RA, Seng DC, Bumpus A, Auffray J (2018) Perceptions of adaptation, resilience and climate knowledge in the Pacific: the cases of Samoa, Fiji and Vanuatu. Int J Clim Chang Strat Manag 10(2):303–322

Westcott L (2016) Tropical cyclone Winston slams Fiji with 200 MPH winds, kills21. Newsweek, 22 February 2016. Online. Accessed 23 May 2020

Chapter 12
Climate Change: Science, People, Religion—Cord of Three Strands

P. H. Havea and A. De Ramon N'Yeurt

Abstract Global climate change is known to have affected government, Churches and people's livelihood, health, well-being and sustainability worldwide. In the literature, although the impacts of climate change on these indicators are clear, no model was developed to address the relationship between them and how to use it to help build better Resilience (Climate Change Adaptation [CCA] and Disaster Risk Reduction [DRR]) cities/communities/planet by 2030 and beyond. Using a Concurrent Convergence Parallel Triangulation Design (CCPTD) of 460 participants from five coastal communities in Tonga, this chapter presents a framework named *Cord of Three Strands*. The first cord represents the science which is the State or the government factor. The second cord represents the people which is the community or the land factor. Then the third cord represents the religion which is the Church or the God factor. Using the insights generated by this threefold cord principle may guide the implementation of Resilience (CCA & DRR) in communities, the Sustainable Development Goals (SDGs), Sendai Framework, United Nations Framework Convention on Climate Change (UNFCCC) and other relevant frameworks towards better outcomes in areas of livelihood, health, well-being, sustainable development and sustainability.

> Hymn 391 verse 5
> "Lau pe he Palofesai (The prophecy said
> Toki 'Olive 'e Ua (The two olives)
> Pule'anga moe Siasi (Government and the Church)
> Kae malohi ha Fonua" (Are the strengths of any nations)
> (Free Wesleyan Church 2006)

P. H. Havea (✉)
Live and Learn Tonga, Tongatapu, Tonga
e-mail: peni.havea@livelearn.org

A. De Ramon N'Yeurt
Pacific Centre for Environmental and Sustainable Development, The University of the South Pacific, Suva, Fiji

J. M. Luetz and P. D. Nunn (eds.), *Beyond Belief*, Climate Change Management,
https://doi.org/10.1007/978-3-030-67602-5_12

The Relationship Between Climate Change, Science, People, Religion, Resilience and Sustainability Communities

The authors of this chapter are writing from a 'pastoral' Pacific Island region worldview orientation that is premised on accepting the Bible as an authoritative document that complements climate scientific and social (people) perspectives. The authors argue that such a research orientation is wholly legitimate in the context of the Pacific islands. For example, in Tonga 99% of the population are self-professed Christians and Christianity remains the predominant worldview (Tonga Department of Statistic & Secretariat of the Pacific Community (SPC) 2008; Tonga Department of Statistic 2013b, 2013a; Statistic Department of Tonga 2016). The authors acknowledge that this paradigm may differ significantly from other religions (e.g., non-Christian) and worldview persuasions.

As a result, it is important to promote closer cooperation between scientists, science, people and religious leaders so that climate change in the Pacific can be more effectively addressed. To bring these three pillars together, the authors used a principle they call the Three Cords God-knot. This is a biblical theory, and this chapter used that Bible concept to argue that God must be factored into adaptation and disaster management, both in Tonga and across the Pacific.

Globally, of the 7.7 billion people currently living on this planet (World Population Review 2019), 7.41 billion (96.23%) belong to some religious denominations (Pew Research Center 2017). Intriguingly, of the 7.41 billion people understood to have some sort of religion, 31.2% (2.3 billion) are understood to be Christians (Pew Research Center 2017). This means that the vast majority of the people in the world believe in the importance of spirituality and religious dimensions in orienting and guiding their lives and caring for the environment.

In the Pacific, using Christian principles to build resilience (climate change adaptation and disaster risk reduction) and sustainable communities is highly favoured in the sense that most of the Pacific Islanders identify as Christian (Mone 1974; Leiserowitz et al. 2015; Schuman et al. 2018; Cox et al. 2018; Handley 2019; Goldberg et al. 2019; Luetz and Nunn 2020). For example, in Tonga because the Head of the State is a devoted Christian, using God to guide adaptation and disaster management by the affected communities is extremely relevant. So, using this form of three strands model (i.e., God, State, People) scientifically, metaphorically and/or academically is acceptable in the Tongan society as people follow the teaching of the Bible diligently.

This is argued here on the basis of scripture: "Two men (i.e., State and Church) can resist an attack that would defeat one man alone. A rope made of three cords (e.g., Church, State, People) is hard to break" (The Bible Society in Australia 1988, p. 729) (Fig. 12.1). As a result, it is against this background that this chapter brings these three pillars together and then uses that conceptual framework to guide a reversal of the negative climate change and disaster impacts caused by natural hazards on people lives towards achieving a sustainability planet.

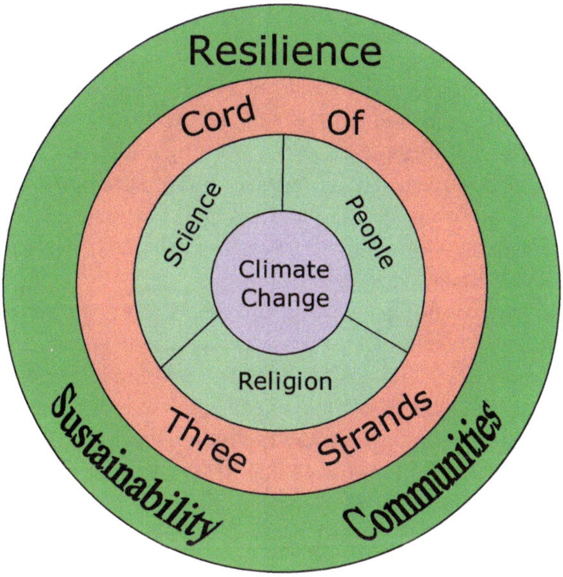

Fig. 12.1 Cord of Three Strands principle: The relationship between climate change, science, people, religion, resilience (CCA & DRR) and sustainability in communities

Why the 'Cord of Three Strands' Framework and 'God-Knot' Approach is Well-Suited in the Pacific

The idea from this chapter emerged from the first author's PhD thesis in which he looked at how climate change affected Tongans aged 15–75 in areas of livelihood, health and well-being in five coastal communities in the main island Tongatapu: Kanokupolu, 'Ahau, Tukutonga, Popua and Manuka (Havea et al. 2018b; Havea et al. 2019a). The study found that people in Tonga used their faith in their Christian God to try and reverse the impacts of climate change and disasters caused by natural hazards from affecting their livelihoods, health and well-being. This was accomplished through Bible reading, fasting and praying. More importantly, because in Tonga the State and the Church are not separated, associating religion/God to science and climate change is highly expected and accepted in society throughout the region and beyond (Hill and Pargament 2003; Nunn 2009; Michaelis 2011; Lata and Nunn 2012; Luetz and Nunn 2020). So, building the SDGs, resilience (CCA & DRR) and sustainability on the basis of faith-based approaches (e.g., Cord of Three Strands theory) is socioculturally promising. More importantly, because the Christian God is integrated into this framework the application of such a God-knot approach is well-suited in the context of the broader Pacific Island region (Nunn et al. 2016; Cox et al. 2018). For the other religions and worldview persuasions, the usage of this particular model may need to be appropriately modified and adjusted.

Climate Change and Science Linkages

The first cord of the strand is about the relationship between climate change and science. From what is known in the literature (IPCC 2014a, 2014b) this relationship is vital and the strongest one in fighting against climate change because science provides the scientific evidence and sets future development goals, targets and objectives to be achieved by 2030 and beyond. For instance, the United Nations Framework Conventions on Climate Change (UNFCCC), Sendai Framework, and Sustainable Development Goals (SDGs).

As a result, this concept represents a broad state-of-the-art of science in climate change. Stakeholder groups may include but are not limited to the State (government), researchers, scientists and relevant stakeholders explaining the scientific assessment on the impacts of climate change and adaptation strategies to the affected communities. For example, in the literature (IPCC 2014a, 2014b), there are four primary impacts of climate change that require science-informed inputs: livelihood impacts, health impacts (WHO 2015; USGCRP 2016), well-being impacts (Nunn et al. 2016; Havea et al. 2018b) and sustainability impacts.

a. Impact on livelihoods and adaptation strategies

The first element of this cord represents the negative impacts of climate change on livelihood. In the literature, these negative impacts include the adverse effects on agriculture, coastal area management, energy and infrastructure, education, fishing and fishery, forestry, tourism and water resources. The theme represents climate adaptation strategies: livelihood adaptation. These includes sustainable farming and agriculture, reforestation and replanting of mangrove and other plant species in coastal areas and zone, the building of seawalls, building of sustainable energy and climate-resilient infrastructure, integrating climate change into the school curriculum, implementation of the Small Management Area (SMA) system in affected communities, conservation and/or turning desert into a forest and improving water resources (IPCC 2014a, 2014b; Ledley et al. 2017; Nagy et al. 2017; Yuan et al. 2017; Wosnitza et al. 2018; Havea et al. 2019b).

b. Impact on health and adaptation strategies

The second element is the impacts of climate change on health. In the literature, the theme used for this category is diseases. Climate-related diseases have been categorised into two major health problems. First, Non-Communicable Diseases (NCD) include several diseases such as pneumonia[1] (Nisar et al. 2007; Walter et al. 2008), asthma, post-traumatic stress disorder, anxiety, depression, injuries, death and others (Friel et al. 2011). Second, Communicable Diseases (CD) include malaria, diarrhoeal diseases, dengue, typhoid fever, Sexually Transmitted Infections (STIs), leptospirosis and others. However, the management of these health problems is mandated under the health adaptation and/or Resilience (CCA & DRR) in health program organised

[1]Contagious when caused by small respiratory droplets, in which it contains Mycoplasma Pneumoniae, the bacteria that get people infected when they inhale them intrinsically.

by the Ministry of Health and relevant organisations in their respective countries (WHO et al. 2013; WHO 2015; USGCRP 2016).

c. Impact on well-being and adaptation strategies

The third one is the well-being impacts in the sense that health is always parallel to well-being dimensions. In the literature, there are three components. First, physical well-being impacts include consequences of climate change and/or disasters caused by natural hazards on the physical world or the environment. These influences include effects on health such as climate-related diseases, coastal erosion, demolition of housing, building environment and infrastructure. Second, mental well-being impacts bring together psychological infirmities or conditions affected by the disorder of the mind due to climate change or disasters (Luetz 2020). The most common mental well-being impacts that affected people in a climate change situation are stress, anxiety and/or depression. For example, a participant from Popua in Tongatapu in Tonga expressed his worries when asked to reflect on what climate change, livelihoods, and well-being mean to him. He said:

> Climate change is a relationship between human and nature, for example, air, sun, winds and water. But today, we are worried about climate change, because what has happened in the past is different from what is happening now. Like the sun is getting hotter and hotter and the rainfall has been usually occurred at once and it waters the whole nations but today, we experienced rainfalls per region around Tonga, that's new to us. The weather, say for example stay at the eastern side or the northern side, usually stays there for a day to a week, but what we've experienced today is different in the sense that it changes almost on a daily basis. Like it can stay on this site, and after 20 minutes it shifted to the other side and vice versa. So, that's how I looked at climate change. (Havea, P. [2013, September 24] Personal Interview)

Third, is the spiritual well-being impact. Although these impacts on spirituality are indirect and real, importantly, this can be anticipated and managed through religious programs in the form of mutual communication and spiritual connection with the creation of nature for divine protection. These programs can range from reading scriptures, fasting and praying to God to reverse the impacts of climate change on disasters from affecting their communities and others (Moore et al. 2005; Havea et al. 2018a; Havea et al. 2018b; Freeman 2019; Havea et al. 2019a).

d. Impact on sustainability and sustainability approaches

The fourth element is the sustainability impacts. Based on a study of the literature (Nichols et al. 2009; WHO 2015; Gaziulusoy and Brezet 2015; USGCRP 2016; Blok et al. 2016; Manno 2017; Bhatasara and Nyamwanza 2018; Luetz et al. 2018; Sánchez García and Díez Sanz 2018; Van Beynen et all. 2018; Fuertes-Camacho et al. 2019), the link between climate change and sustainability rests on seven principles. These principles address both the climate change impacts and adaptation strategies of the affected communities to meet the criteria for a sustainable and climate-resilient planet by 2030 and beyond.

The first concept is linking economic impact and economic sustainability. The second concept is connecting social influence and social sustainability. The third one is associating environmental impact and environmental sustainability. Fourth, is joining livelihood impact and sustainable livelihood. The fifth is correlating health impact and sustainability in health. Six is using the principle of resilience (CCA & DRR) to foster a more sustainable society. Lastly is combining well-being impact and sustainability well-being. This chapter argues that only by allowing the State, Church and affected communities to address these principles wholistically can communities expect to achieve resilience and sustainability.

In the cord of three strands approach, these concepts of sustainability are critical because by using them as a guide to implementing sustainability in communities, it strengthens the relationship between climate change, people, society, religion and their lands, which is the hallmark of doing climate adaptation.

Climate Change and People Linkages

The second cord represents climate change and people. The reason people should help the State is because according to the cord of three strands principle, two workers are better than one as they will produce more effectively and efficiently when they work together collaboratively and cooperatively. There are three reasons the literature discusses the importance of using people, community or their land as part of the solutions for this problem.

First, it is because people's perception of climate change and adaptation strategies of the affected communities is vital. This understanding is essential in the sense that it is pertinent in complementing science via perceived impacts of climate change on livelihood, health, well-being and sustainability (Wasswa 2010; Nunn 2013; Nhemachena et al. 2014; Connell 2016; Havea et al. 2017).

Second, people have a spiritual and environmental connection to their land. Understanding these perceptions better may help people to deal more effectively with relevant issues (Havea et al. 2018b; Luetz and Havea 2018; Luetz et al. 2018).

Third, indigenous people have collective traditional and local knowledge of the land, sky, sea, climate or weather, the surrounding environment and climate change, which offers valuable insights in complementing the scientific data and in verifying critical climate models for community-based adaptation and mitigation actions that foster a sustainable and climate-resilient community by 2030 and beyond (Nyong et al. 2007; Nawrotzki and Kadatska 2010; Callison 2017; David-Chavez and Gavin 2018; Mugambiwa 2018; Kupika et al. 2019). As a result, identification of the links between climate change and people is crucial because it holds the key to bridge the gaps between traditional knowledge and climate science, thus allowing the State and the affected communities to tackle climate change in the form of people-led adaptation approaches. Having a complete understanding of the second strand is also a pathway to inform how cord number three of the strands would anticipate building a world that is more resilient to climate change, now and in the future.

Linkages Between Climate Change and Religion

Now, although State and the affected community make a remarkably good team in fighting against climate change and disasters, the vital question here is that if these two pillars fall, who else can help. In the view of God's knot principle, this is where God and the Church come in. They both enter into support the State and the affected communities emotionally and spiritually. In the Bible, there are incidences where God seemingly reverses climate change and disasters. In the Old Testament, the story of Noah shows God controlling heavy rainfall and flooding, and in the New Testament, Jesus Christ commands a strong storm and calms the turbulent sea (Bible Society of the South Pacific 1966; The Bible Society in Australia 1988; Fair 2018).

Adding to the above state of affairs, the literature also discusses three linkages relating climate change to religion. First is the principle of environmental care (Sonya 2016; Schuman et al. 2018; Goldberg et al. 2019). This teaching is based on the principle that God created this universe and, as a result, humans as his children should be responsible for taking care of this planet and everything in it.

Second, are the impacts of climate change on spirituality. In the literature, there are suggestions that climate change may affect people's spiritual well-being (Havea et al. 2018b; Luetz et al. 2018; Luetz and Nun 2020). For example, when heavy rainfalls and cyclones affected people by demolishing their homes and killing somebody close to them, it affected them spiritually (Havea et al. 2018b).

Third, climate change has been viewed as a punishment from God for their sin (Cox et al. 2018). So, by considering the God factor, the application of the cord of three strands approach can be helpful. Significantly, it is because most people rely on God for safety and security. People do this because God as the third cord cannot be torn or broken from the 'God-knot'. From a climate change perspective, He is the only one who can reverse climate change once and for all.

Linkages Between Climate Change, Resilience and Sustainability

The last linkages are between climate change, resilience and sustainability in communities. In order to achieve better outcomes, the State, Church, affected villages and cities must realise and address the concept of climate change and science, people and religion, accordingly. This cosmological framework is expressed in the form of a 'God-knot' approach and is discussed in details below.

Therefore, by taking all this background into account, this chapter aims to achieve two objectives. The first one is to provide a better understanding of the nexus between climate change, science, people and religion in the form of the cord of three strands state-of-the-art idea. Second, to develop a framework or model to guide the community-based implementation of resilience and sustainability by 2030 and beyond.

Methods

a. Methodology
Information for this study was collected using a mixed method approach (Creswell and Plano Clark 2011; Clark and Ivankova 2016) named Concurrent Convergence Parallel Triangulation Design (CCPTD) of 460 residents from five coastal communities in Tongatapu, the main island of Tonga. The quantitative study used a self-administered questionnaire (N = 460) and climate change data such as sea-level rise, temperature changes, rainfall pattern variation, drought, Extreme Weather Events (EWE) (e.g., cyclones), and seasonal variation. Other impacting factors were included such as flooding and ocean acidification. In complementing this data, the qualitative study used Focus Group Discussion (FGD) in the form of a workshop (n = 28), an In-depth Interview (IDI) (n = 24), Key Informant Interview (KII) (n = 12), photos, video footage and observation in the field. The complete analysis of the quantitative data is presented in the first author's PhD thesis. Here, the focus is on using that information to modelling of the three cords God's knot framework.
b. Data analysis
The strategy used for data analysis was a convergence model. First, the quantitative data were analysed using the frequency analysis, correlation and regression. Second, the qualitative data were then analysed using a thematic analytical strategy. Once this had been done, the data were then analysed using a comparative and networking analysis. This strategy allowed modelling the data into the 'God-knot' framework comprising the cord of three strands to guide the community-based implementation of resilience and sustainability by 2030 and beyond. The software packages used to analyse the data were SPSS, Tableau, NVivo, and R. The information gathered from this analysis was then used to develop the model using Smart Draw. Smart draw allows the researcher to connect variable themes and the outcome of that collaboration between scientists, science, people and religious leaders is the achievement of resilient communities now and in the future.
c. Development of the 'three cords God-knot' framework
Before building this model, the researcher should have some basic conceptual framework or foundation to guide the analysis. In this chapter, the authors have used a Tongan concept/theory: "ko e malohinga ha fonua ko hono pule'anga, siasi mo e fonua" (the strength of a nation is his State, Church and land). In the Bible (Bible Society of the South Pacific 1966), the same theory is advocated by Ecclesiastes but instead in the ideology of a cord of three strands principle. This concept was then borrowed by the authors and used to associate climate change and religion to build the three cords God-knot framework. The first author's PhD thesis impinges on this framework because the people in Tonga indicated strongly their perceived need to integrate the God factor to address issues of climate change and science. This chapter extends that theory to demonstrate that such a model is well-suited to both the context of Tonga and the wider Pacific

Island region. Expressed simply, a closer cooperation between scientists, science, people and religious leaders is vital in building climate resilience by 2030 and beyond.

Results

Based on the analysis carried out for this study, a 'God-knot' framework was developed named cord of three strands (Fig. 12.2).

a. God's knot: A framework for community-based resilience and sustainability by 2030 and beyond
   The study found that climate change can be approached and/or solved in three ways. First, through science and government organisations. Second, by the people. Third, through God-informed solutions. Because this chapter promotes God-based adaptation approaches (e.g., reading the Bible, fasting and praying to God to reverse climate change impacts as a solution), the study findings prompted the conceptual development of the 'three cords God-knot' framework. Given the need to demonstrate why close collaboration between scientists, science, people and religious leaders is significant in solving the problems, and to encapsulate or categorise this idea under a theme, the 'God-knot' idea was conceptualised.
   The 'God-knot' name was chosen to represent this model because the framework is considered a God-centred approach. It sums up the importance of including God in this model. Here, in this model people can approach God through reading the scriptures, fasting and then praying to their God to reverse the effects of climate change or disaster. In the context of Tonga, this work is highly practical and affected communities do this routinely. In this framework approach God is seen as the only stakeholder who can ultimately and enduringly reverse the impacts of climate change and disasters.
   As a result, because God is highly significant in the local context, the authors borrow this leading-edge idea for the purposes of innovating climate change leadership and management. The idea is to use this model to better manage climate change and disasters. By applying this state-of-the-art concept in the context of climate change adaptation in the Pacific, this chapter identifies five steps to guide its implementation.

   Step 1 Identification of the Global Issues for the State, Church and Affected Communities
   In the first step, the Church, State and affected communities identify the most pressing issues affecting them in their society (i.e., input). Here the consultant from the State should assist the people and Church accordingly and also throughout this process. These issues are climate change and disasters caused by natural hazards affecting Church, State and people's livelihood, health, well-being and sustainability in their communities.

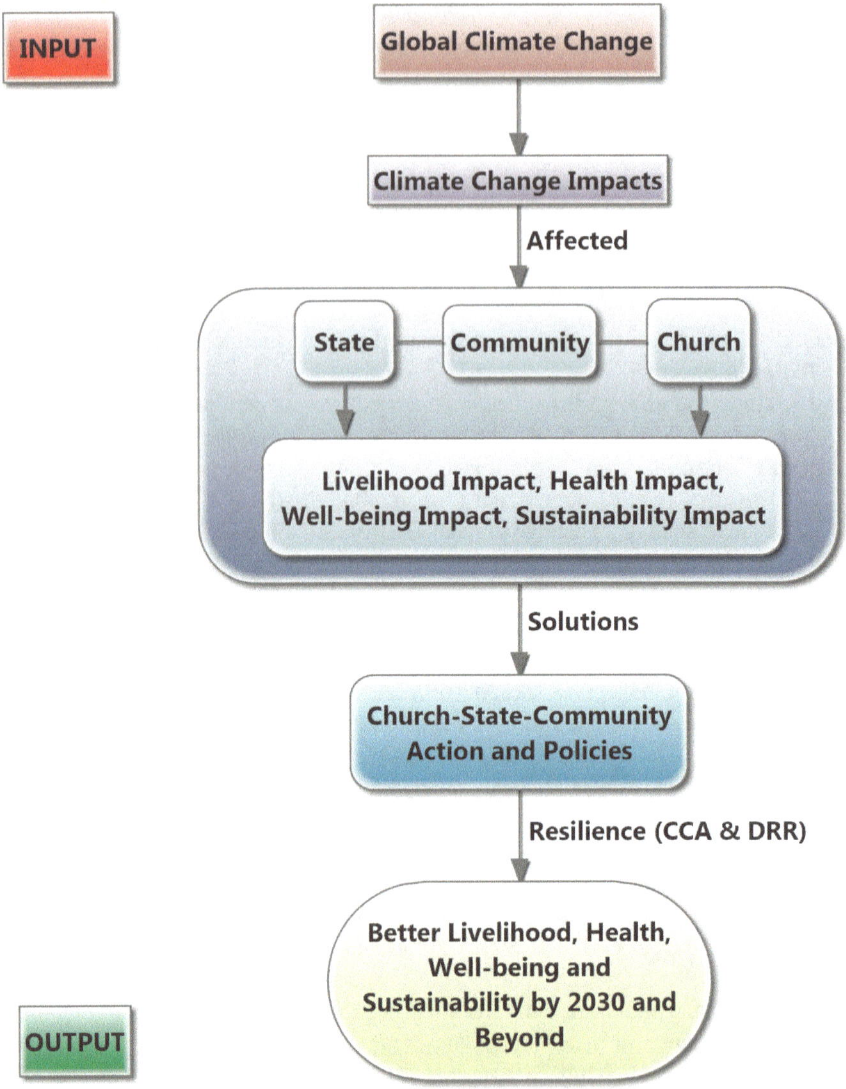

Fig. 12.2 God's knot: A framework of the cord of three strands to guide the implementation of community-based resilience and sustainability by 2030 and beyond

Step 2 Identification of the Climate Change Impacts Affecting State, Church and Communities

The second action is enlisting the State, Church and affected communities to identify concrete climate change impacts. There are two types of climate change impacting factors: climatic and non-climatic. Climatic factors range from sea level rise to temperature changes, rainfall pattern variations,

EWEs (Extreme Weather Events, e.g., cyclones), drought, seasonal variations, flooding, and ocean acidification. Non-climatic factors include environmental pollution (e.g., air, water), deforestation, sand mining, population density, geographical location and other modulating influences.
Step 3 Identification of the State, Church and Community Impacts
The third step is to summon them to identify Church-State-Community impacts. These impacts need to be categorised into four main areas: livelihood impacts, health impacts, well-being impacts and sustainability impacts.
The livelihood impacts may include but are not limited to:

- agricultural impacts;
- coastal management impacts;
- energy and infrastructure impacts;
- educational impacts;
- fishing and fishery impacts;
- forestry impacts;
- tourism impacts;
- water resources management impacts; and
- church service attendance impacts.

The health impacts, on the other hand, will focus on climate-related diseases such as:

- pneumonia;
- asthma;
- dengue;
- typhoid fever;
- malaria and others.

For well-being impacts, there are various types of consequences, which have included impacts in areas of physical well-being, mental well-being and spiritual well-being. The last part is the sustainability impact, which includes issues in areas of social-economic-environmental problems, including poverty, deforestation, sand mining, environmental pollution, waste management, and others.

Step 4 Identification of the Church, State and Communities Solutions
The fourth step is identifying Church-State-Community solutions. There are two approaches. First, taking actions to provide the necessary climate mitigation and adaptation measures such as sustainable farming and agriculture, implementation of the Small Management Area (SMA) or Marine Protected Area (MPA) to the affected communities, the building of seawall, dykes and climate-resilient infrastructure, reforestation initiatives and others.
Second, promoting mitigation and adaptation policies. These policies may include but are not limited to:

developing legal instruments to reduce greenhouse gas (GHG) emissions and carbon footprint;
integrating climate change and sustainable energy topics into the school curriculum or teach them as a stand-alone subjects;
reinforcing the development and use of electric cars and stopping illegal sand mining and deforestation;
engaging the affected communities, Church and State in the reforestation of mangroves and other plant species in coastal areas and zones interalia.
Step 5 Achievements of Better Livelihood, Health, Well-being and Sustainability Church-State-Community by 2030 and Beyond
If the State, Church and the affected communities follow through steps 1–4 diligently, this will result in better livelihood outcomes, health, well-being and sustainability by 2030 and beyond. These expected outcomes are argued here to be quite likely because by partnering with God as a blueprint may lead to a better and more sustainable and climate-resilient future for all. The reason for this expectation is that God is conceptualised in this model as the principal stakeholder. The Bible says: "If my people [or Church and State], which are called by my name, shall humble themselves, and pray, and seek my face, and turn from their wicked ways; then will I hear from heaven, and will forgive their sin, and will heal their land" (2 Chronicles 7: 14). This pledge is understood here as a promise. As shown in Fiji during and after Tropical Cyclone Winston, Christians persevered in their faith and saw the cyclone mainly as 'an act of divine judgment' or as 'punishment on a sinful people' (Cox et al. 2018; Fox and Walsh 2016)

Discussion: Significance of God's Knot Framework for the State, Church and Communities

The development of the God's knot idea in this chapter emerged from mixing the quantitative and qualitative data on spirituality. This proess resulted in a leading-edge framework that now serves as a blueprint for resilience and sustainable development. Ergo, the State, Church and the affected communities should consider implementing it for the following reasons.

First, given that the framework is open, flexible and universal, it allows the model to scope across a spectrum of cultures and traditions locally and regionally, thus enabling the three cords (i.e., Church-State-Community) to reverse the impacts of climate change effectively and efficiently (Cortekar et al. 2016). For example, in Sect. 3 above the framework allows the three cords to model both the climatic and non-climatic issues affecting them, which yields highly relevant results in the context of the Pacific (e.g., Fiji, Tonga, etc.) (Cox et al. 2018; Havea et al. 2018b; Havea et al. 2019a).

Second, given that science, religion and people in affected communities collaborate together in unison, the probability of inflating mutual partnership to make this necessary change happen is higher than by using one pillar alone (e.g., top-down approach only). This closer cooperation between science, scientists, people and religious leaders is of great importance in promoting environmental awareness, action and in producing the most appropriate solutions for problems that could not be tackled by science alone. Instead, the three pillars are needed in the sense that it is the collaboration between these domains of influence that will effectively stem the effects of climate change and disasters caused by hazards (Luetz 2020).

Schuman et al. (2018) support the same theory too and argue that religion, local and indigenous knowledge, and education can help to curb climate change. This is why it is necessary to bring science and religion together and unite them as one because dividing them apart may result in delaying to achieve resilient communities by 2030 and beyond.

Third, including God as the Heavenly Father balances multiple epistemologies and thereby assists in solving adaptation problems once and for all. Protecting God's creation and building capacity in areas of environmental care is "not only important because God has entrusted the world to us men and women, but because human life is itself a gift which must be defended from various forms of debasement" (Pope Francis and Catholic Church 2015). As stated by Pope Francis and Catholic Church (2015) in the encyclical *Laudato si'*, "since everything is closely interrelated, and today's problems call for a vision capable of taking into account every aspect of the global crisis, I suggest that we now consider some elements of an *integral ecology*, one which clearly respects its human and social dimensions" (#137). As a result, "strategies for a solution demand an integrated approach to combating poverty, restoring dignity to the excluded, and at the same time protecting nature" (#139) (Pope Francis & Catholic Church, 2015). God has made a promise treasured in the scripture that "the things impossible with men are possible with God" (Luke 18:27; Matthew 19:26) (Bible Society of the South Pacific 1966; The Bible Society in Australia 1988), so God is theorized in this discussion as being able to effect climate change mitigation and adaptation in areas where humans have so far failed to make a meaningful contribution. For example, in Tonga, relying on the Christian God for support and adaptation is quite normal for people as it is a Christian State. More specifically, because people have called on their God to help them solve their problems before, soliciting support from God for problem solving is both anticipated and accepted in society. This attitude was reflected by a male participant from 'Ahau who elaborated using God as a solution, saying:

> Tupou 1st have already devoted that God and Tonga are my inheritance. As for my own witnessed Jesus has stated to his disciples, when we have this kind of [climate change] problem, we shouldn't be so surprised about it. It is like a metaphor for a woman in labour suffering because something good is coming. The same principle applies this is the sign of the second coming of Jesus. And I hope that this blessing will linger in Tonga forever.

Another woman from Kanokupolu went further to discuss why God is crucial in this framework, saying, "we need God to help us understand nature and solve this [climate change] problem" for all. As shown by this woman, Tongans already refer to God as a saviour and a solution to their problems. So, when people read their Bible, discuss the scripture, fast and then pray to reverse the impacts of climate change and disaster, they are simultaneously implementing the elements of this three cords 'God-knot' framework.

Furthermore, this framework is intended to be implemented in tandem with other models (e.g., UNFCCC, Sendai framework and SDGs) that serve a similar purpose. In addition, this framework includes a collection of indigenous ideology, phenomenology, faith and scientific beliefs. For example, because people in Tonga own their lands, their spiritual connections to these traditional and cultural affairs are vigorous and robust. Therefore, integrating local and indigenous knowledge to adaptation is prominent in addressing spirituality and faith-based approaches. This call to be united in the face of differences is indeed a good sign for better livelihoods, health, well-being and sustainability for the Church, State and the affected communities, both now and into the future.

Moreover, as an integrated methodological approach the model can also merge both the bottom-up and top-down approaches. Combining top-down and bottom-up approaches has already been proven to be valid for effective climate change adaptation and disaster management (Lata and Nunn 2012; Butler et al. 2015; Dittrich 2016; Tra et al. 2018, Luetz 2020). For example, in India, combined top-down and bottom-up approaches were used for hydrological adaptation of the Kangsabati river catchment by increasing forestation and reforestation of the study area which helped to revive the river catchment (Bhave et al. 2014).

Last but not least, when God is called upon to lead and supervise, not only is there a clear pathway about what to achieve, but there is a threefold cord. By working hard and then relying on the Lord to do everything for and with the Church, State and the afflicted communities, God can offer protections that no other stakeholder can guarantee. Because of this, the afflicted communities stand to benefit from implementing this threefold model.

From a climate change and Christian theological, hermeneutical and eschatological perspective, the implementation of this 'God-knot' may promote a better future for all that is not reliant on chance but rather operates in tandem with God's bigger purposes and plans. According to our reading of scripture, this theory of collaborating 'with God' was planned and designed by God a long time ago (Bible Society of the South Pacific 1966; The Bible Society in Australia 1988; Gansle 2007; Howells 2010; Seekingtruth.co.uk 2016; Luetz et al. 2018). For example, in Psalm 23:1-4 (Bible Society of the South Pacific 1966; The Bible Society in Australia 1988), the Bible presents God as a shepherd. By allowing Him to be the shepherd of the Church, State and climate change-affected communities, the framework openly enlists his input. God's promises go beyond human intentions as enshrined, for example, in the global Sustainable Development Goals (SDGs). When God promises to do something, He means it wholeheartedly. In consequence it is important to ensure that God be integrated early in any adaptation planning.

Conclusions

It is clear from this chapter that climate change adaptation is significantly reliant on science, people and religion. The interrelationships between these three assets are revealed quite clearly through the development of our cord of three strands model. This 'God's knot' framework is unique and makes an important contribution to scholarship by turning problems into opportunities (and people away from their sins to please God). Moreover, it recognises that the power of a nation to adapt effectively to climate change depends on three principal stakeholders:

1. Government/state;
2. Church/religion/God;
3. People/communities and their land;

These three contributors are brought together in the three God-knot approach (Fig. 12.3).

The model can be used to solve climate change adaption and faith-based issues concurrently. By investing in developing the relationships and collaborations between key domains and stakeholders (scientists, science, people and religious leaders) the model promises to generate multiple benefits for climate change-affected communities.

Importantly, the following main lessons were learned from this chapter:

1. The government cannot reverse the impacts of climate change on livelihood, health, well-being and sustainability alone, and neither can science nor the people do it singly. Therefore, the model advocates that people should turn to God for help and support.
2. Combining the unique contributions from science, people and religion in the service of environmental care should be naturalised and cultivated. When science, people and religion work together their combined contribution will be much stronger than the sum of their individual parts.
3. God should be acknowledged as a concerned stakeholder of climate change adaptation by both the Church, State and the affected communities. In climate change

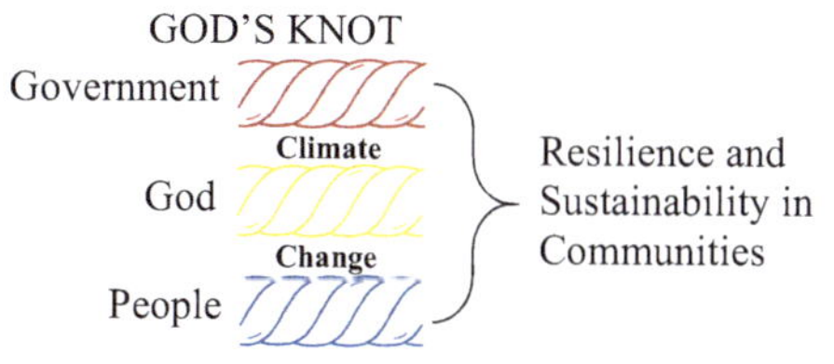

God's Knot: Government (red strand) + People (blue strand) + God (gold strand) = Resilient and Sustainable Communities

Fig. 12.3 Summary of God's knot framework: Cord of three strands approach and disaster situations, humble reliance on God for support acknowledges that the task of adaptation is so formidable that it transcends what is both scientifically required and humanly possible to achieve.

These lessons learned are vital because they hold the keys to unlocking locally meaningful climate change and faith-based adaptation approaches. Most importantly, they may spawn more holistic anticipatory adaptation strategies where preparations for the necessary changes can be made *before* they are needed. The benefits of preparedness in relation to climate change are noted in the literature (Luetz 2008). For the future, this chapter recommends the following further research:

1. Case study research to pilot the implementation of the 'God-knot' framework in local Pacific Island settings;
2. A longitudinal study to evaluate the implementation of the 'cord of three strands' framework in affected communities over time.

Acknowledgements We want to acknowledge the people of Tonga, the government and churches. Special thanks also goes to the University of the South Pacific (USP).

References

Bhatasara S, Nyamwanza A (2018) Sustainability: a missing dimension in climate change adaptation discourse in Africa? J Integr Environ Sci 15(1):83–97
Bhave AG, Mishra A, Raghuwanshi NS (2014) A combined bottom-up and top-down approach for assessment of climate change adaptation options. J Hydrol 518(Part A): 150–161
Bible Society of the South Pacific (1966) Koe Tohi Tapu Katoa (The Holy Bible in Tongan). Suva, Fiji: The Bible Society in the South Pacific
Blok V, Gremmen HGJ, Wesselink R (2016) Dealing with the wicked problem of sustainability: the role of individual virtuous competence. https://doi.org/10.5840/bpej201621737: Bus and Profes Ethics J
Butler JRA, Wise RM, Skewes TD, Bohensky EL, Peterson N, Suadnya W, Yanuartati Y, Handayani T, Habibi P, Puspadi K, Bou N, Vaghelo D, Rochester W (2015) Integrating top-down and bottom-up adaptation planning to build adaptive capacity: a structured learning approach. Coastal Management 43(4):346
Callison C (2017) Climate change communication and indigenous publics. In: Nisbet MC, Ho SS, Markowitz E, O'Neill S, Schäfer MS, Thaker J (eds) The oxford encyclopedia of climate change communication. Oxford University Press, UK
Clark VLP, Ivankova NV (2016) Mixed method research: A guide to the field. Sage, London, United Kingdom
Connell J (2016) Last days in the carteret islands? climate change, livelihoods and migration on coral atolls. Asia Pacific Viewpoint 57(1):3–15
Cortekar J, Bender S, Brune M, Groth M (2016) Why climate change adaptation in cities needs customised and flexible climate services. Climate Services 4:42–51
Cox J, Finau G, Kant R, Tarai J, Titifanue J (2018) Disaster, divine judgment, and original sin: christian interpretations of tropical cyclone winston and climate change in Fiji. University of Hawai'i Press 30(2):380–410
Creswell JW, Plano Clark VL (2011) Designing and conducting mixed methods research (2nd ed.). Los Angeles, United State of America, Sage

David-Chavez DM, Gavin MC (2018) A global assessment of Indigenous community engagement in climate research. Environ. Res. Lett. 13(12):123005
Dittrich R (2016) Top-down and bottom-up decision-making for climate change adaptation. An application to flooding (Doctor of Philosophy in the Economics of Climate Change Adaptation PhD Thesis), University of Edinburgh, Scotland
Fair H (2018) Three stories of noah: navigating religious climate change narratives in the pacific island region. Geo: Geogr Environ 5(2): e00068
Fox L, Walsh M (2016) Cyclone Winston: Fiji death toll reaches 42 with reports entire villages wiped out on remote islands. ABC News: Australia. Retrieved [2017 April 28]; from http://www.abc.net.au/news/2016-02-24/cyclone-winston-entire-villages-wiped-out-on-fijis-koro-island/7195842
Free Wesleyan Church (2006) Ko e Tohi Himi ‘a e Siasi Uesiliana Tau’ataina ‘o Tonga. Nuku’alofa, Tonga, Kar Seng Trading PTE Limited
Freeman BM (2019) Promoting global health and well-being of Indigenous youth through the connection of land and culture-based activism. Glob Health Promot 26(3_suppl): 17–25. https://doi.org/10.1177/1757975919831253
Friel S, Bowen K, CampbellLendrum D, Frumkin H, McMichael AJ, Rasnathan K (2011) Climate change, noncommunicable diseases, and development: The relationships and common policy opportunities. Annu Rev of Public Health 32:133–147
Fuertes-Camacho TM, Graell-Martín M, Fuentes-Loss M, Balaguer-Fàbregas CM (2019) Integrating sustainability into higher education curricula through the project method, a global learning strategy. Sustain 11(3):767
Gansle DJ (2007) Your world, your future and bible prophecy. Infinity, USA
Gaziulusoy AI, Brezet H (2015) Design for system innovations and transitions: A conceptual framework integrating insights from sustainability science and theories of system innovations and transitions. J Clean Prod 108(Part A): 558–568
Goldberg MH, Gustafson A, Ballew MT, Rosenthal SA, Leiserowitz A (2019) A social identity approach to engaging Christians in the issue of climate change. Science Communication 41(4):442–463
Handley GB (2019) Climate scepticism and Christian conservatism in the united States: a transnational ecocritical analysis. In: Garrad G, Goodboy A, Handley G, Posthumus S (eds) Climate change scepticism: a transnational ecocritical analysis. Bloomsbury Academic, London
Havea PH, Hemstock SL, Jacot Des Combes H (2017) Preparing for better livelihoods, health and well-being—a key to climate change adaptation. In: Leal Filho W (ed) Climate change adaptation in pacific countries: fostering resilience and improving the quality of life. Cham, Springer International Publishing, pp 87–99
Havea PH, Hemstock SL, Jacot Des Combes H (2018a) Improving health and well-being through climate change adaptation. In: Duncan LT (ed) Advances in health and disease, vol 4. Hauppauge, NY, Nova Science Publisher, pp 215–230
Havea PH, Hemstock SL, Jacot Des Combes H, Luetz J (2018b) “God and Tonga are my inheritance!”—climate change impact on perceived spirituality, adaptation and lessons learned from Kanokupolu, ‘Ahau, Tukutonga, Popua and Manuka in Tongatapu, Tonga. In: Leal Filho W (ed) Climate change impacts and adaptation strategies for coastal communities—climate change management. Cham, Springer
Havea PH, Hemstock SL, Jacot Des Combe H, Luetz J, Liava‘a L FC (2019a) Impact of climate change on spirituality. Pac Dynamics—Journal Interdiscip Res 3(1): 33–45
Havea PH, Siga A, Rabuatoka T, Tagivetaua Tamani A, Devi P, Senikula R, Hemstock SL, Jacot Des Combes H (2019b) Using vocational education to support development solutions in the Pacific: an emphasis on climate change and health. Applied environmental education & communication, 1–16
Hill PC, Pargament KI (2003) Advances in the conceptualization and measurement of religion and spirituality: implications for physical and mental health research. Am Psychol 58(1):64–74
Howells K (2010) Making sense of bible prophecy. Lulu, European Union

IPCC (2014a) Climate change 2014: impacts, adaptation, and vulnerability. part A: global and sectoral aspects. Contribution of working group II to the fifth assessment report of the intergovernmental panel on climate change. Cambridge, Cambridge University Press, p 1132

IPCC (2014b) Climate change 2014: impacts, adaptation, and vulnerability. Part B: regional aspects. Contribution of Working Group II to the Fifth Assessment Report of the Intergovernmental Panel on Climate Change. Cambridge: Cambridge University Press, p 688

Kupika OL, Gandiwa E, Nhamo G, Kativu S (2019) Local ecological knowledge on climate change and ecosystem-based adaptation strategies promote resilience in the middle Zambezi biosphere reserve Zimbabwe. Scientifica 2019(3069254):15

Lata S, Nunn P (2012) Misperceptions of climate-change risk as barriers to climate-change adaptation: a case study from the Rewa Delta. Fiji. Clim Change 110(1–2):169–186

Ledley TS, Rooney-Varga JN, Niepold F (2017) Addressing climate change through education. Oxford Research Encyclopedia of Environmental Science, Oxford University Press, Oxford, UK

Leiserowitz A, Maibach E, Roser-Renouf C, Feinberg G, Rosenthal S (2015) Climate change in the American Christian mind: March 2015. New Haven, CT, Yale Project on Climate Change Communication: Yale University and George Mason University

Luetz JM (2008) Planet prepare: preparing coastal communities in Asia for future catastrophes. Los Angeles/Singapore, World Vision International. DOI/ISBN: 1-887983-47-3 https://www.wvi.org/asia-pacific/publication/planet-prepare

Luetz J, Havea PH (2018) We're not refugees, and we'll stay here until we die!"—Climate change adaptation and migration experiences gathered from the Tulun and Nissan Atolls of Bougainville, Papua New Guinea. In Leal Filho W (ed) Climate change impacts and adaptation strategies to coastal communities. Switzerland, Springer

Luetz JM, Buxton G, Bangert K (2018) Christian theological, hermeneutical and eschatological perspectives on environmental sustainability and creation care—the role of holistic education. In: Luetz JM, Dowden T, Norsworthy B (eds) Reimagining Christian education—cultivating transformative approaches. Singapore, Springer Nature

Luetz JM (2020) Disaster-resistant schools for disaster-resilient education. In Leal Filho W, Marisa Azul A, Brandli L, Gökcin Özuyar P, Wall T (eds) Quality education: encyclopedia of the UN sustainable development goals. Springer Nature, Switzerland, pp 1–17. https://doi.org/10.1007/978-3-319-69902-8_15-1

Luetz JM, Nunn PD (2020) Climate change adaptation in the pacific islands: a review of faith-engaged approaches and opportunities. In: Filho WL (ed) Managing climate change adaptation in the Pacific region. Springer International Publishing, Heidelberg, Germany

Manno J (2017) Environmental sustainability/sufficiency. In: Denemark RA, Marlin-Bennett R (eds) The international studies encyclopedia. UK, Wiley-Blackwell

Michaelis L (2011) Climate change and spirituality. In: Bouckaert L, Zsolnai L (eds) Handbook of spirituality and business. Palgrave Macmillan UK, London, pp 236–244

Mone, AM (1974) A study of adolescence in the changing Tongan social situation and its implications for Christian education in the Free Wesleyan Church of Tonga (Degree of Bachelor of Divinity Degree of Bachelor of Divinity), The Pacific Theological College, Suva

Moore SE, Leslie HY, Lavis CA (2005) Subjective well-being and life satisfaction in the Kingdom of Tonga. Soc Indic Res 70(3):287–311

Mugambiwa SS (2018) Adaptation measures to sustain indigenous practices and the use of indigenous knowledge systems to adapt to climate change in Mutoko rural district of Zimbabwe. Jamba 10(1):388

Nagy GJ, Cabrera C, Coronel G, Aparicio-Effen M, Arana I, Lairet R, Villamizar A (2017) Addressing climate adaptation in education, research and practice: the CLiVIA-network. Int J Clim Chang Strat Manag 9(4):469–487

Nawrotzki R, Kadatska P (2010) Addressing climate change with indigenous knowledge. Int J Clim Chang: Impacts Responses 2(1):33–48

Nhemachena C, Mano R, Mudombi S, Muwangwa V (2014) Perceptions on climate change and its impact on livelihoods in Hwange district Zimbabwe. J Disaster Risk Stud 6(1):1–6

Nichols A, Maynard V, Goodman B, Richardson J (2009) Health, climate change and sustainability: a systematic review and thematic analysis of the literature. Environ Health Insights 3:63–88
Nisar N, Guleria R, Kumar S, Chawla TC, Biswas NR (2007) Mycoplasma pneumoniae and its role in asthma. Postgrad Med J 83(976):100–104
Nunn PD (2009) Bridging the gulf between science and society: imperatives for minimizing societal disruption from climate change in the Pacific. In: Sumi A, Fukushi K, Hiramatsu A (eds) Adaptation and mitigation strategies for climate change. Springer, Tokyo
Nunn PD (2013) The end of the Pacific? Effects of sea level rise on Pacific Island livelihoods. Singap J Trop Geogr 34(2):143–171
Nunn PD, Mulgrew K, Scott-Parker B, Hine DW, Marks ADG, Mahar D, Maebuta J (2016) Spirituality and attitudes towards nature in the Pacific islands: insights for enabling climate-change adaptation. Clim Change 136(3):477–493
Nyong A, Adesina F, Osman Elasha B (2007) The value of indigenous knowledge in climate change mitigation and adaptation strategies in the African Sahel. Mitig Adapt Strat Gl 12(5):787–797
Pew Research Center (2017) The changing global religious landscape. Pew Research Center, USA
Pope Francis, and Catholic Church (2015) Laudato Si: on care for our common home. Our Sunday Visitor, United States of America
Sánchez García JL, Díez Sanz JM (2018) Climate change, ethics and sustainability: an innovative approach. JIK 3(2):70–75
Schuman S, Dokken J-V, van Niekerk D, Loubser RA (2018) Religious beliefs and climate change adaptation: a study of three rural South African communities. Jamba 10(1):509
Seekingtruth.co.uk (2016) Climate change, extreme weather and bible end time prophecy. Seekingtruth, UK. Retrieved [2016 May 22]; from http://www.seekingtruth.co.uk/weather.htm
Sonya S (2016) Religious identity, beliefs, and views about climate change. Oxford University Press, UK. Retrieved [2019 Sep 5]; from https://oxfordre.com/climatescience/view/10.1093/acrefore/9780190228620.001.0001/acrefore-9780190228620-e-335
Statistic Department of Tonga (2016) Tonga 2011 census of population and housing: Volume 1—basic tables and administrative report. Tonga: Tonga Department of Statistics
The Bible Society in Australia (1988) Good news bible: Australian Edition. Canberra: The Bible Society in Australia
Tonga Department of Statistic (2013a) Tonga 2011 census of population and housing, volume 2: Analytical report. Nuku'alofa, Government of Tonga
Tonga Department of Statistic (2013b) Tonga 2011 census of population and housing: volume 1—basic tables and administrative report. Nuku'alofa: Tonga Department of Statistics
Tonga Department of Statistic, & Secretariat of the Pacific Community (SPC) (2008) vol. 2. New Caledonia, SPC
Tra TV, Thinh NX, Greiving S (2018) Combined top-down and bottom-up climate change impact assessment for the hydrological system in the Vu Gia—Thu Bon River Basin. Sci Total Environ 630:718–727
U.S. Global Change Research Program (USGCRP) (2016) The Impacts of Climate Change on Human Health in the United States: A Scientific Assessment. In: Crimmins A, Balbus J, Gamble JL, Beard CB, Bell JE, Dodgen D, Eisen RJ, Fann N, Hawkins MD, Herring SC, Jantarasami L, Mills DM, Saha S, Sarofim MC, Trtanj J, Ziska L (eds) Washington, DC, U.S. Global Change Research Program, p 132
Van Beynen P, Akiwumi FA, Van Beynen K (2018) A sustainability index for small island developing states. Inter J of Sust Dev & World Ecol 25(2):99–116
Walter ND, Grant GB, Bandy U, Alexander NE, Winchell JM, Jordan HT, Sejvar JJ, Hicks LA, Gifford DR, Alexander NT, Thurman KA, Schwartz SB, Dennehy PH, Khetsuriani N, Fields BS, Dillon MT, Erdman DD, Whitney CG, Moore MR (2008) Community outbreak of Mycoplasma pneumoniae infection: school-based cluster of neurologic disease associated with household transmission of respiratory illness. J Infec Dis 198(9):1365–1374
Wasswa FN (2010) Climate change and livelihoods: implications and evidences from Uganda (Doctor of Philosophy PhD Thesis), University of Pretoria, Pretoria (SA)

WHO, Ministry of Health in Fiji, GEF, & UNDP (2013) Human health vulnerability to climate change in Fiji. Government of Fiji, Suva, Fiji
World Health Organisation (WHO) (2015) Human health and climate change in Pacific Island Countries. World Health Organisation, Geneva
World Population Review (2019) 2019 World Population by Country. World Population Review, USA. Retrieved [2019 September 24]; from http://worldpopulationreview.com/
Wosnitza M, Peixoto F, Beltman S, Mansfield CF (2018) Resilience in education: concepts, contexts and connections. Springer, Switzerland
Yuan K-S, Wu T-J, Chen H-B, Li Y-B (2017) A study on the teachers' professional knowledge and competence in environmental education. EURASIA J Math Sci Tech Ed 13(7):3163–3175

Peni H. Havea is a Tongan Scholar and a training Pastor from the Kingdom of Tonga in the South Pacific.

Antoine de Ramon N'Yeurt obtained his PhD in marine botany from the University of the South Pacific (USP, Suva) in 1998, and now works as a Research Fellow in climate-change issue at the Pacific Centre for Environment and Sustainable Development (PACE-SD) of the University of the South Pacific in Fiji. For the last 20 years he devoted his activities to the study of marine algae of the Pacific Islands such as Fiji, Rotuma, the Cook Islands, and has been involved in surveys of the marine floras of French Polynesia, Wallis, the Solomon Islands, Santo (Vanuatu), the Seychelles and Clipperton. He is the author of numerous publications, books and floras on taxonomy and described several new genera and species of marine algae. More recently, he has been involved in the topics of Ocean Algal Afforestation (OAA) and renewable energy, ocean acidification and its effects on coral reefs, and population-level climate change adaptation in the South Pacific.

Chapter 13
Influence of Spiritual Beliefs on Autonomous Climate-Change Adaptation: A Case Study from Ono Island, Southern Fiji

Jalasayi Atkinson-Nolte, Patrick D. Nunn, and Prudence Millear

Abstract Most climate-change adaptation in the Pacific Islands has been framed in secular terms yet applied in contexts where routine decision-making is spiritually informed. This may be a major reason why such externally-sponsored adaptation has met with very limited success to date. This highlights an opportunity for reframing adaptation in ways that are more acceptable to target communities. Using a case study of traditional communities on a remote island in southern Fiji, information was collected using focus groups and informal interviews about people's attitudes towards climate change, its causes, effects and remedies. Particular attention is given to understanding the ways in which people's spiritual beliefs, incorporating both Christian and pre-Christian understandings, influence their perceptions of climate risk and determine their preferred adaptive responses. Given that this study is likely to be representative of rural, faith-engaged, contexts across the Pacific Islands region, it is argued that the results are widely applicable and should inform a re-evaluation of international/donor approaches to enabling climate-change adaptation in this region.

Introduction

It is widely acknowledged that islands such as those in the Pacific Ocean are more exposed to climate change than most other inhabited land masses. The reasons for this are mostly geographical, particularly the lengthy coastlines of islands (compared to their land areas) and the highly skewed distribution of people and infrastructure which means that coasts are the most valued parts of most islands (Kumar et al. 2020; Robinson 2019).

J. Atkinson-Nolte · P. Millear
School of Health and Behavioural Sciences, University of the Sunshine Coast, Maroochydore, QLD 4558, Australia

P. D. Nunn (✉)
School of Law and Society, University of the Sunshine Coast, Maroochydore, QLD 4558, Australia
e-mail: pnunn@usc.edu.au

J. M. Luetz and P. D. Nunn (eds.), *Beyond Belief*, Climate Change Management,
https://doi.org/10.1007/978-3-030-67602-5_13

People have lived on islands in the west Pacific Ocean for more than three thousand years (Carson 2018). Their prolonged and unbroken occupation of most Pacific islands required the development of considerable societal resilience, the remains of which are still apparent in many island cultures today in both tangible and intangible forms (Nunn et al. 2017; McMillen et al. 2017).

Yet the global perception of the comparatively high vulnerability of islands to climate change led to the development and implementation of external adaptive solutions that naïvely disregarded or undervalued such culturally-grounded resilience. Calls have been growing recently for such resilience to be re-valued and even used as the foundation of future adaptation initiatives in the Pacific Islands (McNamara et al. 2020; Westoby et al. 2019). Yet achieving this goal requires that external agencies, who might help fund future initiatives, are adequately persuaded of the efficacy of cultural resilience in such contexts. This applies especially to the adaptive capacity of communities and their ability to use cultural resilience to cope with future climate change, which is expected to occur at unprecedented rates relative to past climate change.

With this in mind, this chapter reports on a study into the perceptions of climate change, especially its drivers, on a small remote island in the Fiji group. The people occupying this island are strongly spiritually engaged and largely disengaged from secular science, meaning that their understanding of environmental changes and their responses to it are mostly driven by spiritual beliefs and rationalizations together with elements of longer-standing cultural resilience. These sometimes fatalistic views may not provide a perception of climate change which supports the necessary adaptive response. This chapter first discusses the study area and describes relevant aspects of its inhabitants' livelihoods and beliefs, before a short account of the methods used, then results and analysis. There follows a discussion in which several insights into the role of spiritual beliefs in climate-change decision-making here are explained, before a brief conclusion.

Study Area and Context

The island of Ono (Fig. 13.1) is located within the eastern part of the Kadavu Island group in southern Fiji.[1] Although there is a small airstrip in the west of the group, most people from the main island of Fiji (Viti Levu) where the national capital (Suva) is located reach Ono by an 8–10-h boat ride. There are six villages on Ono, all coastal and on lands that are owned by resident extended family groups (*mataqali*). Most people on Ono routinely subsist on foods obtained from local environments, including ocean and reefal ones and mangroves. Like all islands in the Kadavu group, Ono is an old volcanic island with fertile well-watered soils in which many crops are grown. While the people of Ono sell some surplus fish locally, most cash income

[1] Note that there is another Ono Island in Fiji, at the southern end of the Lau Island chain east of this Ono.

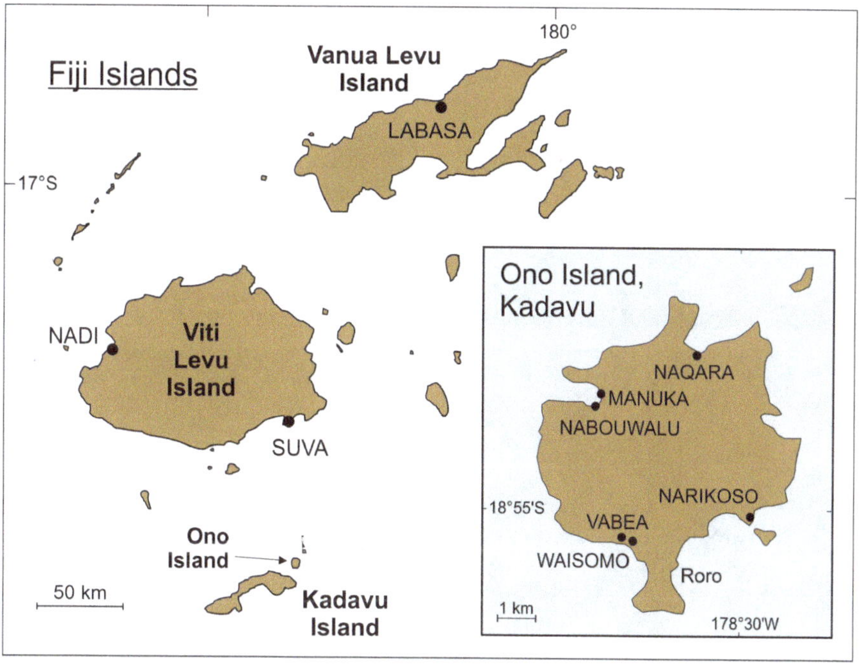

Fig. 13.1 Map of the Fiji archipelago showing the location of Ono Island (details in inset)

comes from sales of *kava* (*Piper methysticum*), used to make a non-alcoholic drink consumed both ceremonially and socially in the Pacific Islands and increasingly in demand elsewhere in the world.

For this study, we focused on the villages of Vabea and Waisomo (where we stayed for two weeks in June–July 2019) and Narikoso which we visited several times. We also visited the other three villages and became familiar with much of the environment of this part of eastern Kadavu. We obtained agreement from local leaders to observe environmental impacts attributable to climate change and to discuss these with a range of local residents.

Climate change has been affecting Ono Island with increasingly severe and diverse impacts over the last few decades, although few of these impacts have been formally documented. As on many smaller islands, the most common expression of climate change on Ono is the effect of sea-level rise, which is currently occurring in this part of the world at some 4 mm/year, slightly above the global average (Wu et al. 2017). Sea-level rise here causes shoreline erosion, increases the incidence and extent of lowland flooding, and causes the spread of groundwater salinization. Shoreline erosion is common along most sandy coasts on Ono (Fig. 13.2a,b) while lowland flooding has reportedly been occurring more frequently in all six coastal settlements on Ono; in one of these, Vabea, the Fiji government, adjudging the problem to be the river (not the rising sea level) has recently constructed a river-bank wall (Fig. 13.2c). Groundwater salinization appears less of a problem on Ono because

Fig. 13.2 Views of Ono Island. **a** Eroding shoreline at Vabea-Waisomo. **b** Old seawall at Narikoso. **c** Government-built river-bank wall at Vabea. **d** Methodist church at Narikoso. **e** Methodist church at end of village green (*rara*) in Vabea-Waisomo. Photos by authors

most intentionally-planted food crops are grown inland and upslope while lowland food plants largely grow wild.

Another aspect of climate change affecting Ono is the rise in temperature which here, as elsewhere in Fiji and the tropical Pacific Islands (Nunn and Kumar 2019b; Scott-Parker and Kumar 2018), means people are less able to work as long as customary during the hottest times of the day. Conversely, Ono residents report that warming temperatures on the island over the past decade have enabled them to plant *kava* whereas this was not possible before.

Temperature rise is also implicated in recent changes in the nature of West Pacific tropical cyclones, to which all of Fiji is exposed. An observed drop in the frequency of tropical cyclones in the region over the past decade has been accompanied by an increase in the intensity of the strongest, something in line with projections (Knutson et al. 2010; Sugi et al. 2017). This change was memorably manifested in Fiji in 2016 when Severe Tropical Cyclone Winston, packing the strongest winds ever recorded in the southern hemisphere, passed through the centre of the group (Cox et al. 2018).

The people of Ono are exclusively Christian, each village having at least one church (Fig. 13.2d,e) and Vabea-Waisomo, which are contiguous, having four churches of different Christian denominations at the time of our visit—One Nation, Seventh-Day Adventist, Catholic and Methodist. For a long time, Methodism was the majority 'orthodox' Christian denomination in Fiji but in recent decades a number of newer denominations, typically funded from overseas congregations and supported by vigorous missionary training, have gained footholds throughout Fiji (Morgain 2015). But it is also clear that, as in other Pacific Island countries, Christianity in Fiji never wholly replaced pre-Christian belief systems but subsumed their most strongly-held tenets, a key to its success (Winch-Dummett 2010). And even today, there are aspects of Christian worship in Fiji that can be considered culturally-grounded, respectful of traditional beliefs and worldviews. These are also expressed in the way that people live, especially in rural more traditional communities like those on Ono Island, ranging from the hierarchical hereditary decision-making framework to the existence of cultural taboos that govern the gamut of everyday decision-making including agricultural practices, food distribution, even homework and childbirth (Granderson 2017; Rubow and Bird 2016; Lukere and Jolly 2001; Walid 2017).

Methods

The results reported in this chapter came from three sources.

The first was enabled by our residence within communities, our participation in routine community activities, and our familiarization with the environment of Ono, especially those coastal parts where the evidence for environmental changes attributable to climate change is most readily apparent.

The second involved interviews, effectively one-on-one discussions, with key persons on Ono, including traditional chiefs, chiefs' spokespersons (*matanivanua*), religious (Christian) leaders in the community (*talatala* or ministers, *vakatawa* or lay preachers), school teachers (*qasenivuli*) and the administrative leaders in each community (*turaga ni koro*). These people speak not just for themselves but also reflect community views, aspirations and preferred ways of doing things, all of which were considered key to understanding the research context.

The third utilized focus-group sessions which are an appropriate unit for acquisition of data in such communal societies (Guest et al. 2017). A focus group guide was used which consisting of open ended questions about climate-attributable environmental changes they had observed. The words "climate change" were not mentioned by the researchers during this questioning as this has previously been linked to biased responding (Rudiak-Gould 2012). Pacific Island societies are communal so focus groups are a widely-accepted way of obtaining information of the kind sought in this study (Warrick 2009). Yet for the same reason, it is not always easy to control the composition and size of focus groups as people invariably come and go at will. Our survey gathered information from a number of less-formal, more fluid focus groups of this kind as well as from eight focus groups separated by age and gender (young female, mature female, young male, mature male) in the villages of Narikoso and Vabea where participants were selected using maximum variation purposive sampling. Twenty-seven adults were engaged in this manner and this is the key source of data used in this chapter.

Discussions took place in places and contexts in which interviewees and community participants all felt comfortable, typically on the floors of houses, where *kava*, tea or snacks might be consumed while questions were asked. All formal interactions were guided by the researchers' University's ethics requirements, participation was informed and voluntary and unremunerated. Questions were generally asked in the Fijian language (*iTaukei* Bauan) and responses translated in summary as necessary. All discussions were audio recorded and transcribed verbatim as needed.

Results and Analysis

Since most of the data collected was qualitative, it is presented in this section alongside a discussion about its meaning and implications for understanding and addressing climate-change impacts on Ono Island and in comparable contexts elsewhere.

This section is divided into four parts. The first (section "Observations of Climate Change by the People of Ono") explains Ono residents' observations of environmental change that are attributable to climate change. The second (section "Ono Residents' Views on the Causes of Climate Change") explains residents' views about the causes, both proximate and ultimate, of observed changes while the third (section "Responses to Climate-Change Impacts on Ono") focuses on their responses, past and planned, to these changes. Finally in section "Views About Future Coastal-Environmental Changes on Ono", we report on what people on Ono feel might happen in the future and how they expect to be affected and to respond.

Observations of Climate Change by the People of Ono

Like most Pacific Island peoples, those of Ono Island understand that the part of the Earth's surface they occupy is dynamic and changeable (Galipaud 2002) but consider that it is usual for this to happen within a fixed framework of land and sea. In other words, while accepting that landslides sometimes occur or even that superficial islands (sand cays) may alternately form or vanish after storms, the idea that the configuration of land and sea might change profoundly is generally unthinkable.

That this is a largely recent idea informed by Christian beliefs, to which the overwhelming majority of Pacific Islanders today subscribe (Nunn et al. 2016), can be demonstrated by the fact that most who live in coastal areas consider their right to occupy these as God-given. This explains why there has traditionally been so much resistance by coastal communities in the Pacific Islands region (and similar) to the idea of inland (even off-island) relocation forced by climate change (Marino 2015; Farbotko 2019; Piggott-McKellar et al. 2019b). Justifications of this resistance commonly invoke the idea that coastal lands are God-given and that he does not want his people to abandon them; indeed to do so would be a betrayal of religious belief.[2] Relocation is almost unthinkable, "barely conceivable" (Kempf 2012, p. 250).

Yet when Pacific Islanders become more detached from their religious belief systems, as invariably occurs when they discuss their (culturally-recalled) histories, a contradiction often emerges. For instance, while claiming the present (vulnerable coastal) sites of their villages to be God-given, many residents when asked about where their distant ancestors used to live will point elsewhere. Indeed many oral histories in the Pacific region corroborate the archaeological evidence of migration within the last millennium. In the past 700 years this typically involved moving from coasts to hilltops to coastal hinterlands to coasts, the latter move commonly required when people and communities became Christianized more than a century ago in most places (Nunn 2007). Yet some oral histories recall the latter move as ill-advised. Examples appear widespread but few are documented, an exception being the recent history of relocation in the Torres Islands of northern Vanuatu (Siméoni and Ballu 2012). In January 2019, an elderly informant on Ono Island told the second author that

> My grandfather always used to say that *his* grandfather was forever saying it was a mistake for our people to move to the coast. That was a time when we lived in the hills, safe from the waves, but then you people [meaning Europeans] came along and forced us all down to the water's edge. The old ones, they knew this was a dangerous place, they said not to go, but we had no choice. Here we are and *dina saraga* – too true – we now discover it is a dangerous place! We should listen to the past.

There is a clear contradiction between acceptance of Christian views about the entitlement to coastal land (where nineteenth-century missionaries told the community to settle) and the culturally-grounded acceptance of a history of (largely) pre-Christian mobility (Campbell 2011; Nunn and Campbell 2020). A similar contradiction often arises when Pacific Islanders talk about the cause of particular impactful phenomena like climate change (see following section), simultaneously attributing it to God and human-enhanced global warming. We return to this issue, which we characterize as 'flexible attribution', in the Discussion in section "Discussion: Spiritual Beliefs and Climate Change".

---

[2]Note we use the words 'he' and 'his' in respect of God in the Pacific Islands, reflecting its exclusive gender usage in the region.

The importance of all this to Ono residents' observations of climate change is that their religious views (and the messaging of local church leaders) allow people to downplay (occasionally exaggerate) observed effects. If land is deemed God-given, then its existence must be under divine protection so that, if this land is damaged, then God must have been angered and the most-commonly favoured solution (as discussed in section "Responses to Climate-Change Impacts on Ono" below) may be spiritual rather than secular, immanent rather than pragmatic.

There is also a gender dimension to the understanding of climate-change impacts that reflects a woman's role in traditional societies such as those on Ono (du Bray et al. 2019). The expression of this view is also gendered, most women reluctant to talk about such matters when men are present but often voluble on the subject when men are absent. From our female-only focus groups, we deduce that many women understand perfectly well what is causing increased lowland flooding and shoreline erosion. More than any other groups, mature women appear more resigned to a future requiring upslope relocation, an expression of the greater responsibility that many feel for the welfare of the community compared with some males. But what was unexpected was that most younger women we interviewed had little interest in observing environmental changes on Ono, something we attribute both to their being preoccupied with more-immediate matters (like childcare or tending sickly family elders) and to the fact that many come from elsewhere in Fiji—most newly-wed couples settle in the husband's village—so had comparatively little familiarity with the Ono environment.

Ono Residents' Views on the Causes of Climate Change

Most people we spoke to on Ono Island had heard about 'climate change' but their understanding of its nature and its scientific causes was variable. In terms of proximal causes, many residents were keen to blame their actions and those of others for unwanted environmental changes like coastal erosion and flooding. Some residents, more commonly male, argued that people's interference with coastal processes—by chopping down mangroves, for instance—had aggravated shoreline erosion. Others, more commonly female, often argued that their neighbours or their community had become insufficiently devout or had strayed from the true faith, for which God was punishing them. Perceptions of this kind clearly influence the responses favoured for climate-change adaptation (see section "Responses to Climate-Change Impacts on Ono").

In terms of ultimate causes, we found a similar divide. Younger residents in particular were ready to blame global human activities (like fossil-fuel combustion and deforestation) for causing climate change, a situation that is a microcosm of the entire Pacific Islands region from which some of the loudest protests come from younger people like the Pacific Climate Warriors (see Ledderucci, this volume; Titifanue et al. 2017; Steiner 2015). On Ono Island, part of this awareness stems from the presence in Narikoso Village of Kelepi Saukitoga, an articulate and well-informed person who was the community voice of the Pacific Islands at the Global Climate Action Summit in 2018 in San Francisco.

Yet ranged against this belief in secular ultimate causes is one that regards climate change, like other socio-environmental stressors, as an expression of divine anger by a jealous God (Cox et al. 2018). This view also finds expression in other ways, notably through anger directed at churches (and churchgoers) that are new and involve novel systems of worship. For example, among the 140 residents of Vabea and Waisomo Villages there are four Christian churches (see Fig. 13.2e), each of a different denomination, and each resident generally attends only one of these. And although we never witnessed any signs of conflict between these churches on Ono, such conflict is becoming more common in similar communities elsewhere and even at national level, driven in part by disillusionment with the Methodist Church, the most widespread and longest-established in Fiji (Ryle 2010; Tomlinson 2012).

Our research shows that many people on Ono, especially those in community leadership roles and those younger people who recently finished high school, understand much about climate change that coincides with western scientific explanations. Members of the former group may express different views in different social contexts. A leader on Ono conveyed to the second author that they are well aware of global warming and its relation to sea-level rise but that they cannot communicate this to members of their community as it would be met with disbelief due to their strong religious convictions.

Younger people are often knowledgeable about climate change and recognize its impacts in many parts of Ono Island yet, because Fiji society is largely hierarchical, the views of younger adults are generally unheard within the community, especially if these views conflict with those of their elders. This is a common source of frustration among younger people in Fiji and a source of growing anxiety (Scott-Parker and Kumar 2018). Elsewhere in the Pacific Islands, for example on Butaritari Island in Kiribati, so frustrated have younger adults become at the reluctance of their elders to respond to climate-change impacts that the traditional hierarchy itself is under threat (Nunn et al. 2014). One explanation for this is that younger people were found to have more formal education and awareness of climate change issues compared to traditional decision makers. The knowledge held by the latter and the framework within which their decision-making is made may both be insufficient to cope with the long-term and unprecedentedly complex challenges to livelihoods posed by climate change (Laursen et al. 2018; Foley 2018).

Responses to Climate-Change Impacts on Ono

Responses to unwelcome environmental changes, such as those likely to be driven by climate change in Fiji, are framed within the understanding of their cause/s. And since on Ono Island, causes are generally divided between the spiritual and the secular (section "Ono Residents' Views on the Causes of Climate Change"), occasionally encompassing both, responses can be similarly divided.

Our surveys suggest the most widespread response to climate change among the people of Ono was a spiritual one, but we are aware this might be a default judgement; it is simpler and cheaper to pray than just about anything else. In its simplest form, most Ono residents explain climate change as a consequence of an insufficiency of piety and—more pragmatically—tithing. This situation angers God but can be reversed by 'praying more and paying more' in the muttered words of one commentator on Ono. The role of religious leaders in developing and promulgating this view often appears quite substantial. Research indicates that environmental ills may be capitalized on by such people to boost church coffers and to ensure that churches recover quickly in the aftermath of disaster (Gard and Veitayaki 2017; Cox et al. 2018).

In contrast (but not alternatively), we found that many Ono residents, typically younger males, were driving or contemplating much more pragmatic solutions to climate-change impacts, such as (enhanced) seawall construction, tree/creeper planting along the backs of eroding shorelines, mangrove replanting offshore, and even the incremental relocation of settlements inland and upslope. The people who expressed such views most volubly were tolerant of others who favoured spiritual responses but appeared unconvinced of their efficacy.

A third element has entered the mix of climate-change responses on Ono Island, which was one reason it was selected as a site for this study in the first place. This element involves the imposition of externally-designed and externally-funded adaptation solutions, typically by agencies with secular agendas and often little understanding why local residents might privilege spiritual ones. On Ono, there are two main interventions.

The first is the relocation of low-lying Narikoso Village (110 residents), something first planned in 2011 but which, following years of inaction and the downscaling of promises by aid donors, was only starting to happen at the times of our visits in 2019 (Bertana 2020; McMichael and Katonivualiku 2020). In January 2019, we were shown raw materials for two new houses (to be constructed at the new uphill village site) that had been dumped on the beach adjoining the village, none of its residents sure what they should do with them. In June–July 2019, two government workers were assembling these two new houses; more than 20 had been promised by the aid donor in 2016–2017. Narikoso residents were unsure whether more houses would ever be built at the new site which was showing signs of degradation resulting from clearance of the forest and the associated exposure of the regolith/soil to sometimes heavy rain.

The second was the Government-built river-bank wall, started and completed in mid-2019, in Vabea Village (see Fig. 13.2c). In 2016–2017, the river that runs through Vabea broke its banks and flooded adjacent houses for several days. The issue was reported upwards through the local subnational decision-making hierarchy and the Government, we were told, unilaterally and without consulting the Vabea community, decided to construct a solid river-bank wall along the southern bank of the lower (tidal) part of the river. When we were on Ono in June–July 2019, fifteen Government workers were billeted in Vabea and, using imported rocks, bulldozers and mechanical diggers, were in the final stages of completing the river-bank wall.

The decision to build this wall appears to have been premised on the belief that river flows had increased to the point that (unprecedented) flooding started to occur, whereas it is more likely that rising sea level—especially during high tide—has radically reduced the ability of river water to be dispersed in the ocean. It is unlikely, as many local residents told us, that the river-bank wall is going to prevent flooding in the future, but they were grateful nonetheless for the Government's investment.

Views About Future Coastal-Environmental Changes on Ono

Many informants on Ono, largely irrespective of age and gender, told us in 2019 that they would need in the future to move their dwellings and communal buildings, especially churches (*valenilotu*) and community halls (*valenisoqo*), to higher ground to escape the increasingly-visible effects of rising sea level. Religion-informed climate change denial was not evident in our surveys of Ono despite this being apparent in other parts of the Pacific Islands. Previous research has documented the local people's expressed belief that current period of sea-level rise will cease in about a decade (Lata and Nunn 2012) or they completely deny sea-level rise, given (the Christian) God's promise in the aftermath of the Deluge to never again flood the Earth (Fair 2018; Kempf 2017). For although people emphasized spiritual causes and the efficacy of spiritual responses, many were also pragmatic, stating that they were prepared to eventually move upslope.

One reason for this view may be the attention, nationally and globally, that the Narikoso relocation has attracted over the past five years or so, something which may have normalized the idea of coastal-settlement relocation for communities in this part of Fiji more than elsewhere. All coastal settlements on Ono Island have land at their rear that is higher, beyond the reach of storm waves, yet not so steep that buildings could not be readily constructed there. Indeed in all villages except Narikoso and Waisomo, there are already buildings in these higher areas, perhaps signalling some degree of autonomous adaptation to sea-level rise (Piggott-McKellar et al. 2019b).

Some of the most dramatic cases of shoreline recession around Ono have occurred where mangroves are once said to have existed but were cleared decades ago for various reasons; a similar study of this in Fiji comes from the islands of Ovalau and Moturiki where coastal communities with an uncleared mangrove fringe are experiencing considerably fewer problems than those where this was cleared decades earlier (Nunn 2000). On Ono, the beach named Roro (location shown in Fig. 13.1) has experienced shoreline recession of as much as 80 m in some 30 years, largely because the intertidal mangrove zone has been cleared, exposing the sandy substrate to wave erosion. This situation has sparked community replanting of mangroves here, something that local residents believe might halt erosion. In contrast, the village of Nabouwalu has retained a mangrove fringe which they consider a barrier to shoreline erosion and are forbidden to cut. Like other 'green' or 'nature-based' solutions, mangrove reforestation was once a popular remedy for shoreline erosion in Fiji and other Pacific Island countries (Walton et al. 2006) but now its limitations may be starting to be recognized, especially in a situation where the rate of sea-level rise is accelerating (Morris et al. 2018; Pearson et al. 2020).

There is also a temporal dimension to this, which has been detected in studies of relocation elsewhere in Fiji (Piggott-McKellar et al. 2019b, c), that involves the overcoming of community leaders' resistance to relocation. In some situations outside Ono, it has been claimed that only the passing of elderly leaders who refused to contemplate relocation could open the door to this relocation. Although we did not hear such sentiments openly expressed on Ono Island, we did hear numerous complaints about the difficulties of persuading elderly decision-makers of the growing need to relocate and the advantages of beginning this process sooner rather than at the last minute, when the need would be clear to everyone.

Discussion: Spiritual Beliefs and Climate Change

In culturally-grounded societies like those on Ono Island, extraneous livelihood stressors like climate change are viewed through three lenses, sometimes separately, sometimes in combination. These lenses are religious, secular-scientific, and traditional-cultural. To different degrees, each has a role in the development and implementation of responses.

Religious Beliefs

Like other Pacific Island contexts, especially rural ones, the majority of people on Ono Island consider that climate change has a divine cause and that an important response is to pray and become more pious. We were part of several conversations in which island residents discussed what might have caused God's anger in this regard and how this might be lessened. For example,

> Not respecting the religious belief and that's how the God is cursing the land. With the changing weather, the climate change that's happening around, we really need to make sure all children are looked after well, especially with the sea level rising. (Narikoso, young male)

It is important to note that climate change is not a unique stressor in this regard; for most of the time that Christian belief systems have influenced Pacific Islander worldviews, undesired environmental changes have been attributed to divine anger and solutions developed and implemented accordingly (Rubow and Bird 2016; Haluza-DeLay 2014).

Two recent studies of the (Christian) Biblical story of Noah and the way it has influenced climate-change perceptions in the Pacific Islands are illuminating in this regard (Fair 2018; Kempf 2017). While diverse and nuanced in different Pacific Island contexts, the essence of the story of Noah—that he and his family survived a global flood, in the aftermath of which God promised never to flood the Earth again—in most such contexts focuses on the denial of (climate-driven) sea-level rise, or at least its continuation in the future, because of God's unretracted promise. In some situations, the acceptance of sea-level rise and its associated effects (like shoreline erosion and increased lowland flooding) has been regarded as almost heretical, leading to a refusal to respond/adapt. For example, a former Kiribati politician, Harry Tong, stated in 2010 that "climate change is an all-natural phenomenon. You can't really do much about that, unless you talk to God and negotiate with God ... He made a promise to Noah that he will never again destroy this earth with flood" (quoted in Fair 2018, p. 5). We did not encounter such views in our surveys over nearly four weeks on Ono Island (which does not mean no-one there believes this) but it is anecdotally one reason why many other communities in the Kadavu Island group are reluctant to relocate from their low-coastal locations. Yet we did find similar sentiments expressed, for example—

> So it's respecting the land and meeting the obligations of being a faithful servant or custodian of this land. You will bear many fruits or you will have many blessings from the gods or from their faith as a being loyal to their gods or to their teachings, to their Christian teaching. (Narikoso, mature female)

We also find the influence of religious leaders, both nationally and within communities, to be significant in the development of ideas among community residents on Ono concerning climate change and its attribution: a comparable situation to that reported on Lombok Island, Indonesia, where the influence of religious leaders (*tuan guru*) extends far beyond people's spiritual lives (Kingsley 2014). In communal societies such as those found on Ono, it is common for 'persons of influence' to drive community members' understandings of various phenomena and direct their responses to these—often to a degree that more individualistic western societies might consider unusual. In fact, it has been suggested that climate-change adaptation strategies in the Pacific Islands region (and in developing-country contexts elsewhere) might focus less on funding national governments to drive adaptation and more on empowering such persons of influence within communities to do so (Nunn 2009).

Secular-Scientific Beliefs

Many scientists, representing western/global scientific enterprise, have viewed people's religious beliefs in regions of the world like the Pacific Islands as 'barriers' to climate-change adaptation (Kuruppu and Liverman 2011; Paton and Fairbairn-Dunlop 2010), one even characterizing faith in such a situation as a 'touching silliness' (Haraway 2016). Such views are often as myopic as they are uninformed. People like those on Ono are not going to abandon their religious views just because they are told these are worthless, least of all by an outsider. So if science-based future projections of climate change are to feature in decision-making for environmental sustainability on Ono, scientists must learn how to work with community leaders.

The first step is to acknowledge their contrasting worldviews and to incorporate these where possible within community-driven and owned adaptation strategies (Donner 2007; McNamara et al. 2020).

Our surveys on Ono Island showed that more people than might be expected in a rural peripheral Pacific Island context demonstrated an understanding of climate science and, as a result, were considering adaptation strategies which were aligned with those that western science would likely prioritize (such as coastal reforestation and upslope relocation). This highlights the needed and likely future (a decade or so hence) where religious beliefs are maintained whilst also responding to climate change in an adaptive manner. We attribute the comparatively 'enlightened' situation on Ono to the presence of Narikoso Village, which has achieved iconic status as a community requiring (and accepting the need for) relocation, plausibly as a result of sea-level rise. Yet the notable awareness evident on Ono should not be overstated as it is clear that this could also be partly due to the inclusion of climate change information within high-school curricula within the Pacific Islands region for more than a decade (Nakayama et al. 2019; Scott-Parker et al. 2017). Accordingly, younger adults on Ono have acquired a science-based understanding of climate change that appears to already be influencing (planned) community responses.

Traditional-Cultural Beliefs

People have probably lived on Ono Island for more than two thousand years (Burley and Balenaivalu 2012), developing an intimate knowledge of its natural environment (marine and terrestrial), its possibilities and limitations. Navigating changes in the climate drivers of environmental sustainability over this time, especially within the last millennium (Nunn 2007), has also led to the evolution of culturally-grounded protocols for helping ensure sustainability. While undoubtedly grounded in pre-Christian (pre-1860) worldviews about which we know little today, these protocols include the conservation/restoration of overutilized resources, an intimate understanding of seasons (when to plant and harvest), and the routine production of surpluses (and their long-term preservation) to sustain communities in times of exigency (Johannes 2002; Lefale 2010; Aalbersberg et al. 1988).

Many such practices continue on Ono Island today and are considered by its Christian residents as cultural (rather than religious and pre-Christian). These practices can be considered culture-defining and some appear as important in decision-making, especially for environmental futures, as faith-informed beliefs. Rather than being in competition with Nature, Pacific Islanders view themselves as highly connected and embedded within their natural environment (Nunn et al. 2016). This may be expressed as respect for places, their boundaries, and their history and relation to revered ancestors. Such places exist on Ono Island, whereby cultural practices must be performed and outsiders may not be granted access. This is of relevance as many ceremonies, such as those considering changes to land, invoke the spirits of former inhabitants and even ancient 'gods' who are believed to have created the environment and made it hospitable.

We are unlikely to have a complete picture of such protocols and practices on Ono Island, both because there is a considerable diversity of these, varying between households and *mataqali*, but also because such knowledge is often not considered proper to share with outsiders; the issues that may arise when such knowledge becomes public are well documented (Farran 2014).

Flexible Attribution and Community Responses

In addition to a range of contemporary religious and cultural influences on residents' understanding of climate change, modern Ono is also touched by a growing understanding of western-scientific explanations for climate change. We consider cultural views (outlined in section "Traditional-Cultural Beliefs") as basal or immanent, whereas the religious and scientific explanations of climate changes (sections "Religious Beliefs" and "Secular-Scientific Beliefs" respectively) have the potential to be competing frames of understanding. One view or the other was generally favoured by every Ono resident we spoke with in 2019 but it became increasingly clear that many residents also held both views together, sometimes expressing exclusively one for particular audiences or in particular social situations. This dualism extends to responses, most residents apparently favouring principally spiritual responses, others secular ones, yet others both simultaneously. This is something we label 'flexible attribution', the idea that people can simultaneously hold contradictory explanations for the same phenomenon.

This is not the first time that the dualistic beliefs underpinning flexible attribution have been identified in a Pacific Islands context. Arising from her fieldwork in Vanuatu which found a similar dichotomy between religious and scientific explanations of climate change, Fair (2018) saw the potential for placing these side-by-side in order to explore "the possible convergences and tensions between different knowledges of climate change" (p. 2), an approach she termed *tufala save* which means 'double knowledge' in Bislama. Our research also focuses on dualistic understandings of climate change but goes further to see how these might inform the responses of people on Ono Island to climate-change impacts, both experienced and anticipated.

How flexible attribution influences adaptive behavioural decisions is interesting. It is not clear from our study that flexible attribution confuses responses to climate-change impacts on Ono but the issue of response determination is one that is different in communal societies (like Ono) compared to individualistic societies like most western democracies. Decisions involving environmental governance are rarely made by individual family units (*tokatoka*) or even extended family units (*mataqali*) unless the issue is confined to their land. Recognizing that everyone on Ono is largely sustained by the natural environment and that its health is thus key to community sustainability, most such decisions are communal. On Ono, as elsewhere in Fiji and other Pacific Island countries, an environmental issue will be discussed among all adult community members before a decision is made on the community response (Nunn et al. 2014). The hereditary leader of the whole-island community, advised by (typically) elderly leaders, will reach decisions which the community will not normally question and will implement. Since many responses to climate change, like building seawalls or replanting vegetation (like mangroves) along eroding shorelines, are labour intensive, the community works together cooperatively to achieve them. This situation, named *solesolevaki* in Fiji, remains commonplace in rural communities like those on Ono where every adult male is required to devote 1–2 days each week to communal activities, which may include house repair/construction and maintenance of communal food gardens as well as responses to climate change impacts. In less-rural, often peri-urban, Fiji communities, *solesolevaki* has largely disappeared because residents are busy with other things, often in wage employment (Movono and Becken 2018). Recently there have been calls to restore such practices as a way of adapting to climate change without any or little need for cash (Nunn and Kumar 2019a).

Conclusion and Future Research

This study shows how non-western scientific beliefs (religious or cultural) influence people's understanding of climate change in the Pacific Islands, a conclusion similar to that of others who have studied this region (Kempf 2012; Haluza-DeLay 2014; Nunn et al. 2016). Yet this study also finds, as did Fair (2018) and others, that residents of rural remote communities in the Pacific Islands may comfortably hold contrasting beliefs about climate change. Our study shows that this dualism does not hinder responses to climate change on Ono Island; in fact, it may drive them more than might be the case if people adhered to a single explanation.

Dualistic responses to climate-change impacts are similar to those that are likely to underpin recent ecumenical efforts to develop guidance for people of faith in the Pacific Islands region to address climate-change concerns (Rowland 2019). The way forward that is most likely to succeed in terms of minimizing livelihood disruption from future climate change is one that is "likely to lie in a mixture of traditional knowledge and scientific information filtered through worldviews of Pacific Island decisionmakers" (Nunn et al. 2016, p. 478). With this end in mind and as previously synthesised, "it remains an important priority for climate change adaptation in the Pacific Island region to continue building consilience in local communities through a heightened emphasis on science-faith integration" (Luetz and Nunn 2020, p. 304).

As the pace of climate change accelerates and its impacts on human livelihoods in places like Ono Island diversify and increase in severity, the need for effective and sustainable adaptation becomes ever more exigent. At present, much adaptation in the Pacific Islands region is funded by external agencies, typically bilateral donors or international agencies, who uncritically privilege their own (supposedly global) adaptive solutions, attempt to impose these on 'others', then wonder why they do not produce the intended outcomes. The study presented here, like others recently which highlight reasons for past intervention failure (Nunn and McNamara 2019; Piggott-McKellar et al. 2019a), shows that the onus is on these external agencies to reappraise their models of assistance, to take a step back and become facilitators rather than doers of adaptation in others' places (McNamara et al. 2020).

Note

Since our research on Ono Island in mid 2019, Severe Tropical Cyclone Harold (April 2020) tore across the island, flattening many of the buildings in the villages we visited and damaging livelihoods. We expect that the impact of this may have changed some of the views about climate change on which we report here.

Acknowledgements JAN was funded by a New Colombo Plan mobility grant from the Australian Government, PN was funded by an Australian Research Council grant (LP160100941). For help with community liaison, we thank Elia Nakoro and Meli Nanuku from the Fiji Museum. For their many kindnesses and hospitality, we thank the people of Ono Island, especially Bulou Katarina Rarasea, the chief of Narikoso, who was generous with her time and knowledge.

References

Aalbersberg WGL, Lovelace CEA, Madhoji K, Parkinson SV (1988) *Davuke*, the traditional Fijian method of pit preservation of staple carbohydrate foods. Ecol Food Nutr 21:173–180

Bertana A (2020) The role of power in community participation: relocation as climate change adaptation in Fiji. Environ Plan C-Polit Space 28:902–919

Burley DV, Balenaivalu J (2012) Kadavu archaeology: first insights from a preliminary survey. Domodomo 25:13–36

Campbell IC (2011) Worlds apart: a history of the Pacific Islands. Canterbury University Press, Christchurch

Carson MT (2018) Archaeology of Pacific Oceania: inhabiting a sea of islands. Routledge, London

Cox J, Finau G, Kant R, Tarai J, Titifanue J (2018) Disaster, divine judgment, and original sin: Christian interpretations of tropical cyclone Winston and climate change in Fiji. Contemp Pac 30:380–411

Donner SD (2007) Domain of the Gods: an editorial essay. Climatic Change 85:231–236

du Bray M, Wutich A, Larson KL, White DD, Brewis A (2019) Anger and sadness: gendered emotional responses to climate threats in four island nations. Cross-Cult Res 53:58–86

Fair H (2018) Three stories of Noah: navigating religious climate change narratives in the Pacific Island region. Geo-Geogr Environ 5:e00068

Farbotko C (2019) Climate change displacement: towards ontological security. In: Klöck C, Fink M. (eds) Dealing with climate change on small islands: towards effective and sustainable adaptation? Gottingen University Press, Gottingen

Farran S (2014) 'That plant is my ancestor': dilemmas for intellectual property in developing countries, food security and Pacific Island countries. Queen Mary J Intellect Prop 4:277–296

Foley AM (2018) Climate impact assessment and "islandness": challenges and opportunities of knowledge production and decision-making for Small Island Developing States. Int J Clim Chang Strat Manag 10:289–302

Galipaud JC (2002) Under the volcano: Ni-Vanuatu and their environment. In: Torrence R, Grattan J (eds) Natural disasters and cultural change. London: Routledge.

Gard AR, Veitayaki J (2017) In the wake of Winston—climate change, mobility and resiliency in Fiji. Int J Saf Secur Eng 7:157–168

Granderson AA (2017) The role of traditional knowledge in building adaptive capacity for climate change: perspectives from Vanuatu. Weather Clim Soc 9:545–561

Guest G, Namey E, Taylor J, Eley N, McKenna K (2017) Comparing focus groups and individual interviews: findings from a randomized study. Int J Soc Res Methodol 20:693–708

Haluza-Delay R (2014) Religion and climate change: varieties in viewpoints and practices. Wiley Interdiscip Rev-Clim Chang 5:261–279

Haraway D (2016) Staying with the trouble: anthropocene, capitalocene, chthulucene. In: Moore JW (ed) Anthropocene or capitalocene? Nature, history and the crisis of capitalism. PM Press, Oakland, CA

Johannes RE (2002) The renaissance of community-based marine resource management in Oceania. Annu Rev Ecol Syst 33:317–340

Kempf W (2012) Climate change, migration, and Christianity in Oceania. In: Hastrup K, Olwig KF (eds) Climate change and human mobility: challenges to the social sciences. Cambridge University Press, Cambridge

Kempf W (2017) Climate change, Christian religion and songs: revisiting the Noah story in the central Pacific. In: Dürr E, Pascht A (eds) Environmental transformation and cultural responses. Palgrave Macmillan, New York

Kingsley JJ (2014) Redrawing lines of religious authority in Lombok, Indonesia. Asian J Soc Sci 42:657–677

Knutson TR, McBride JL, Chan J, Emanuel K, Holland G, Landsea C, Held I, Kossin JP, Srivastava AK, Sugi M (2010) Tropical cyclones and climate change. Nat Geosci 3:157–163

Kumar L, Eliot I, Nunn PD, Stul T, McLean RF (2020) Comparison of the physical susceptibility of Pacific islands to risks potentially associated with variability in weather and climate. In: Kumar L (ed) Climate change and impacts in the Pacific. Springer, Berlin

Kuruppu N, Liverman D (2011) Mental preparation for climate adaptation: the role of cognition and culture in enhancing adaptive capacity of water management in Kiribati. Glob Environ Chang 21:657–669

Lata S, Nunn P (2012) Misperceptions of climate-change risk as barriers to climate-change adaptation: a case study from the Rewa Delta, Fiji. Climatic Change 110:169–186

Laursen S, Puniwai N, Genz AS, Nash SAB, Canale LK, Ziegler-Chong S (2018) Collaboration Across worldviews: managers and scientists on Hawai'i Island utilize knowledge coproduction to facilitate climate change adaptation. Environ Manage 62:619–630

Lefale PF (2010) *Ua 'afa le Aso*, stormy weather today: traditional ecological knowledge of weather and climate. The Samoa experience. The Samoa experience. Climatic Change 100:317–335

Luetz JM, Nunn PD (2020) Climate change adaptation in the Pacific Islands: a review of faith-engaged approaches and opportunities. In: Leal Filho W (ed) Managing climate change adaptation in the Pacific region. Climate Change Management. Springer Nature, Cham, pp 293–311. https://doi.org/10.1007/978-3-030-40552-6_15

Lukere V, Jolly M (eds) (2001) Birthing in the Pacific: beyond tradition and modernity. University of Hawai'i Press, Honolulu

Marino E (2015) Fierce climate, sacred ground: an ethnography of climate change in Shishmaref, Alaska. University of Alaska Press, Anchorage

McMichael C, Katonivualiku M (2020) Thick temporalities of planned relocation in Fiji. Geoforum 108:286–294

McMillen HL, Ticktin T, Springer HK (2017) The future is behind us: traditional ecological knowledge and resilience over time on Hawai'i Island. Reg Environ Change 17:579–592

McNamara KE, Clissold R, Westoby R, Piggott-McKellar A, Kumar R, Clarke T, Namoumou F, Areki F, Joseph E, Warrick O, Nunn PD (2020) An assessment of community-based adaptation initiatives in the Pacific Islands. Nat Clim Chang 10:628–639

Morgain R (2015) 'Break down these walls': space, relations, and hierarchy in Fijian evangelical Christianity. Oceania 85:105–118

Morris RL, Konlechner TM, Ghisalberti M, Swearer SE (2018) From grey to green: efficacy of eco-engineering solutions for nature-based coastal defence. Glob Change Biol 24:1827–1842

Movono A, Becken S (2018) Solesolevaki as social capital: a tale of a village, two tribes, and a resort in Fiji. Asia Pac J Tour Res 23:146–157

Nakayama M, Taafaki I, Uakeia T, Seru J, McKay Y, Lajar H (2019) Influence of religion, culture and education on perception of climate change, and its implications. J Disaster Res 14:1297–1302

Nunn PD (2000) Coastal changes over the past 200 years around Ovalau and Moturiki Islands, Fiji: implications for coastal zone management. Aust Geogr 31:21–39

Nunn PD (2007) Climate environment and society in the Pacific during the last millennium. Elsevier, Amsterdam

Nunn PD (2009) Responding to the challenges of climate change in the Pacific Islands: management and technological imperatives. Climate Res 40:211–231

Nunn PD, Campbell JR (2020) Rediscovering the past to negotiate the future: how knowledge about settlement history on high tropical Pacific islands might facilitate future relocations. Environ Dev 35:100546

Nunn PD, Kumar R (2019a) Cashless adaptation to climate change in developing countries: unwelcome yet unavoidable? One Earth 1:31–34

Nunn PD, Kumar R (2019b) Measuring peripherality as a proxy for autonomous community coping capacity: a case study from Bua Province, Fiji Islands, for improving climate change adaptation. Soc Sci 8:#225.

Nunn PD, McNamara KE (2019) Failing adaptation in island contexts: the growing need for transformational change. In: Klock C, Fink M (eds) Dealing with climate change on small islands: towards effective and sustainable adaptation. Gottingen University Press, Gottingen

Nunn PD, Aalbersberg W, Lata S, Gwilliam M (2014) Beyond the core: community governance for climate-change adaptation in peripheral parts of Pacific Island Countries. Reg Environ Change 14:221–235

Nunn PD, Mulgrew K, Scott-Parker B, Hine DW, Marks ADG, Mahar D, Maebuta J (2016) Spirituality and attitudes towards Nature in the Pacific Islands: insights for enabling climate-change adaptation. Climatic Change 136:477–493

Nunn PD, Runman J, Falanruw M, Kumar R (2017) Culturally grounded responses to coastal change on islands in the Federated States of Micronesia, northwest Pacific Ocean. Reg Environ Change 17:959–971

Paton K, Fairbairn-Dunlop P (2010) Listening to local voices: Tuvaluans respond to climate change. Local Environment 15:687–696

Pearson J, McNamara KE, Nunn PD (2020) iTaukei ways of knowing and managing mangroves for ecoystem-based adaptation. In: Filho WL (ed.) Managing climate change adaptation in the Pacific region. Springer Nature, Cham, Switzerland

Piggott-McKellar A, McNamara K, Nunn PD, Watson J (2019a) What are the barriers to successful community-based climate change adaptation? A review of grey literature. Local Environment 24:374–390

Piggott-McKellar A, McNamara KE, Nunn PD, Sekinini S (2019b) Moving people in a changing climate: lessons from two case studies in Fiji. Soc Sci 8:#133.

Piggott-McKellar A, Pearson J, McNamara KE, Nunn PD (2019c) A livelihood analysis of resettlement outcomes: lessons for climate-induced relocations. Ambio 49:1474–1489

Robinson SA (2019) Mainstreaming climate change adaptation in small island developing states. Climate Dev 11:47–59

Rowland P (2019) Ecosystem services and faith communities in Oceania. Ecosyst Serv 39

Rubow C, Bird C (2016) Eco-theological responses to climate change in Oceania. Worldviews-Glob Relig Cult Ecol 20:150–168

Rudiak-Gould P (2012) Promiscuous corroboration and climate change translation: a case study from the Marshall Islands. Glob Environ Chang-Hum Policy Dimens 22:46–54

Ryle J (2010) My god, my land: interwoven paths of Christianity and tradition in Fiji. Ashgate, Burlington, VT
Scott-Parker B, Kumar R (2018) Fijian adolescents' understanding and evaluation of climate change: implications for enabling effective future adaptation. Asia Pac Viewp 59:47–59
Scott-Parker B, Nunn PD, Mulgrew K, Hine D, Marks A, Mahar D, Tiko L (2017) Pacific Islanders' understanding of climate change: where do they source information and to what extent do they trust it? Reg Environ Change 17:1005–1015
Siméoni P, Ballu V (2012) Le mythe des premiers réfugiés climatiques: mouvements de populations et changements environnementaux aux îles Torrès (Vanouatou, Mélanésie). Ann Géographie 3:219–241
Steiner CE (2015) A sea of warriors: performing an identity of resilience and empowerment in the face of climate change in the Pacific. Contemp Pac 27:147–180
Sugi M, Murakami H, Yoshida K (2017) Projection of future changes in the frequency of intense tropical cyclones. Clim Dyn 49:619–632
Titifanue J, Kant R, Finau G, Tarai J (2017) Climate change advocacy in the Pacific the role of information and communication technologies. Pac Journal Rev 23:133–149
Tomlinson M (2012) The generation of the now: denominational politics in Fijian Christianity. In: Tomlinson M, McDougall D (eds.) Christian politics in Oceania. Berghahn Books, New York
Walid M (2017) Towards a mega-Pacific Islands education curriculum for climate adaptation blending traditional knowledge in modern curriculum. In: Filho WL (ed.) Climate change adaptation in Pacific countries: fostering resilience and improving the quality of life. Springer, Cham
Walton MEM, Samonte-Tan GPB, Primavera JH, Edwards-Jones G, le Vay L (2006) Are mangroves worth replanting? The direct economic benefits of a community-based reforestation project. Environ Conserv 33:335–343
Warrick O (2009) Ethics and methods in research for community-based adaptation: reflections from rural Vanuatu. In: Ashley H, Kenton N, Milligan A (eds.) Participatory learning and action: community-based adaptation to climate change. IIED, London
Westoby R, McNamara KE, Kumar R, Nunn PD (2019) From community-based to locally-led adaptation: evidence from Vanuatu. Ambio 49:1466–1473
Winch-Dummett C (2010) Christianity and cultural reconfigurations in south West Pentecost. Oceania 80:78–101
Wu QR, Zhang XB, Church JA, Hu JY (2017) Variability and change of sea level and its components in the Indo-Pacific region during the altimetry era. J Geophys Res-Ocean 122:1862–1881

Chapter 14
For Kin, God and Other Beings: Mixtures of Conservation Practice in Raja Ampat, West Papua

Ian N. Parker

Abstract This chapter describes the emergence of hybrid conservation practices in the Raja Ampat islands of Indonesian New Guinea in the context of significant socio-economic and ecological change. On West Papua's coast fringe, different values motivate different types of geographic resource management systems with varying stakes for the people who support them. Here, a patchwork of interlinked regimes of land and sea-based resource governance have contributed to a composite approach to adaptive governance, rather than an inherently conflicting set of practices or norms. I describe how resident Beteo and Ma'ya people revived a limited type of seasonal harvest prohibition and taboo called *sasi* to incorporate Christian ethics and the significance of ancestral sites of nonhuman spirits. On Waigeo island, a center point of international conservation programs and ecotourism, two distinct forms of *sasi* are currently practiced: *sasi gereja*, a type of Christian village-based resource protection and *sasi mon*, a set of clan-mediated rules for areas beyond villages inhabited by ancestors or nonhuman spirit beings. In the past decade, the practices have become formalized through engagement of West Papuan communities with international non-governmental organizations. The varieties of conservation practices in coastal West Papua reflect distinct but perhaps commensurable ethical norms and values. Engagements with valued places highlights how conservation in Raja Ampat is consequential to people's understanding of themselves and others, amidst ongoing resource degradation, the denial of Christian social virtue or economic marginalization.

Introduction

As with societies elsewhere, Pacific Island communities have incorporated beliefs, adapted values, and engaged across boundaries of social difference in times of socio-economic and ecological change. Interdisciplinary research on climate change adaptation has identified how cultural values shape human–environment relations, particularly how religious or spiritual practices provide mechanisms to support cultural

I. N. Parker (✉)
Department of Anthropology, University of California San Diego, 9500 Gilman Drive, La Jolla, CA 0532, USA

J. M. Luetz and P. D. Nunn (eds.), *Beyond Belief*, Climate Change Management,
https://doi.org/10.1007/978-3-030-67602-5_14

and ecological resilience (Adger et al. 2013; Murphy et al. 2016: 102). In many Melanesian and Pacific societies, beliefs about proper interactions between people and nature are reflected in the organized stewardship of resources and habitats that reflect social values in ways that also support ecosystem services (Berkes and Folke 2003; McMillen et al. 2014).

Research into cultural ecosystem services, values and local worldviews about people's relationship to the environment has mobilized a search for examples of strategies or institutional approaches that incorporate religious belief and scientific principles (Chan et al. 2016). Luetz and Nunn (2020) identify how religious worldviews frame how people think about and engage with the natural world. Misunderstandings about the intentions and goals for environmental protection have led to conflicts between indigenous groups and outsiders. Yet in some Pacific island communities and elsewhere, Christianity, local perceptions about nonhuman beings and scientific management have fostered new institutions for supporting livelihoods and habitats (Awuah-Nyamekye 2019; Scoville-Simonds 2018; Tarusarira 2017).

As a way of dealing with social and ecological change West Papuan communities in the Raja Ampat islands of Indonesian New Guinea have adapted Christianity and local spiritual practices as forms of interlinked conservation of marine reef and forest zones. Communities here have transposed beliefs about human–environment relations into new types of protected zones that are sometimes commensurate with—and also distinct from—environmental programs introduced from abroad. Environmental protection includes marine protected areas (MPAs) or forest reserves (*cagar alam*) established by international conservation organizations or Indonesian government agencies, ritually-closed sea or land zones known as *sasi* (McLeod et al. 2009), and zones of nonhuman spirits managed by clans. In Raja Ampat, different values support resource management practices, with different stakes involved for the people who support them.

Around Waigeo island, a center of environmental programs in Raja Ampat, *sasi* is a genre of locally-inflected conservation that has evolved in unexpected ways following large-scale interventions in the 2000s to protect the region's coral reef ecosystems. There are two distinct forms of *sasi* concurrently enforced among resident Beteo and Ma'ya communities: *sasi gereja*, a type of Christian village-based resource protection common in Beteo areas and *sasi mon*, a set of rules and regulations for territories inhabited by Ma'ya ancestors or nonhuman spirits. These zones have become interlinked with donor-funded MPAs throughout Raja Ampat, contributing to governance that comingles endogenous spiritual practices with exogenous ecosystem-based management.

I first provide an overview of studies that describe local mechanisms of resource management, before introducing Indonesia's Raja Ampat islands as a meeting place of people and species. I then discuss how Papuan evangelical Christianity, beliefs in non-human spirits and newly introduced conservation programs have become interlinked in ways that have adapted belief to support natural resource management and socio-ecological resilience.

Pacific Island Communities' Environmental Practices at the Juncture of Custom and Science

The designers, funders and implementers of environmental projects tend to make assumptions about which type of knowledge matters to beneficiary communities. In Asia and elsewhere, donors and governments tend to privilege quantifiable metrics over qualitative responses or knowledge practices of resource-dependent populations (Agrawal 2005). Similarly, nongovernmental organizations (NGOs), scientists and development agencies often promote technocratic solutions to improve the social development of indigenous groups (Li 2007; Dove 2011). Misunderstandings about human–environment relations have frustrated the development of holistic community-engaged responses to threats (Klöck and Nunn 2019; Nunn and McNamara 2019; Luetz and Havea 2018).

Anthropological research has documented how indigenous stewardship of ecological zones has supported biodiversity and the resilience of people who make a living from the sea or forest. Such practices typically entail the establishment of community-based spatial and temporal boundaries, the prescription of acceptable gear and catch restrictions to maintain valued biota in specific ecological habitats (Hviding 1996; Cinner and Aswani 2007), and the creation of forms of cooperative governance that prevent exploitation of common-pool resources (Berkes et al. 2000). People, clans, or factions tend to compete for control of forests, reefs and fishing grounds (Adhuri 2013; Thorburn 2000). Misunderstandings have occasionally led to conflicts over ownership and access rights to coastal fishing locations and marine harvest sites (Pollnac and Pomeroy 2005). Traditional environmental practices may also be focused on optimizing harvests for yearly celebrations rather than being driven by an intrinsic environmentalist interest for protecting turtles, sharks or rays (Foale and Manele 2004). Research also highlights that indigenous communities often compete over resources, with different interests (Brosius 1999; Sillitoe 1998; Nazarea 2006; Buggy and McNamara 2016).

Locals and outsiders with stakes in natural resources occasionally form alliances with NGOs or outside agencies to protect species or clash over the management plans for a specific site (Brosius 1999; Li 2007). Relatedly, a society's interactions with the surrounding environment—including beliefs about nonhuman spirits—can change in relation to outside pressures, incentives or forces (Jacka 2010). People rely on cultural values and beliefs to develop new mechanisms for coping with change (Sasaoka and Laumonier 2012). Adger et al. (2013: 114) note that "developing new narratives, alternative meanings or strategies to lead meaningful lives" can help societies adapt to change.

Framing Environmental Adaptations in the Raja Ampat, at the Juncture of Oceania and Southeast Asia

The Raja Ampat islands, an archipelago off the northwestern coast of Indonesia's West Papua Province, it is a zone with a long history of human settlement and contact (O'Connell and Allen 2015). The islands are also embedded in a unique biogeographical seascape of globally recognized high species diversity known as the Coral Triangle (Polhemus 2007). The major islands are Waigeo, Salawati, Batanta and Misool, along with many smaller islets of limestone, coral and atolls. Raja Ampat is also a political unit or Regency (*Kabupaten*), with seventeen subdistricts (*kecamatan*), with a combined population of 46,613 (2016) living in 89 villages—the largest town is Waisai.[1] In the villages they speak a Halmahera-West New Guinea language—Beser, Ambel, Ma'ya—alongside Papuan Malay (Kamholz 2014).

The people of Raja Ampat have long been at a crossroads of ocean currents, trade circuits and politics. Traders, pirates, European voyagers, explorers and sultans have interacted with resident populations in the Raja Ampat area since the sixteenth century (Andaya 1993; Ellen 2003).[2] Waves of settlers from islands to both the east and west have shaped the region's hybrid linguistic and social character. Goods and gifts have facilitated marriage alliances, slave raiding, inter-regional trade networks, global spice monopolies, and contests over access rights (Ploeg 2002: 89; Timmer 2011).

While there are several distinct ethnic communities living among the islands and bays of Raja Ampat, this study focuses on interactions with Ma'ya, Ambel and Beteo settlements in and around Waigeo Island.[3] The Ma'ya people are considered to be original inhabitants of Waigeo Island (Mansoben 1995; Leeden 1989). In the past, they were intermediaries in the exchange of natural resources for precious goods (and people) between Moluccan Sultans, Biak raiders and people from the New Guinea mainland. Today, they make their living from fishing, subsistence gardening and selling selected products. In every Ma'ya community, different clans (*gelet*) have rights to control access and use of particular land and sea zones (Mansoben 1995: 255). The Ambel (a Ma'ya subgroup that speaks a distinct language) live in several small villages along the shores of Waigeo's Ma'yalibit Bay, fishing, collecting sea products, and trading garden produce or sago. The Beteo are the most numerous ethnolinguistic group in the Raja Ampat region generally. They arrived from Biak at least 200 years ago on large canoes (*karures*), and established coconut reserves on outer islands (Ploeg 2002: 79). Once feared as pirates (*bajak laut*), the Beteo later

---

[1]https://rajaampatkab.bps.go.id/linkTableDinamis/view/id/44, accessed 20 April 2018.

[2]The name of Raja Ampat—the islands of the four (Indo. empat) Rajas—highlights a history of loose-knit relational governance known regionally as *sosolot*, "based on intense trade and migration among autonomous kin and trade friends" (Goodman 2006: 10).

[3]It is unclear whether the Ambel should be considered a separate people or a Ma'ya sub-group that speaks the Ambel language. For purposes of this chapter, Ambel refer to those people (whether self-identified as Ma'ya or not) who live among the seven predominantly Ambel-speaking settlements in Ma'yalibit Bay.

became important traders, employees of commercial fishing companies, and more recently, pioneers of village-based marine tourism.

While parts of Indonesia have been Christian since sixteenth century Jesuit voyages, along New Guinea's coastal zone a key moment occurred in 1855 when German missionaries Carl Ottow and Johann Geissler introduced the Gospel to people living near the modern city of Manokwari. Conversion signified a rupture from a mobile past towards settled village life. It hailed the end to tribal wars (*perang suku*) and roving witches. Many Papuans became defined as members of a congregation (*jemaat*), rather than through dispersed family units. Today, most residents identify with the Evangelical Christian Church (*Gereja Keristen Injili de Tanah Papua,* GKI), followed by the Roman Catholic and Evangelical Tabernacle Church (*Gereja Kemah Injil di Indonesia,* GKII) (Aritonang and Steenbrink 2008: 377–380).

The implementation of large-scale marine conservation programs in the late 2000s increased the frequency and variety of cross-cultural interactions, particularly between Papuan communities, conservation scientists, environmental NGOs and ecotourists. Following surveys that identified Raja Ampat as a global refuge for coral reef and forest habitats, the World Wildlife Fund for Nature (WWF), The Nature Conservancy (TNC) and Conservation International (CI) coordinated the establishment of twelve MPAs to safeguard the region's coral reef and mangrove habitats mining. Conservationists view Raja Ampat as a refuge for important marine and terrestrial species; a buffer against global degradation of tropical ecosystems from anthropogenic change (Atmodjo et al. 2019; Cox and Bright 2017) (Fig. 14.1).

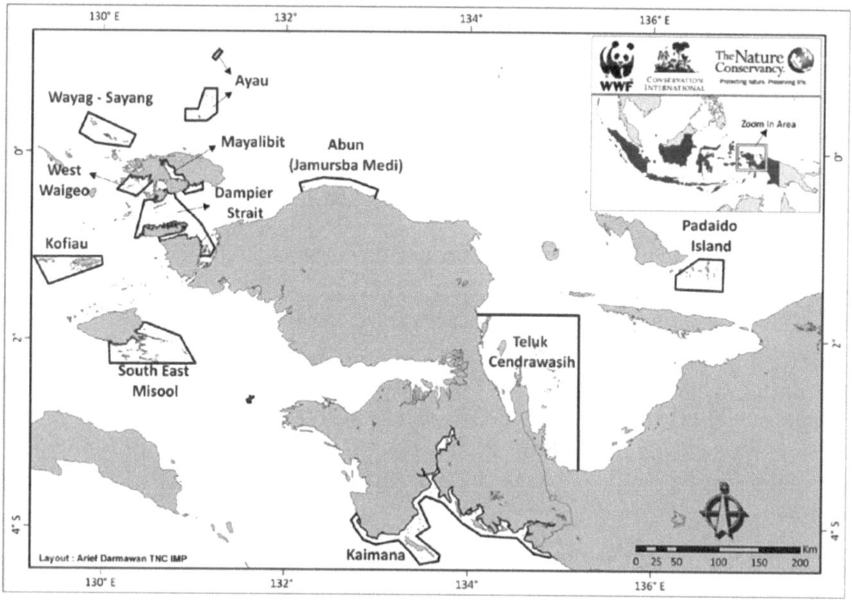

Fig. 14.1 Marine Protected Areas in Raja Ampat, adapted from Wilson et al. (2010)

Amidst these conflicting interests, many Papuans in Raja Ampat view environmental projects as a means to assert cultural rights to landownership. People have re-sacralized reef or forest areas by involving church leaders or cultural mediators who advise environmental organizations about customary rules and practices (*adat*) and places of spiritual significance, and then help resolve resource disputes. The Evangelical Church of West Papua has played a key role in the emergence of new forms of successful conservation: people have noted the return of manta rays to sandy shoals, larger fish aggregations and bigger sea cucumbers for sale abroad. Protection of reefscapes close to villages has in turn fostered the rise of an endogenous village-based tourism market, which has become an engine for economic development and a fulcrum of local politics.

In Raja Ampat, as with Melanesian societies in the Solomon Islands (see Hviding 1996; Leal Filho et al. 2020), clan-managed zones extend from land to sea and include marked fishing sites, bays and lagoons. First arrivals and families who established a garden, coconut grove, or house can claim rights to compensation for use of forest products, hunting, gardening, fishing or collecting marine animals. Through marriage alliance, the passage of time and collective recognition, use rights can eventually become ownership claims. These issues have become more important as the spread of tourism developments has led to several inter-village conflicts.

Study Sites and Methods

This study is based on three research trips to the Raja Ampat archipelago of 12 months duration between July 2012 and October 2015. The main research populations include Beteo and Ambel-speaking West Papuan informants and kin-networks on Waigeo, Gam, Mansuar Besar and Kri Islands. Other informants included Papuan, Indonesian and non-local staff of CI, WWF and TNC as well as the Flora and Fauna International and Yayasan Nazaret Papua, a local Christian charity who advises conservation groups.

Participant observation related to conservation practices, beliefs, values about the nonhuman world and ecotourism, fisheries surveys, and structured interviews, in Bahasa Indonesia and Papuan Malay forms the basis of this study (Creswell and Creswell 2017). Data collected includes two dozen key informant and life history interviews, conversations and recorded talks with 200 cultural intermediaries, 12 group discussions, taxonomies of 300 important locally-named species, a mixed-methods protocol for small-scale fisheries research (n = 95); oral transcription of narratives about resource conflicts, and clan histories; and notes from conservation-focused workshops.[4]

[4]Research Incorporated Institutional Review Board (IRB) protections for human research via University of Califoria San Diego IRB (Project 130737S) that included three primary consent regimes and two corollary consent procedures for audio and photo recordings. Research in Indonesia

Fig. 14.2 Land-based *sasi* for coconuts, Kri Island, 2012

Christianity and Ancestral Beliefs as Hybrid Environmental Practices in Raja Ampat

Throughout the Raja Ampat island zone, different communities have specific rules for overseeing culturally important species of animals and plants, whether gathered for subsistence, shared at feasts, or sold to markets. Fallow gardens, clan managed land and sea areas, and ritually closed zones support the capacity to adapt to unpredictable rains and rising populations. In recent years, Beteo and Ma'ya people have institutionalized hybrid forms of conservation in cooperation with international NGOs. Environmental protection includes no-take marine protected areas, village-based customary rules, restricted zones of nonhuman spirits, and Christian or Islamic prohibitions referred to as *sasi*. *Sasi* is a type of time-bound prohibition (or taboo) that restricts harvesting forest products or marine animals (Fig. 14.2). They are social-environmental practices that communicate ethical values about peoples' ideal relations with their natural surrounds amidst a context of socio-political transformation, as well as a set of marine conservation practices calibrated to different habitats (e.g. mangrove, reef slope, shoal, open sea) and target species.[5] Here I contrast two forms

was approved by the Indonesian government's ministry of research (RISTEK) who facilitated local residency status (KITAS) and travel papers(*surat jalan*) to West Papua.

[5]Equivalent terms for 'sasi' in Indonesian and Matlol are *tempat keramat* (sacred place), *zona inti* (a forbidden zone), *daerah yang dilarang* (a prohibited area), *tempat pamali* (hallowed site), *kabuis* (Matlol, spirit) and *saum samom* (Matlol, protected area).

of *sasi,* their origins in clan history, their connections to external conservation efforts and recent successes.

Near Waigeo, *sasi* takes the form of spatial boundaries and seasonal restrictions on harvesting marine animals such as sea cucumbers and lobster, sago or coconut (Hviding 1996; Thorburn 2000). Some ritually closed zones aim to ensure adequate supplies of prestige foods, commercial shell (*Tectus niloticus*) for sale abroad, while other areas are for subsistence fishing or gathering. Most harvest closures follow a seasonal pattern of winds, tides or ceremonial events. These restrictions also prohibit the use of compressors, commercial-scale nets and large vessels while permitting wooden spears, hand nets, paddle canoes and diving glasses. *Sasi* also incorporates specific norms for engaging in appropriate ethical action to protected areas or species. Breaking the rules will lead to sanctions, illness or death from divine retribution or enraged spirits. "Other penalties can be imposed on *sasi* violators such as reprimands, monetary fines, confiscation of fishing gear and boats, and exile from the village" (Boli et al. 2014: 137). Fear of social ostracism, beliefs about *sasi* as a moral practice, and goals for improved biological targets have contributed to its effectiveness.

In addition to Christian-oriented conservation, sacred sites throughout Ma'yalibit Bay on Waigeo Island are associated with the protection of nonhuman domains. These places, named for the spirit entities (*mon*) or ancestors who inhabit them, are different from Church *sasi,* but are important buffer zones that support several threatened species and habitats. Among the Ambel, each kin network (*gelet*) is responsible for managing ecological resources as well as nonhuman forces within its domain. People perceive the landscape as places of stories, sacred sites (Indo: *pamali*), guardian spirits (Ambel: *mútum*), malevolent forces (Ambel: *kábyo*) and cannibal witches (*suanggi*). A Ma'ya conservation adviser said that *mon* sites are "good in terms of ecology, because people never go [there], so there are more animals and fish there. Forest people leave those places alone." If not, nature responds. For example, if a person enters Kabui Bay near Kaliraja—a place the Ma'ya consider their origin source—without giving adequate respect, "great winds will blow anyone who arrives off course."[6]

The various types of ecological stewardship currently practiced in Raja Ampat can be viewed as examples of traditional ecological knowledge: locally meaningful practices of environmental management maintained through stories and authority figures and memory (Berkes et al. 2000). Yet if they are tokens of indigenous ecologies, they also represent reinventions or adaptations of socio-ecological knowledge practices rather than examples of practices that are distinct from larger-scale initiatives overseen by NGOs or state agencies. And just like these other endeavors, Raja Ampat conservation practices involve politics and conflict over land rights claims, access and control. I the sections below, I consider in more detail the two sorts of *sasi: sasi gereja* and *sasi mon.*

[6]Conversation with Yohanes Gaman, Yayasan Nazaret, Sorong, West Papua, August, 2012.

Conservation as a Faith-Based Adaptation to Change

In the 2000s, marine-oriented Church *sasi* zones expanded near Waigeo island. Originally a way to mark valued fruiting or palm trees or garden sites, they later became a means to protect egg-laying fish aggregating during spawning. Consultants from Jakarta and elsewhere held workshops in different villages as part of a World Bank coral reef management initiative (COREMAP). Over several meetings, Beteo-speaking villages near Waigeo island agreed to support the banning of trawling nets (*jaring harimau*), curtail shark fishing and to restrict hunting sea turtles. West Papuan churches became involved by linking conservation to a claim that care for the environment provided a vehicle for redemption. Soon afterwards, CI and TNC began working with Papuan church-affiliated advisors who promoted *sasi gereja* alongside larger-scale marine zoning projects. In extending to the sea over the past decade, such practices have also broadened from family-managed to village-level oversight.

In Raja Ampat today, Church *sasi* has become a measure of West Papuan virtue as well as communal shortcomings. To the Beteo people of Deer village off Kofiau island, *sasi gereja* signals that Raja Ampat's marine realm is an integral part of people's spiritual lives. People speak affectionately about manta rays, nudibranchs, and reef sharks. Young men mourn seeing blast craters of dead elkhorn coral. On Waigeo island, moral responsibility for the land and sea is connected to the desire to abide by Christian values that are important to people's sense of identity in a Muslim-majority country.

In survey responses and group discussions, tribal advisers, congregants and conservation staff commented how *sasi gereja* has become a way to act rightly towards nature and among others. One informant said, "It is easy to organize people through the church…Everyone prays to stop people from harvesting sea cucumber…in certain places they say, 'leave it there.'" At a church in Saporkren village (Fig. 14.3), the priest asked congregants to thank Jesus for providing abundant fish and bountiful gardens. On Manyaifun island another pastor viewed *sasi* as an integral practice to his church as well as a means to ensure a community's livelihood. "We bring [offerings] to the church so that the Lord protects them; so that the Lord guards our things…we are grateful for the results that he provides us." On Rutum Island on the Ayau atoll north of Waigeo near Palau in November 2014, a marine ranger explained that *sasi* is "primarily for protection of the marine environment and particularly to ensure that species that have decreased will be maintained for future generations." He said, "We respect *sasi* because of our faith in God." He distinguished such practices from conservation-sponsored management, which he says, "is a government thing."

When asked about the differences between village regulations and *sasi gereja*, an advisor at Misool's Nature Conservancy office said that *sasi gereja* is considered to be the most effective because people are afraid to enter closed *sasi gereja* zones: they are fearful of punishment (*kutuk*) from God or nature spirits. At a fisheries discussion in the nearby settlement of Go, participants remarked that God would have animals—including crocodiles, wild boar, and certain venomous snakes—act

Fig. 14.3 Church service, Saporkren village, Waigeo, 2015

to punish greedy people from taking too much without sharing. A man in Saporkren village recalled that in 2015, crocodiles attacked four men who had trespassed into *sasi gereja* zones that were under protection. Winds and currents could also draw someone off course, or they could drown. To them, *sasi* is a divine law (*hukum dari Tuhan*): "Someone who enters and violates a *sasi* place will be subject to punishment from God. The person will become sick or he will be a victim from nature—bitten by a fish, or a tree can fall on them." An elder on Ayau atoll said that if a fisherman steals, his wife's body would become barren. By taking from a place whose abundance is a sign of divine grace, God would 'steal' his wife's life-giving fertility. Only until he confesses to a member of the clan with land and sea rights will his wife be capable of giving birth. According to a Beteo conversation advisor on Ayau, "Only people of bad character would take from a *sasi* place, in contrast with the majority who desire goodness."

To protect oneself from future retribution after trespassing into a protected *sasi* zone, a person must first apologize to the landowning clans or kin groups who have management oversight (*pemilik*). Most then pay a fine (*denda*) to the owners or church congregation of up to 10 million Indonesian Rupiah (about 700 USD), a percentage of future catch, or fishing gear (Boli et al. 2014: 134.)[7] Others give gifts such as processed sago flour as compensation. Taken together, these excerpts underscore how *sasi gereja* has become an effective mechanism to limit degradation, promote surveillance, and communicate moral values.

Sasi gereja provides for a set of moral prescriptions that aims to promote a sense of mutual responsibility and cooperative sharing in a fluid seascape. As one priest argued, "Rascals who use chainsaws to cut valuable trees or steal fish reflect the temptations of greed." Those who transgress are punished for negating the moral code of social solidarity that underpins Christian practice against a backdrop of cultural erasure and economic uncertainty. By engaging with *sasi* alongside new environmental practices, Papuans in Raja Ampat communicate moral claims about the value of communal management of natural resources amidst increasing tension over the spread of private tourism developments.

And the *sasi gereja* are effective: since closing areas off to large-scale fishing and restricting activities to occasional harvests, Ayau's lagoon has seen greater abundance: in August 2014, grouper gathered to spawn in huge aggregations. Lobsters are bigger and sea cucumber sales have increased. Several men expressed hope that the fishing catch will increase even more, from five to eight tons per month. Alongside village regulations *(peraturan kampung)* and continued government support, they are optimistic about the future.

Nonhuman Zones as Unintended Ecological Refuges

Among the ten Ambel villages of Ma'yalibit Bay, a different type of local conservation called *sasi mon* restricts the building of houses, making of gardens, or fishing in areas where ancestral spirits or agents of the forest dwell. *Mon* places include caves, natural springs, uninhabited karst islands, inaccessible mountains, riparian zones and areas of primary rainforest. They also include burial sites of ancient dead, sites of remembered events and clan origins, or places associated with cannibal witches.

Each Ambel village has specific areas that are taboo (off limits) because of associations with spirit beings or malevolent forces. Ambel believe that in the past people lived alongside trans-human beings such as heroic giants *(mambri)*, shape-shifting entities and sorcerers (*mawi*). Beyond the village space are forest and sea areas, and higher up on mountain peaks loom the realm of dragons (*kórben*). These beliefs underscore the ways human connections with the natural world are expressed as intimate social relations. Even today, the Ambel believe in nonhuman realms marked by boundary stones, shells or rivers. Entering into these places requires caution, and a guide from the right *gelet* to prevent being eaten by cannibalistic witches (*kábyo*—Ambel, *suanggi*—Indo.) or bitten by a poisonous animal. Intermediaries (*sadaká*)

[7]Conversation with Ali Oherenan, The Nature Conservency, Harapan Jaya, Misool, 2 December 2014; talk with Niko Sauyai, Sawinggrai, Gam island, 26 January 2015. Monetary fines vary but are meant to be a powerful deterrent.

give offerings (*kakes*) of food or gifts before embarking on any treks into Waigeo's spirit domains (Arnold 2018: 648).

Christian conversion in the 1930s diminished the power of the *mambri* and *mawi*, ensured that the *suanggi* no longer walked as men during the day, and promoted interclan unity. Ambel began living in villages near the shore with increased trade and travel, and aligned with Indonesian government settlement plans, but still maintained bush houses to avoid gossip and to tend to their inland food gardens. But Christian conversion did not significantly change pre-existing beliefs about the nonhuman realm. Almost all Ambel speak about *mon* spirits who inhabit primary forest, caves, natural springs or trees. *Mon* are generally tolerant but sometimes annoyed with human meddling or inappropriate talk. They are said to cause blindness, crop loss or drought, illness or animal attack if their area is disrespected. The spatial topography of *mon* zones reflects a belief that human–environment relations are inherently unstable: people's intentions are opaque, and the surrounding forest is a place of affordance, but also of danger.

And yet zones of nonhuman entities are associated with increased biodiversity and species richness. In the sea, they encompass habitats of the Indonesian speckled carpetshark (*Hemiscyllium freycineti*), juvenile blacktip reef sharks, eels, mangroves and saltwater crocodiles. On land, *mon* areas include New Guinea lowland forest mosaics with stands of threatened Indonesian eaglewood (*Aquilaria Sp.)*, Moluccan ironwood (*Intsia spp.*) and sub-montane forest stands on karst peaks. *Sasi mon* areas appear to have positive ecological spillover effects for certain habitats and species, even if this was not the intended goal. In this sense it aligns with forms of species and habitat prohibitions such as *tapu* (Tonga) and *fady* (Madagascar) (Colding and Folke 2001; Cinner 2007).

Across the landscape there is a persistent belief among Ambel people that nature has power in *mon* areas similar to how Beteo perceive that God has power in *sasi gereja* zones. *Mon* sites, as protected zones, are not only nature preserves but are anthropomorphized sites of collective memory that communicate their views about appropriate conduct with the environment to others. For example, behind Kalitoko village, the Dawa and Nok families speak about a place along the Kakit river which is a repository of stories and a refuge for nonhuman entities. Despite the lifegiving and sustaining properties associated with this place, Kakit is also a place of warning. In one story, two sisters, Lorena and Dubul, became blind when spirits became angry after they ate a pig hunted from the wrong side of the river (and without permission). People from Warimak say the sisters remain in large boulders by the river, reminding others to pay proper respects to the forest spirits.

Conservation groups have supported the expansion of *mon* zones. In August 2012, CI, the Raja Ampat government and members of the Ma'ya customary council (*dewan adat*) declared a *sasi mon* area near Lopintol village (Fig. 14.4). Participants led a ritual procession to the sea where they left food offerings, and placed tree branches, clove cigarettes, areca nut, and yellow rice in secret places. Clan elders chanted special prayers and expressed their desire to safeguard their lands.

Fig. 14.4 Sign marking *sasi mon* zone near Warsambin village, Ma'yalibit Bay, 2012

Linked Resource Management in Raja Ampat as a Buffer to Increased Threats

Sasi gereja and *mon* are two different but analogous types of resource management that have become intertwined with large-scale environmental projects such as the Coral Triangle Initiative on Coral Reefs, Fisheries and Food Security. Engagements with strangers have shaped new behaviors—in the past few years fishermen in Waigeo's Dampier Straits marine protected area have stopped killing sharks for export to Chinese markets, following meetings with conservation scientists who relayed their importance as keystone species, and new rules banning shark finning. Beteo residents living in villages in and near these protected zones increasingly value manta rays and nudibranch species for attracting divers. While there are still disagreements between kin networks about the long-term benefits of working with governmental institutions and NGOs, many residents say that attention from outsiders has helped to sooth rivalries over fishing rights and ensure better harvests.

While development programs routinely failed to address underlying socio-economic disparities, people said that linked conservation projects have facilitated trust between Waigeo's heterogeneous social groups, and even among families. For instance, in the Muslim Lopintol village nearby, cooperation to manage local fisheries has led to increased numbers of Spanish mackerel (*Rastrelliger brachysoma, R. kanagunta*), the village's main economic livelihood. Residents on Arborek Island view different types of conservation areas working together to support fishing and tourism. Today, Beteo people at Arborek village live within four conservation zones: a

Fig. 14.5 Preparing to sell fish at Waisai market, Saporkren, August 2015

no-take zone near a manta aggregation site; *sasi gereja* areas; an area with village gear restrictions; and the encompassing Dampier Straits MPA NGO-sponsored community marine patrols have been critical to the return of reef sharks, and have reduced fishing raids from gangsters near Sorong city.

While increasing numbers of tourist visitors have created some tensions over trash, fishermen reported that they now catch much closer to shore than ten years ago (Fig. 14.5). Funds from larger *sasi* harvest revenues have also supported church construction and holiday feasts. In February 2017, reflecting increased cooperation between Beteo residents and conservationists, nine residents from local villages in the area were recognized as manta guardians to protect pregnant females from harassment by over-eager tourists at a mating and feeding site on a shoal near Arborek village. Such actions are also designed to reinforce institutional mechanisms to continue protecting important species.

Sasi and no-take marine protected areas draw their moral foundations from different beliefs and values but together have formed a lattice of norms, institutional practices, and effective controls for supporting key habitats in several island zones across Raja Ampat. This is notable because conservationists and locals reported different motivations for supporting the same environmental protections: conservation biologists and NGOs promoted the intrinsic value of biodiversity or specific-based protections, while many Beteo and Ma'ya communities cooperated with environmental programs as a means to greater cultural recognition for land rights, or to support better harvests or fisheries landings, or to attract foreign eco-tourists to village homestays and dive at nearby reefs.

West Papuans across island settlements of Raja Ampat think about their role in protecting or making use of their nonhuman surrounds in diverse ways. They take shape as a moral grammar of religious edicts with *sasi gereja*, an ethical stance of 'do no harm' when interacting with manta rays in tourist zones, a guarded caution towards *suanggi* caves, a reverent attitude towards ancestors in *mon* areas, and an economic view toward future profitability in garden plots. According to a Ma'ya elder and conservation advisor (see Fig. 14.6):

> We are very much in tune with the NGOs [to] safeguard natural resources. Because from the perspective of the Ma'ya *adat* council, these resources need to be managed well…So there are areas such as 'no take zones' [that] must exist for the benefit of indigenous peoples, and to ensure the sustainability of fisheries. This can't be accomplished alone. We cannot walk alone – [we must work] with NGOs and the community.

He identified a congruence between traditional spirit practices and modern conservation goals: "We are talking to folks again using the language of conservation

Fig. 14.6 Ma'ya elder Kris Thebu (in headdress, right) oversees a new conservation agreement 2018

– conservation is indeed the same as sacred places. This means that conservation is regulated in similar ways." Other informants believe that West Papuans must lead future efforts to support socio-ecological resilience. A Beteo leader acknowledges that NGOs have provided some benefits, but he argues that "they don't know what conservation is." He says conservation must first arise from West Papuan communities, who have long safeguarded their resources, "without *sasi*, without Conservation International rules, without The Nature Conservancy conservation rules, without the rules of the government." Another Papuan advisor believes that government, custom (*adat*) and religion should be involved "[b]ecause all three have the power to make decisions, to protect the people and natural resources." He harbors hopes for future political freedom, but also for a rejuvenated indigenous governance through the *Ma'ya* Traditional Council (*Dewan Adat*).

Many West Papuan children now speak about *konservasi* at home with their parents following visits by educators who lead environment-focused workshops. For many Beteo, *sasi gereja* has become a mechanism for enacting Christian virtues of communal harmony while showcasing their conservation values. Among the Ma'ya, *sasi mon* is a type of spiritual wilderness ethic for respecting places where nature spirits dwell, as well as a way to map clan-managed territory at a time of increased competition over resources.

Local forms of conservation have also become a political-economic lever: they support claims by Biak and Ma'ya villages for official Indonesian government recognition and channel claims for increased material support from conservation organizations to villages. Relatedly, in Ma'yalibit Bay, *sasi mon* areas have become important sites for negotiating access and control of valued patches of primary forest or fish aggregation sites. Sandra Pannell (1997: 298) suggests that *sasi* practices are relevant not because of their potential fit with contemporary marine management regimes, but because they express local understandings about what is sacred, about ownership and about spatial differentiation. To Nils Bubandt (2005), *sasi* is a dialectic between local cultural values and outside political forces. Such practices are not merely a practical method of ensuring bountiful harvests but also a mechanism for ensuring the continuity of values about people and nature.

By operationalizing belief and ethical norms, West Papuan environmental practices express a desire to have greater leverage in negotiating over land rights. They hope for greater partnership with environmental NGOs and outsiders. By representing *sasi gereja* and *mon* prohibitions as analogous types of land and sea-based conservation, Ma'ya and Beteo residents have forged new alliances to confront looming threats.

West Papuans and other Pacific island communities are navigating increased threats from overfishing, ecological shocks and habitat fragmentation. As with other island peoples, they confront species loss, unpredictable rains and ocean acidification. The sardines and anchovy are gone, but green turtles and manta rays have returned. Increased migration of traders from Java and Sulawesi islands has created new problems with trash, decreased water quality, and competition for formal jobs. While tourism seems to be mitigating commercial exploitation of reef ecosystems, nickel mining, road construction, and illegal logging is transforming the terrestrial landscape, with negative consequences from coastal runoff. Awareness of such hazards have motivated Beteo and Ma'ya communities to link with outsiders to promote environmental protection. For now, *sasi* and other conservation regimes provide ways to limit resource degradation, as well as an institutional structure for transmitting cultural values about people and the places they cherish.

Acknowledgements The author would like to thank Dedi Adhuri, Hamid Toha and Tony Rudyansjah for logistical and research support. In Raja Ampat, Reuben Sauyai, Yohanes Goram, Abraham Gaman, Kris Thebu, Yudas Sauyai, Enggelina Dimara, Yeskel Dawa, Wolter Gaman, Martin Makusi, Anache Goram, Angela Beer, Salman Wiyai and Mety Mongdong provided critical insights and connections. Research was supported by Fulbright-Hays Doctoral Dissertation Research Abroad (P022A1300-(20)-007), National Science Foundation Integrated Graduate Education and Research Training (IGERT) Fellowship, "Global Change, Marine Ecosystems, and Society" (Award number 0903551), Waitt Foundation, Scripps Institution of Oceanography, United States-Indonesia Society, and UC San Diego Department of Anthropology.

References

Adger W, Barnett J et al (2013) Cultural dimensions of climate change impacts and adaptation. Nature Clim Change 3:112–117. https://doi.org/10.1038/nclimate1666

Adhuri DS (2013) Selling the sea: a study of conflict over marine tenure in Kei Islands, Eastern Indonesia. ANU Press, Canberra

Agrawal A (2005) Environmentality: community, intimate government, and the making of environmental subjects in Kumaon, India. Curr Anthropol 46(2):161–190. https://doi.org/10.1086/427122

Andaya L (1993) The world of Maluku: Eastern Indonesia in the early modern period. University of Hawaii Press, Honolulu

Aritonang JS, Steenbrink, KA (eds) (2008) A history of Christianity in Indonesia. Studies in Christian Mission, vol. 35. Brill, Leiden

Arnold LM (2018) Grammar of Ambel, an Austronesian language of Raja Ampat, west New Guinea. Dissertation, University of Edinburgh

Atmodjo E, Lamers M, Mol APJ (2019) Governing dynamics in marine conservation tourism in Raja Ampat, Indonesia. Tour Plan & Dev. https://doi.org/10.1080/21568316.2019.1686652

Awuah-Nyamekye S (2019) Climate change and indigenous Akan religio-cultural practices lessons for policy-makers and implementers in environmental conservation in Ghana. Worldviews-Glob Relig Cult Ecol 23(1):59–86

Berkes F, Colding J, Folke C (2000) Rediscovery of traditional ecological knowledge as adaptive management. Ecol Appl 10(5):1251–1262

Berkes F, Folke C (2003) Navigating social-ecological systems: building resilience for complexity and change. Cambridge University Press, Cambridge

Boli P, Yulianda F et al (2014) Benefits of sasi for conservation of marine resources in Raja Ampat, Papua. J Manaj Hutan Trop 20(2):131–139

Brosius JP (1999) Analyses and interventions: anthropological engagements with environmentalism. Curr Anthropol 40(3):277–310. https://doi.org/10.1086/200019

Bubandt N (2005) On the genealogy of Sasi: the transformations of an imagined tradition in Eastern Indonesia. In: Otto T, Pedersen P (eds) Tradition and agency: tracing cultural continuity and invention. Aarhus Universitetsforlag, pp 193–232

Buggy L, McNamara KE (2016) The need to reinterpret "community" for climate change adaptation: a case study of Pele Island, Vanuatu. Clim Dev 8(3):270–280. https://doi.org/10.1080/17565529.2015.1041445

Chan KM, Balvanera P et al (2016) Opinion: why protect nature? Rethinking values and the environment. Proc Natl Acad Sci 113(6):1462–1465

Cinner JE (2007) The role of taboos in conserving coastal resources in Madagascar. SPC Trad Mar Res Manage Knowl Inf Bull 22:15–23

Cinner J, Aswani S (2007) Integrating customary management into marine conservation. Biol Conserv 140(3):201–216

Colding J, Folke C (2001) Social taboos: "invisible" systems of local resource management and biological conservation. Ecol Appl 11(2):584–600

Cox K, Bright J (2017) Raja Ampat: a biodiversity hot spot and the future of marine conservation. Fish 42(9):462–467

Creswell J, Creswell D (2017) Research design: qualitative, quantitative, and mixed methods approaches. Sage Publications, Thousand Oaks, CA

Dove M (2011) The banana tree at the gate: a history of marginal peoples and global markets in Borneo. Yale University Press, New Haven

Ellen RF (2003) On the edge of the Banda zone: past and present in the social organization of a Moluccan trading network. University of Hawaii Press, Honolulu

Foale S, Manele B (2004) Social and political barriers to the use of marine protected areas for conservation and fishery management in Melanesia. Asia Pac Viewp 45(3):373–386

Goodman TE (2006) The sosolot: an eighteenth century East Indonesian trade network. Dissertation, University of Hawaii at Manoa

Hviding E (1996) Guardians of Marovo Lagoon: practice, place, and politics in maritime Melanesia, vol 14. University of Hawaii Press, Manoa

Jacka J (2010) The spirits of conservation: ecology, Christianity, and resource management in highlands Papua New Guinea. J Study Religion Nat Cult 4(1):24–47

Kamholz D (2014) Austronesians in Papua: diversification and change in South-Halmahera-West New Guinea. Dissertation, University of California, Berkeley

Klöck C, Nunn PD (2019) Adaptation to climate change in Small Island Developing States: a systematic literature review of academic research. J Environ Dev 28(2):196–218. https://doi.org/10.1177/1070496519835895

Leal Filho W (2020) Climate change adaptation as a development challenge to small Island states: a case study from the Solomon Islands. Environ Sci Policy 107:179–187. https://doi.org/10.1016/j.envsci.2020.03.008

Leeden A (1989) The Myth of Wawage. In: Haenen P, Pouwer J (eds) Peoples on the move: current themes of anthropological research in New Guinea. Centre for Australian and Oceanic Studies, University of Nijmegen, Nijmegen, pp 102–136

Li TM (2007) The will to improve: governmentality, development, and the practice of politics. Duke University Press, Durham

Luetz JM, Havea PH (2018) "We're not refugees, we'll stay here until we die!"—climate change adaptation and migration experiences gathered from the Tulun and Nissan Atolls of Bougainville, Papua New Guinea. In: Leal Filho W (ed) Climate change impacts and adaptation strategies for coastal communities. Springer Nature, Cham, pp 3–29

Luetz JM, Nunn PD (2020) Climate change adaptation in the Pacific Islands: a review of faith-engaged approaches and opportunities. In: Leal Filho W (ed) Managing climate change adaptation in the Pacific region. Climate Change Management. Springer, Cham, pp 293–311. https://doi.org/10.1007/978-3-030-40552-6_15

Mansoben JR (1995) Sistem Politik Tradisional di Irian Jaya, Indonesia: Studi Perbandingan [Traditional political systems in Irian Jaya, Indonesia: a comparative study]. Leiden University and the Indonesian Institute of Sciences, Leiden and Jakarta

McLeod E, Szuster B, Salm R (2009) Sasi and marine conservation in Raja Ampat, Indonesia. Coast Manag 37(6):656–676. https://doi.org/10.1080/08920750903244143

McMillen HL, Ticktin T et al (2014) Small islands, valuable insights: systems of customary resource use and resilience to climate change in the Pacific. Ecol Soc 19(4):44. https://doi.org/10.5751/ES-06937-190444
Murphy C, Tembo M et al (2016) Adapting to climate change in shifting landscapes of belief. Climatic Change 134:101–114. https://doi.org/10.1007/s10584-015-1498-8
Nazarea VD (2006) Local knowledge and memory in biodiversity conservation. Annu Rev Anthropol 35:317–335. https://doi.org/10.1146/annurev.anthro.35.081705.123252
Nunn PD, McNamara KE (2019) Failing adaptation in island contexts: the growing need for transformational change. In: Klock C, Fink M (eds) Dealing with climate change on small islands: towards effective and sustainable adaptation. Gottingen University Press, Gottingen, pp 19–44
O'Connell JF, Allen J (2015) The process, biotic impact, and global implications of the human colonization of Sahul about 47,000 years ago. J Archaeol Sci 56:73–84. https://doi.org/10.1016/j.jas.2015.02.020
Pannell S (1997) Managing the discourse of resource management: The case of sasi from 'Southeast' Maluku, Indonesia. Oceania 67(4):289–307
Polhemus D (2007) Tectonic geology of Papua. In: Marshall AJ, Beehler BM (eds) The ecology of Papua. Part One. Periplus Editions, Singapore, pp 137–164
Pollnac RB, Pomeroy RS (2005) Factors influencing the sustainability of integrated coastal management projects in the Philippines and Indonesia. Ocean & Coast Manag 48(3–6):233–251
Ploeg A (2002) De Papoea: what's in a name? Asia Pac J Anthropol 3(1):75–101
Sasaoka M, Laumonier, Y (2012) Suitability of local resource management practices based on supernatural enforcement mechanisms in the local social-cultural context. Ecol Soc 17(4)
Scoville-Simonds M (2018) Climate, the Earth, and God—entangled narratives of cultural and climatic change in the Peruvian Andes. World Dev 110:345–359
Sillitoe P (1998) The development of indigenous knowledge: a new applied anthropology. Curr Anthropol 39(2):223–252
Tarusarira J (2017) African religion, climate change, and knowledge systems. Ecumenical Review 69(3):398–410
Thorburn CC (2000) Changing customary marine resource management practice and institutions: the case of Sasi Lola in the Kei Islands, Indonesia. World Dev 28(8):1461–1479. https://doi.org/10.1016/S0305-750X(00)00039-5
Timmer J (2011) Cloths of civilisation: Kain Timur in the bird's head of West Papua. Asia Pac J Anthropol 12(4):383–401. https://doi.org/10.1080/14442213.2011.587020
Wilson J, Rhodes K, Rotinsulu C (2010) Aggregation fishing and local management within a marine protected area in Indonesia. SPC Live Reef Fish Inf Bull 19:7–13

Part V
Belief and Action: A Way Forward?

Chapter 15
Pacific Climate Warriors and Local Narratives on Climate Change: An Analysis of a Faith-Informed Indigenous Rhetoric

Claudia Ledderucci

Abstract Climate change is probably among the most urgent issues today. Its growing relevance and presence in the media, together with the evidence of a critical unfolding situation, has led to the emergence of a wide array of movements contesting western policies and raising awareness among the public. Specifically, it is evident that climate change and related sea-level rise are threatening some of the Pacific Island States and their livelihood. Several local associations are strongly objecting to the dominant view which asserts that Pacific Islanders are victims waiting for the imminent catastrophe. By using campaigns, actions and call for prayers they raise awareness encouraging participation and resilience. In this chapter, the focus is on the narrative of one of these associations, namely the Pacific Climate Warriors. On one hand, the narrative is heavily based on the dominant rhetoric about climate change demanding a better future and a safe planet. On the other hand, it calls upon the concepts of connection, tradition, and faith to contest the reductionist view that Pacific Islanders are hopeless victims.

Introduction

Climate change is one of the most pressing issues of our time. The effects of global warming such as ice melting, sea-level rising, flooding, drought, and desertification are already observable. Yet, if the current model of energy exploitation is not changed, they might soon unfold into catastrophic scenarios. According to the Intergovernmental Panel on Climate Change (IPCC):

> Climate change refers to a change in the state of the *climate* that can be identified (e.g., by using statistical tests) by changes in the mean and/or the variability of its properties and that persists for an extended period, typically decades or longer. Climate change may be due to natural internal processes or external *forcings* such as modulations of the solar

C. Ledderucci (✉)
Dipartimento di Culture, Politica e Società, Università degli Studi di Torino, Lungo Dora Siena, 100, 10153 Turin, Italy
e-mail: claudia.ledderucci@unito.it

J. M. Luetz and P. D. Nunn (eds.), *Beyond Belief*, Climate Change Management,
https://doi.org/10.1007/978-3-030-67602-5_15

> cycles, volcanic eruptions and persistent *anthropogenic* changes in the composition of the atmosphere or in the land use [...]. (IPCC 2018, p. 544)

Alongside with the scientific definition and findings related to the phenomenon, other aspects, like cultural and local perception and faith-related and symbolic issues, need to be included in the discourse as they significantly impact a wider understanding of climate change (Crate 2011; Hulme 2009).

Firstly, this chapter aims to highlight the influence of the political debate and the historical context on the indigenous narratives posed by a particular grassroots movement, the Pacific Climate Warriors. Such influence can work at various levels and plays a role in guiding both the approach and the findings of the scientific community, as different scholars have shown (Demeritt 2001, 2006; Jasanoff 2010; Lahsen 2010; Latour 2010; Latour and Woolgar 1979; Miller 2004; Sarewitz 2000).

Secondly, the chapter aims to explain the singular way through which the spokespeople of this movement reject the dominant rhetoric and resist the 'doomed fate' that has been attributed to them (McNamara and Farbotko 2017).

The magnitude of the environmental issue in the Pacific Islands region has been investigated by numerous scholars (see Brunois [2005] for ethnoecological approaches or Trichet and Leblic [2008] for an extensive discussion about the environment). The projected anthropogenic changes caused by the emissions of greenhouse gases are likely to modify the delicate relationship humans maintain with nature, at the same time impacting the cultural values and local cosmologies of the indigenous communities (Crate 2011; Weir et al. 2017). Indeed, climate change is thought to be a real threat to the Islanders' livelihoods (Connell 2016; Nunn et al. 2020), as shown by the alarming results of the IPCC (Nurse and McLean 2014). During the last decade, the Pacific islands have become one of the global symbols of the environmental crisis in the eyes of the Western and eco-conscious public (Farbotko 2010a). This reality, as widely publicized by the mass media, increases the resonance of the problem, presenting such places as highly vulnerable while others get little attention (for a discussion on the uneven attention attributed to climate change see Orlove et al. 2014).

For a deeper understanding of the issue, I propose to examine the following questions: What are the understandings of the climate-change phenomenon by selected Pacific indigenous communities? Are the major narratives on climate change, such as those transmitted by the IPCC, compatible with the local ones? How can indigenous narratives enrich the scientific understanding of the environmental crisis? And finally, what is the role of faith in this dense interplay?

Such questions have been raised by other scholars. Among many, Lazrus (2012) posed interesting questions about island communities facing sea-level rise; Rudiak-Gould (2014) examined science communication and indigenous climate-change perceptions; Kempf and Hermann (2014) explored the importance of place-making, identity and mobility and how these categories may vary because of climate change; Farbotko (2005, 2010a, b) examined the construction of the category of 'climate refugees' and how this is used by Western media and NGOs to corroborate their narratives; Luetz and Havea (2018) analyzed the complex situation in the Carteret Islands;

Worliczek (2013, 2014) studied climate change and its understandings among the local communities in Rangiroa (French Polynesia) and Wallis (Wallis and Futuna), both French territories; lastly, Kempf (2012) analyzed climate-induced migration and the role of Christianity in Oceania.

Before answering these questions and investigating the Pacific Climate Warriors' rhetoric about environmental issues, different discourses, which consider climate change as both a scientific and a social object, will be briefly analyzed.

Climate Change: One Set of Facts, Multiple Values

Climate change was first conceptualized by the IPCC in 1990 and 1992 (IPCC 1992). The Panel stated in its report that starting from the Industrial Revolution, the growing emission of carbon dioxide in the atmosphere has led to a warming of the global climate. As observed by Latour and Woolgar (1979), the publication of a scientific paper—the IPCC First Assessment Report in this case—allows laboratory results to be accepted as objective facts. Hence, the publication of this report marked the birth of climate change as the familiar scientific object we know today. It is assumed to be true by the lay public because of the social and academic prestige attributed to it by the scientists who have authority on the matter, and its reinforcement by the media. As shown by Rudiak-Gould (2013), much of its credibility is the result of the invisibility of the phenomenon itself: only the scientists can reveal the reality of the changing Earth system and, as specialists of the matter, they can determine the degree of urgency it carries. Latour and Woolgar stress that "to say that [hypothesis] is constructed is not to deny its solidity as a fact. Rather, it is to emphasize how, where, and why it was created" (1979, p. 127).

Yet, the results published by the IPCC are based on Global Climate Models (GCMs)—scientific reproductions that can simulate climate trends during decades by using statistical models which have limitations and involve uncertainties. Such models approximate the climate system and are based on an imperfect understanding of complex interactions between the components of the earth-system (the sun, the atmosphere, the ocean, etc.). They are used by scientists to understand how human activities may alter the global climate and its associated natural and social processes. Even if GCMs are "central tools in areas of global climate science" (Lahsen 2005, p. 897) they nonetheless entail some critical points. An objective assessment of these models is very hard to obtain, especially because scientists often lack reliable and consistent global data against which to check them. Indeed, this is the reason why GCMs could be criticized for their reliability.

In addition to this, the socially constructed origin of climate change (Hulme 2009) suggests considering it more as a 'discourse' and not as the truth, as an ongoing issue and not as something given. The particular way we perceive and react to the scientific debate that highlights the dangers caused by climate change depends on the degree of trust we have in scientists and science, as well as on the cultural context with which we ascribe meaning and value to such a phenomenon (Roncoli et al. 2009).

Moreover, decision-makers' actions are influenced and embedded in socio-cultural and political values, as well as scientific contexts in which 'facts' are created (Lahsen 2010).

At this point, it is important to mention the role of mass media in the creation of climate change as one of the most discussed realities in the twenty-first century. The scientific issue of global warming was popularized by mass media during the mid-1980s, strengthening the idea that climate change is an undeniable scientific truth. Over the last decades, media coverage has shaped the public perception on the theme of climate change by utilizing what Boykoff and Boykoff (2007) identified as *first-order journalistic norms* exploiting personalization, dramatization and novelty. This has been possible because of an accurate choice of news coverage influencing an interplay between scientists, policy actors and the public. More recently, mass media played a key role in suggesting the controversy over climate change and its reality, promoting popular debate regarding the issue (Swain 2017).

Social sciences and humanities play a key role in this debate for two main reasons. Through a critical analysis of natural sciences, they can help the decision-making process to be more effective, and they can improve the knowledge about climate change and its cultural meaning. Thus, social scientists are well-positioned to explain consensus and controversy and their 'creation' (or construction). They can unravel the interplay between science, power, and politics, highlighting those particular processes that transform norms into habits (Latour and Woolgar 1979). They can, therefore, help find solutions to climate change-related political issues examining social and cultural representations. For this reason, Crate (2011) proposed a new methodology, namely that of 'climate ethnography'—that is a multi-sited, critical collaborative ethnography able to adopt cross-scale, multi-stakeholder, and interdisciplinary approaches into research and practice—to investigate and understand, on a local level, how indigenous peoples perceive and conceptualize the climate and its changing nature. On the other hand, Kempf and Hermann (2014) criticized this approach because of the supposed cultural essentialism[1] it entails. Moreover, Crook and Rudiak-Gould (2018) suggest the importance of taking into account indigenous ecologies and philosophies to better understand the reception of climate change in regions of the world like the Pacific Islands.

Finally, social scientists can interpret and observe the role of images and the role of the Internet, nowadays a fundamental means of communication strategically positioned to influence trust towards specific discursive constructions (Scott-Parker et al. 2017). The Internet and social media are a key component of many climate activists' campaigns. Indeed, one of the characteristics I first noted was their activity on the web. This was a privileged point of observation, explicitly used by the activists to explain their positions and views on climate change.

The data used in the article was collected between February and June 2018 in New Caledonia. In Nouméa, I met and interviewed Billy Cava, the coordinator of *350 Nouvelle-Calédonie*, the Caledonian branch of the movement *350 Pacific*. Following

[1]Cultural identity is not a static concept; rather, it is both historically created and the result of interpretation.

this, I moved to the island of Ouvéa to carry out a participant observation on the steps of the project conducted by Le Duff and Dumas on coastal erosion monitoring (Le Duff et al. 2017).

The methodology used in this study is that of discourse analysis combined with interviews with strategical respondents and activists to examine and understand indigenous strategies and narratives regarding climate change in the Pacific Islands region. It is also supported by bibliographical research in New Caledonia and by information gathered through online sources such as 350.org, the 350 Pacific official website, as well as Facebook and Instagram. Social activism and its peculiarities in Oceania are now analyzed, focusing attention on the international movement 350.org and its regional branch, 350 Pacific which gave birth to the Pacific Climate Warriors.

Social Movements and Indigenous Activism in Oceania

The political inaction in addressing climate change in the past few years has led to significant growth in grassroots involvement, especially amongst indigenous peoples. These movements have been acknowledged by different sectors, from governments to institutions and local associations and have acquired significant social relevance. Obtaining wider attention has enabled the different communities strongly impacted by the environmental change to strengthen their connection.

In the recent past, several newly founded associations have given their views about climate-driven environmental changes, drawing the attention of the public by using the media and social networks. New technologies are used as megaphones to make local stories heard by the public, especially in times of crisis, through what has been called digital storytelling (Cunsolo Willox et al. 2012; Otto 2017).

Instead of dissolving local peculiarities as feared by many, globalization has caused the emergence of grassroots 'social forms' that are questioning and contesting the reality they are facing through social mobilization. As stated by Appadurai (2000), this new dimension is called 'grassroots globalization', an idea that requires domestication of modernity and its languages. Such demands of political acknowledgements are daily claimed by grassroots movements which are strongly (but not exclusively) linked to climate justice and environmental welfare protests. As observed by Dhillon and Allooloo (2016) regarding the Canadian and American reality, "indigenous peoples are leading the struggle for environmental justice worldwide". These kinds of protests unravel the unequal power and impact that indigenous communities and multinational corporations have on the environment and highlight the discrepancy of views that these two actors have with regards to human domination over nature. The connection between nature and spirituality found in the tradition of local communities cuts to the heart of the diverse indigenous responses to climate change in regions like the Pacific Islands (Fair 2018; Schlehe 2010).

Social mobilization for the climate and the environment is invariably inspired by a political claim, the targets of protests being the fossil fuel industry and the production models that the current economic system reproduce, such as the centre-periphery relation at the core of colonialism and capitalism. As shown below, a good example comes from the international organization 350.org.

350.org

350.org is an international environmental movement founded in 2008 in the United States. Today its network extends across 188 countries. Among the numerous initiatives organized by the movement, the 'global days of action' lie at the core of the activists' campaign, consisting of protests linking individuals and organizations all over the world. Other fundamental actions are worldwide campaigns to promote awareness about climate change among people and sit-ins to block energy and industrial projects which use fossil fuels and are therefore a source of carbon pollution. The association aims to organize bottom-up protests and actions on a local scale, working with grassroots movements and communities which can influence the power and the decision-making processes of the fossil fuel industry. The importance given to transnational networks of grassroots organizations highlights the emergence of new local social forms that, as demonstrated by 350.org, present themselves as informal networks, founded by 'friends' (persons with shared values and aspirations) wishing to spread the environmental message and save the planet. Arguably, this kind of movement could be identified as a green ecologist imagined community,[2] that is a solidarity group whose members share the same values and ideas (Gamson 2004).

The style of the narrative employed by this organization highlights the focal points of their campaigns and the ways these are conducted. The emphasis here is put on the concept of environmental justice: a *safe* climate and a *better* future can and must be claimed by ordinary people. This discourse assumes even more importance when confronted with the perception of 'environment' and the idea of nature. These ideas are culturally and historically determined; they vary from place to place, differing from the universal definition proposed by 350.org. Although some of the statements on their website are vague and idealistic, for example: "help 350.org build a powerful climate movement. We believe in a safe climate and a better future - a just, prosperous, and equitable world built with the power of ordinary people. Help us get there!", they touch on ideas and themes regarding the environment that are recognized worldwide. Nonetheless, as shown below, activists from indigenous worlds in Oceania (and elsewhere) have appropriated this universal discourse, making it 'local' through the process of indigenization (Appadurai 1996).

[2]As defined by Anderson (1983, p. 49) referring to the nation-state, "[A community] is imagined because the members of even the smallest nation will never know most of their fellow-members, meet them, or even hear of them, yet in the minds of each lives the image of their communion."

350 Pacific and the Pacific Climate Warriors

The need to amplify the global movement's transnational connections found fertile ground in the Pacific Islands region, where local activists decided to join 350.org, founding a regional branch of the organization with the name of 350 Pacific, and referring to themselves as the Pacific Climate Warriors (see also McNamara and Farbotko 2017; Steiner 2015; Titifanue et al. 2017). This new movement has led popular fights against climate change, organizing and coordinating a lot of initiatives in Oceania.

The network is composed of young indigenous adults who want to create awareness about the vulnerabilities of Pacific islands among a growing number of people, both locally and internationally. At the same time, 350 Pacific highlights the power and resilience of local communities. In fact, to increase the mobilization indigenous protesters made an online call to all Pacific Warriors:

> Every morning, we wake up and the ocean is there, surrounding our islands. But now the ocean, driven by climate change is creeping ever closer. Unless something changes, many of our Pacific Islands face losing everything to sea-level rise. For twenty years we've asked world leaders to take action to stop polluting the atmosphere. We couldn't wait any longer. Now, warriors of the Pacific are rising peacefully to protect the Pacific Islands from climate change. Our message: we are not drowning. We are fighting! (350 Pacific 2014a)

The emphasis here is put on politics: as members of Pacific indigenous communities, the Warriors asked world leaders to take action for climate change. Since the answer never arrived, these young fighters decided to intervene on their own, making known the urgency of the crisis.

But who are these Warriors? As observed by Rowan Gard (2018) the concept of 'warrior' as a classic male figure in the Pacific is today re-imagined for all genders (see also George 2019; McNamara and Farbotko 2017). A warrior is anyone who will stand beside these Pacific Islanders environmental activists, as 'warriors' defending their homes and world (Gard 2018). As we can read on their website, there are principles by which a warrior must abide:

> A warrior is resilient. A warrior is not aggressive or violent but is assertive. A warrior serves to protect their community, culture, land, and ocean. A warrior is always learning. A warrior responds to the need of those around them and of the greater good. A warrior non-violently stands their ground against an adversary, against injustice, and against oppression. A warrior respectfully embodies their local culture and traditions. A warrior is accountable for their actions and words. A warrior serves those who cannot fight for themselves - future generations, animals and plants, environments. (350 Pacific 2014b)

This assertion shows parallels between the use of such 'universal' principles (non-aggressive or violent attitude, protection of the environment, injustice, and oppression) by the Warriors and other international environmental NGOs. For instance, while Greenpeace aims to use symbols that can trigger an emotional involvement in the public to raise awareness among people and mitigate climate change (Doyle 2007), the Warriors' narrative seems slightly different for it aims to protect the Pacific islands from climate change.

The Pacific islands and their indigenous communities, popularly considered to be isolated and underdeveloped compared to the Western world (see Hau'ofa 1994; Nunn and Kumar 2019 for a critique of this narrative), are the perfect representation of the contemporary environmental crisis: they embody Nature's fragility, smallness, remoteness both in geographic space and time. Nonetheless, the Pacific Climate Warriors also assert local ways to express their dissent.

> Five years ago, our network of young Pacific Islanders began to form under the name of 350 Pacific, to join with the global climate change movement, 350.org. Active in fifteen of the Pacific Island Nations, we have a unique approach of empowering young people to understand the issue of climate change and to take action to protect and enrich our islands, cultures, and oceans. (350 Pacific 2014a)

As shown below, the Pacific Climate Warriors claim a particular and therefore different strategy to approach the environmental problem, which is the empowerment of local communities instead of top-down development programs, usually delivered by international NGOs.

> We know we cannot deliver this project on our own. We don't have big bank accounts, big funders, or big email lists. That's why we are asking the international community to continue to stand in support with the Pacific Warriors. (350 Pacific 2014a)

Here stands a recurring paradox in the Warriors' rhetoric. On the one hand, they are trying to empower indigenous local communities, by fostering resilience in them. On the other, they affirm their peripheral position compared to the Western discourse and seek global help to fight this battle. This shows they are aware of the idea of smallness and remoteness of the Pacific Islands, usually attached to them by the Western public, and furthermore, they can use such a narrative to gain attention from other countries. Climate change becomes then a political instrument used by indigenous communities to assert and confirm their agency and their right to speak.

> What we do have is a network of courageous Pacific Islanders - from Niue to Tuvalu - that have been learning new skills and campaign tactics to take on the challenge of achieving global action on climate change - the future of our islands depends on it. (350 Pacific 2014a)

Their goal is to foster resilience and to empower the local communities with whom they live and work to create awareness among people about climate change.

In this context, the metaphors proposed by Farbotko (2010a) of the 'canary in the coal mine' and that of 'wishful sinking' are helpful. She posits that Tuvalu—allegory of the entire world—is expendable as the canary, usually carried inside the mine by the miners to check the safety of its air. The possibility of Tuvalu's disappearance makes it a great moral symbol in the eye of the Western public. In the words of the author: "only after they disappear are the islands useful as an absolute truth of the urgency of climate change, and thus a prompt to save the rest of the planet" (Farbotko 2010a, p. 47).

The Warriors and their communities have a completely different view about themselves, their cultures and traditions. As observed also by Hau'ofa (1994), Pacific Islanders do not think about themselves as marginal. This is made clear in the final lines of his paper:

> Oceania is vast, Oceania is expanding, Oceania is hospitable and generous, Oceania is humanity rising from the depths of brine and regions of fire deeper still, Oceania is us. We are the sea, we are the ocean, we must wake up to this ancient truth and together use it to overturn all hegemonic views that aim ultimately to confine us again, physically and psychologically, in the tiny spaces which we have resisted accepting as our sole appointed place, and from which we have recently liberated ourselves. We must not allow anyone to belittle us again, and take away our freedom. (Hau'ofa 1994, p. 16)

Nonetheless, I argue that the activists strategically use a 'negative' rhetoric and the attention given to them by the Western media and organizations to reach their aim (in the same way that other organizations like Greenpeace do). This aim consists in creating awareness among Pacific and non-Pacific people about climate change and environmental problems in the Pacific Islands region which threaten their livelihoods. Or, put differently, the Pacific Climate Warriors actively utilize this newly achieved attention as a launchpad to make their voices heard in international fora and specialized meetings. They travel all around the world and often take part in world fora such as COPs (Conference of the Parties) to improve international public understanding of the local impacts of climate change.

It is very interesting to note that all these activists mostly come from decolonized countries within the Pacific Islands region. In fact, it is due to their political status that they can communicate widely and effectively and coordinate with other English-speaking communities. On the other hand, as observed by Ferdinand (2018), it is hard for the French Overseas Territories[3]—which stand in a political 'in-between', neither able to become independent States nor to be a full part of France—to access international funding allotted for Small Islands Developing States (SIDS).

For instance, according to my experience in New Caledonia—which continues to be a French colony—it is the lack of activists and mediatic power which weakens communication with other communities involved.

In general, the linguistic barrier is certainly an issue. English and French, as old colonial heritage, are still among the official languages of numerous Pacific Island States and facilitate communication among groups who speak different vernaculars. Yet, very often they are a hurdle between English-speaking and French-speaking communities, further hindering the capacity for transnational network building.

Pacific Climate Warriors can be identified as *passeurs culturels* (Favole and Aria 2010) or social actors on the edge of two worlds that can move through different borders and different universes of meaning. The majority of the Warriors now lives in Australia and New Zealand, even if their country of origin is a 'large ocean state'—such as Samoa or Tokelau—as in the term coined by Ronny Jumeau (2013). These multiple identifications (Kempf and Hermann 2014) of the activists make evident their capacity to behave and act as Pacific Islanders and at the same time their ability to appropriate exogenous discourses and images and putting them in a new form. Their exposure in Western countries, such as Australia, New Zealand or even the west coast of the United States, contributes to the making of their consciousness and activism, not to mention the importance these links have in reconnecting old transnational

[3]For the same reason, there are no 350 Pacific local branches neither in Wallis and Futuna nor in French Polynesia, other French Territories in the Pacific Islands region.

relations interrupted by colonialism. Such connections place the Warriors in the flow of globalization and enable a more informed discourse about climate change and global politics. Reciprocity, as stated by Crook and Rudiak-Gould (2018, p. 2), "is an important principle and social mode: it involves a sophisticated aptitude for discovering new possibilities in resources of all kinds, and an equally adept skill in accepting external interests and turning them into home-grown initiatives".

As recently observed by Jolly (2018, 2019), resistance (instead of resilience[4]) can be found not only in political protests, such as those organized by the Pacific Climate Warriors but also in everyday life. As observed by Miriam Kahn (2011, p. 182) regarding the Tahitian context, "challenges to conventional behavior do have the abilities to effect profound change. These spaces of resistance can occur in myriad ways, both brazenly provocative and quietly subtle".

I argue that the use of traditional clothes and dances, foregrounded by the Warriors as part of the 'Pacific culture', is a space of resistance. By claiming their way to protest and organize social mobilization, using traditional clothes and dancing, the activists re-signify content and objects giving them a different and new meaning.

Billy Cava, the coordinator of the 350 Pacific New Caledonian branch explains how the movement is organized and how their campaigns are arranged.

> What happens is that every year we have… our year always begins in March and ends in November/December and… and our year is about to begin now, this means that next week… next month, in April, we will have our first retreat, on this occasion we will meet with the team 350 Pacific, that is the coordinator of New Caledonia, coordinator of Tuvalu, Kiribati, Fiji, we are about fifteen from the Pacific islands and we will gather in Samoa, in April, to organize the campaigns we will perform this year in the Pacific region. […] We will… develop the regional campaign but at a local level, this means that we will evaluate which actions we can do relating to… but I cannot say it because it is… it is still secret, this year's campaigns are still confidential. (Billy Cava, interviewed on March 15th, 2018 in Nouméa, author's translation from French)

Cava underlines the importance of the annual meeting in Samoa at the beginning of the protest season. This meeting is central to maintain social relationships between activists and especially between islands, and also to create new networks able to cooperate to achieve their common goal: lobbying big polluting industries and companies and expressing dissent on how their lucrative businesses are impacting Pacific Islands States and indigenous livelihoods.

The Warriors' campaigns and actions are developed and conducted regionally. Every Warrior organizes a local initiative, involving local communities, to help the bigger Pacific campaign gain visibility. These actions are intended as key in making local and indigenous people aware of the climate crisis and aim to empower them while fostering resilience.

[4]Resilience is understood as the capacity of a person to recover from a critical event, that is—in this case—the capacity to recover from the impacts and disasters exacerbated by climate change. Resistance, instead, is the capacity to resist and to contest the conventional way to behave and think.

Campaigning in Australia and the Vatican City

Two cases are now analyzed representing the different forms in which the Pacific Climate Warriors have demonstrated their dissent against Western immobility towards climate change and its impacts in the Pacific Islands region. The first case explores the Newcastle canoe blockade, while the second one examines the Warriors' journey to the Vatican.

One of the first protests took place in Newcastle, Australia, in 2014. On October 17th, numerous activists aboard traditional canoes blocked the entrance of the wharf, the world's largest coal export port. This action drew hundreds of people who flocked to the beaches near the wharf, to show their support (see also Fair 2015; McNamara and Farbotko 2017).

Mikaele Maiava, Warrior from Tokelau, spoke at the conference "Interactive conversation on advocacy in the Pacific", that took place in Wellington, New Zealand, in February 2018. He argued that during Newcastle's protest in 2014, the Australian Police destroyed one of the traditional canoes used by the activists. The Warriors' resistance, in the words of Maiava (2018), "was a symbol of strength, that was a symbol of resilience, that was a symbol of who we are. We are resilient! We are people of the land; we are people of the ocean!"

The stress is put on the relation between Pacific communities and two fundamental elements in Islanders' lives: on one hand, the land—in which they are rooted; on the other, the ocean linking the Pacific Islands. During the same conference Maiava stated:

> The land is everything to us. They ask us if we are climate refugees. We are not climate refugees: we are people. We are people with a homeland just as much as you people from up there in the North. What makes you think that we will agree to say that we are climate refugees? No, we don't. Our land is our everything. (Maiava 2018)

Once again the focus is put on the land and the rootedness of indigenous communities. Yet the Pacific Islands region is also well known for the high mobility of indigenous societies (Nunn 2007). To what extent is climate change forcing people to move? And do these people really want to relocate somewhere else? Similar questions were raised by numerous scholars who have investigated topics as mobility, migration, and attachment to place in the Pacific Islands region (Farbotko and Lazrus 2012; Kelman 2010; McMichel and Katonivualiku 2020; Mortreux et al. 2018; Piggott-McKellar et al. 2019). Mobility and attachment to place are two fundamental issues that must be widely explored and in which further research is required to better understand the local perspectives. These issues appear to be opposite, at first sight, yet they are linked to one another in assuming the inevitability of climate change and its consequences. Pacific Islanders envision mobility and migration in the worst-case scenario (Kempf and Hermann 2014) but they are also raising people's awareness and promoting environmentally-conscious behaviours affirming their strong agency (Steiner 2015).

Campaigning in Australia continued after the canoe blockade. Further initiatives were set in motion: *Frontline Truths* aimed to collect stories by activists from different

Pacific Island countries to show the impact of climate disasters to a wider public. In the *#HaveYourSei* campaign, activists asked their sustainers to show support by wearing *sei* (the traditional Polynesian flower). More recent campaigns are the *Stop Adani* marches and sit-ins against the building of the largest coal mine in Queensland; *Pacific Pawa* and the homonymous hashtag campaign seeking to empower local communities, highlight peculiarities and encourage resilience; more recently, *Matagi Mālohi—Strong Winds*, inspired by a Tokelauan poem, uses an evocative photo shooting as storytelling "to weave together young people across the region, and the world, in order to create a collective sense of place, people and purpose" (350 Pacific 2019; see also George 2019 for an analysis of their visual campaigns).

This sense of connection resonates with the words of Koreti Tiumalu, former 350.org Pacific coordinator, who states that "we cannot build a Pacific Climate Movement without engaging our faith communities. Faith is pivotal to our people, and like the ocean, it connects us. In the face of the climate crisis, we need prayer to carry our people and faith to build resilience" (350 Pacific 2016a). *#PrayForOurPacific* has been an important campaign engaging church communities to pray to empower the movement. "We aren't looking to our faith as a way to solve the issue of climate change, but we are looking to our faith to find the strength to continue the fight to protect the planet that was gifted to us. Not a planet that we have dominion over, but a planet we are stewards of" (350 Pacific 2016a).

In 2015, 12 representatives of the Pacific Climate Warriors, including Catholic and non-Catholic members, travelled to Rome to celebrate the Pope's encyclical *Laudato Si'*.[5] "Entering St Peter's Square for the first time [...] was overwhelming for us all. We were all truly in awe of the place we had only seen in images or tv. For some of us, it was the spiritual home our faith. For those who were non-Catholics, the feeling of being in such a Holy place was very strong" (350 Pacific 2016b). The Warriors laid out their traditional hand-woven mats on St. Peter's Square "going to each of the mats individually to silently reflect on the journey of that mat – where it had come from, who had woven it and the people of that Island [...] suffering the impacts of climate change" (350 Pacific 2016b). Religion and tradition were interwoven as the 3-day vigil began: "On each day, four Islands would be given an opportunity to share the story of their mat and stories of impacts with the group. We would then collectively as a group offer up prayers for that Island, prayers for the leaders of that Island, share a moment of silence and where appropriate for the mat, an act of contrition. We finished each Island presentation with a hymn – songs we had been practicing all week" (350 Pacific 2016b).

In June 2018, a delegation of Warriors attended the International Conference for the 3rd anniversary of the encyclical *Laudato Si'*.

It is not possible here to quote the entire dialogue although it is interesting to analyze the participation of a Samoan Warrior, Joseph Moeono-Kolio:

> The Pacific is our home. We are on the frontlines of climate change, it is real and it is our reality now. Climate change is more than just cold data and cold economic models. It

---

[5]Laudato Si' is the second encyclical of Pope Francis. In it, he criticizes consumerism and irresponsible development, common causes of environmental degradation and global warming.

> is a human issue first and foremost. Climate change is cyclone Winston and cyclone Gita threatening Fiji and Tonga and destroying the infrastructures of our island nations. Climate change is the washing away of the graveyards of our ancestors and the future of our children. We must remember that in these gatherings and in the noise of political arguments and ideological narratives we do not forget our humanity and become desensitized of the human faces at the center of this crisis. We flew all the way here because we are fighting not just for our children but yours as well, and your grandchildren.
>
> We in the Pacific cannot wait. We need to understand that 1.5 is the only acceptable outcome. The difference between 1.5 and 2 degrees is the difference between our islands surviving or drowning. […] Remember the human faces of climate change. See us. Partner with us. We are more than a current threat through climate change. We are families, we are communities, we are the memories of our ancestors, we are the saltwater that runs through our veins. We are more than drowning islands, we are fighting warriors. (350 Pacific 2018)

With this talk, Moeno-Kolio is trying to make visible and graspable a phenomenon that is usually hard to understand and experience firsthand (Rudiak-Gould 2013). Climate change is, in his words, personified by cyclones Winston and Gita (in and of themselves personified extreme natural events); it is menacing the past ("the washing away of the graveyards of our ancestors") as well as the future ("and the future of our children"), undermining the very existence of Pacific lifestyles and likely to cause a cultural crisis.

Also, a lot of effort is put in projecting an emotional connection between the Pacific communities and the Western public, while enforcing a strong imagery on how it feels to be threatened by climate change ("we are families, we are communities"), to touch people's sensitivity and trigger a sense of responsibility for their global neighbours.

In the first case—Newcastle—we are confronted with a protest which took place in Oceania. Local popular participation was a very important part of the action: the activists tried to block the wharf riding traditional canoes; traditional dances and demonstrations took place afterwards on the beach. These demonstrations raised the Australian public's attention and attracted other Australian activists.

In the second case—Vatican City—only a few Warriors were selected as spokespersons for the majority of the Pacific communities and the sessions of the conference were not accessible to the lay public. The particular way the movement presents the environmental issue to the Western (and Australian) public is representative of the capacity of the Warriors to speak different languages in different contexts, to move seamlessly between different universes of meaning. The power and agency of these social actors lie exactly here: in the capacity to translate heterogeneous worlds and render them understandable for both Pacific Islanders and the rest of the world.

The Role of the Internet

The Internet plays a fundamental role in these campaigns; in fact, every initiative is routinely eternalized through pictures and videos posted online by the Warriors. The majority of the excerpts cited in this paper were found on the Internet, albeit on different platforms (mostly the 350 Pacific Official website and their Facebook page). The use of social media as a means to tell their stories is called digital storytelling (Cunsolo Willox et al. 2012) and can be considered a new way of revealing oneself, a new form of personal expression that is overriding the older ones. Initially, the web was considered a virtual world, parallel to the real one, where anonymity was feared as a dilution of the self and the identity. It is clear today that online social networks, instead of being a parallel world, are now an extension of the real one (Miller 2016): a useful instrument used to expand interpersonal connections. The exponential increase in the use of online platforms led to a transformation in their status; social media were understood in the past as 'places' where connections with friends and strangers were made possible. Nowadays, they are instead places of unremitting interaction (Miller 2016) used 'to be there' especially across long distances as in Pacific Islander diasporas.

In non-hegemonic anthropology, a working concept is that of 'globalization of the communication' (Lévy 2012). In the context of the present discussion, this refers to the possibility for indigenous and peripheral actors to be present through the web, using new methods of writing and transforming the old ones, building new powerful international networks. From an anthropological perspective, it is therefore relevant to notice the growing use of such platforms in specific communities, where they function as megaphones for local stories to be heard. In fact, "the wide-scale democratization of the basic technologies that allow connection to the network, its permanent availability, and the simplification of the procedures for loading materials and information, make each site a place of expression, exchange, and influence, as well as the formation of international networks" (Lévy 2012, pp. 89–90, author's translation from Italian).

In the case of the Pacific Climate Warriors, social networks such as Facebook and Instagram are not only used to document and make visible climate change and to sensitize the public but also as a way to aggregate people (Titifanue et al. 2017). The important media coverage reached by these initiatives is strongly tied with the use of new technologies which play a fundamental role in making visible what is usually invisible to the eye of Western people, particularly regarding climate change (Doyle 2007; Rudiak-Gould 2013).

This technique is widely used by another environmental organization, Greenpeace, which in the early '70s established a new way to conceive social protest. Through unexpected actions made famous by videos posted on the Internet, and through pictures documenting the environmental changes exacerbated by human-induced climate change, the movement has raised public awareness of these issues. Such a method is very effective and therefore useful for these movements (Doyle 2007) but it is unilinear in spreading a universal message, mostly the Western and eco-conscious ideology proposed by the major environmental organizations. Yet the Pacific Climate Warriors' campaigns are not based exclusively on social media; they also count on local initiatives.

Pacific Activism

There are differences between Western and Pacific activisms which are evident and confirmed by the Warriors as well as Billy Cava:

> We were raising people's awareness and talking, you know, about activism, but Pacific activism, what does it look like because it is different, it is not like the activism… in the West. Western activism is rougher. We will not get yelled at. In the Pacific, we have different methods, different strategies that might be less violent but… which will still give the same results. (Billy Cava, interviewed on March 15th, 2018 in Nouméa, author's translation from French)

These 'other methods' mentioned by Cava are commonly used strategically by the Warriors. Such protests are locally organized and employ a particular strategy to involve indigenous and grassroots communities. This strategy is built on two parallel axes, as Cava stated:

> We have two axes here in the Pacific, and our two axes are the two… it is the focal point, namely our culture and traditions, and our faith, that is religion, Catholic or Protestant […]. Because we are very traditional, we are very spiritual. We use these two axes to touch the indigenous populations' sensibility. Pacific people are very familiar with these two axes. That is why we often use them to communicate our messages, to sensitize people about global warming, and we always use our culture, we organize cultural events, we wear our traditional clothes… and this, this speaks to people, we are also very proud of our cultures, you know, here in the Pacific, that is why when we use things like… a canoe and we sail in the ocean, everybody is interested, they want to come, to participate and at that moment we feel… there is an energy that rises, you know, and we are connected to this energy. This is… we in the Pacific call this energy *mana*, it is a kind of energy that we have in the Pacific. We have this energy and once we are connected to it we have the impression that we can change everything: we can change the world, we can change the country, we can change people's visions because we know that people are listening to us, when we are connected, connected heart to heart, connected to people's hearts. So culture and religion. (Billy Cava, interviewed on March 15th, 2018 in Nouméa, author's translation from French)

According to the Pacific Climate Warriors, tradition and faith are the cornerstones on which Pacific societies are built. These concepts can translate and therefore convey exogenous messages and discourses—such as climate change—through a local Pacific language. In this way, indigenous communities can appropriate these objects and use them as a political claim to exercise their right to speak and fight against climate change. The Warriors often refer to tradition as a supposed 'Pacific culture', an umbrella term which can unite customs and traditions coming from different communities and heritage, through time and space, that nonetheless share some common aspects. This also echoes the regional identity proposed and promoted by Epeli Hau'ofa (1998; see also Lazrus 2012), considered to be a useful means to unite Pacific Islanders and to make Oceania prosperous and able to act when necessary as a united body, bound by cultural ties: the same as the concept of 'grassroots regionalism' as proposed by Titifanue et al. (2017).

Religion is a key component of contemporary Pacific identities (Nunn et al. 2016). The Warriors refer to an ecumenical Christian faith, both Catholic and Protestant, as an endogenous characteristic (Steiner 2015) although Christianity was imposed by Western settlers. Despite being a colonial artefact, religion is so important and fundamental in Pacific societies today that Gnecchi Ruscone stated: "it is not possible to talk about cultural life in Melanesia without considering Christianity" (2012, p. 183, author's translation from Italian). Some other aspects of this common identity proposed by the activists include *mana* (see Tomlinson and Tengan 2016) and the ocean (see Hau'ofa 1998) and the attachment to a place, as mentioned above.

Along with religion, the land and the ocean are fundamental elements of Pacific societies. The articulation between roots and routes proposed by Clifford (2001) and the metaphor of the tree and the canoe (to exemplify traditional identity) suggested by Bonnemaison (1985) can help explain the connection between land and ocean.

The land is usually a symbol of attachment to place in Pacific Island cultures, the homeland in which indigenous people are rooted (Bonnemaison 1985; Clifford 2001; Farbotko et al. 2018; Kempf and Hermann 2014; Kim 2020). The ancestors, the cornerstones of Pacific societies, are buried in it: land in this sense symbolizes 'where one comes from', strengthening the idea of belonging. Yet at the same time, although significant, land does not delimit Pacific spatial, political and cultural values (Farbotko et al. 2015).

The ocean is what makes Pacific societies proud of themselves: their cultures are all about voyaging and navigation (Farbotko 2012; Farbotko et al. 2018). Thus canoe-building is an important event in village life but also a community-building moment (see Steiner 2015 for the canoe-building day of action promoted by the Warriors). Canoes sailing the ocean, following itineraries that connect the 'sea of islands' are a strong symbol of Pacific mobility.

As observed by Hau'ofa (1994, 1998) the land and the ocean are fundamental to place oneself in the structure of Pacific Islands societies. For example, in New Caledonia, everyone can go fishing and has land to cultivate but only a few clans and families have the right to provide crops and fish for different ceremonies, affirming and maintaining in this way their social status (Sabinot and Juncker 2018). Sabinot and Juncker (2018) also underline that climate changes and new fishing techniques have led over the years to a decrease in fish stock and of certain species which to the Kanak community are a means to renew the spiritual connection with the ancestors through rituals. This threatens the customary and spiritual organization of society. The land and the ocean are essential to Pacific societies to place themselves in the world because they organize the local cosmologies (Crook and Rudiak-Gould 2018). Thus it is important to hold together these two concepts in an open and more productive dialogue (Farbotko et al. 2015).

Land and ocean, faith, attachment to place and mobility are used by the Warriors as key notions to contest the inevitability of migration and resettlement, often proposed by NGOs and international institutions. This stark contrast (forced migration vs.

attachment to place) is well expressed by Kathy Jetnil-Kijiner (2017, pp. 70–73), a Marshallese poet and climate activist. In her words:

> […] men say that one day that lagoon will devour you
>
> they say it will gnaw at the shoreline chew at the roots of your breadfruit trees
> gulp down rows of your seawalls
> and crunch your island's shattered bones
>
> they say you, your daughter and your granddaughter too
>
> will wander rootless
> with only a passport to call home […]

This sense of urgency is juxtaposed to a reassuring future on low-lying islands, as mirrored in the following lines of "Dear Matafele Peinam" by Jetnil-Kijiner:

> […] no one
> will come and devour you
>
> no greedy whale of a company sharking through political seas no backwater
> bullying of businesses with broken morals
> no blindfolded bureaucracies gonna push this mother ocean over
> the edge
>
> no one's drowning, baby
> no one's moving no one's losing their homeland
> no one's gonna become a climate change refugee
>
> […] because, baby, we are going to fight!

The use of indigenous concepts and the process of their revitalization, to spread a clear message to fight climate change, places social actors like the Warriors in the flow of globalization, making them protagonists of protests and social movements. As *passeurs culturels* they can cross borders and make sense of the phenomena that are constantly running over them. They can translate exogenous concepts and make them understandable to local communities. They can legitimately appropriate Western narratives that are then indigenized.

Conclusions

After exploring the Pacific Climate Warriors' rhetoric and narratives, the agency of indigenous advocates, which is usually hidden by Western discourse, emerges clearly. Yet, further research could investigate the reception of these kinds of protests among local communities. The Warriors' transnational and globalized communication conveys messages that could not be fully and deeply understood and embraced by their audience. Such messages risk being comprehensible only to a small local elite. Furthermore, the influence of the activists' campaigns on governmental adaptation and mitigation plans need to be evaluated with care.

On the one hand, all the campaigns and the protests organized by the activists are founded on the basic assumption of the scientific proof of climate change, as they also experience firsthand episodes of extreme weather events, such as cyclones and flooding (Rudiak-Gould 2012). Yet it is necessary to distinguish between the scientific fact of climate change and its social discourse and the political claims it brings with it.

On the other hand, the Warriors strongly contest the Western narrative which assumes that Pacific communities, who are at the front lines of climate change, are helpless victims. This victimization has travelled through time and space: Oceania was always thought of as the 'invisible continent' (Le Clézio 2006), composed of tiny islands usually neglected by the continental public. In Western narratives, Oceania is known for its comparative remoteness and smallness, sometimes for its poverty and conspicuous inaction facing modernization and globalization (Hau'ofa 1994). This rhetoric is now updated with climate change: the idea that one day small islands will disappear underneath the ocean and that local populations will be forced to leave their lands and move to Western nations such as Australia and New Zealand. All this negative imaginary is contested by climate activists through their protests and campaigns, highlighting Pacific cultural singularities such as the high mobility of Pacific societies and their spiritual relationship with the environment. This renders climate-related migration an exogenous concept: Pacific Islanders were used to migrating long before the environmental crisis that today is likely to force people to move out of their islands (Kelman et al. 2015; Nunn 2016, 2017; Suliman et al. 2019). Moreover, as observed by McNamara and Farbotko (2017), the Warriors' narrative does not deny the worst-case scenario (which implies mobility and migration), merely its inevitability. The Warriors' motto is, therefore, "we are not drowning, we are fighting!" (350 Pacific 2013). This local rhetoric echoes what Epeli Hau'ofa (1994) has called 'a sea of islands', contesting the reductionist view of Pacific islands as just tiny dots in a vast ocean.

Other than just contesting the Western narrative and raising awareness among local communities, the Warriors' movement is also trying to reconnect traditional inter-island links through a transnational network, renewing the economic, political, and kinship relations which suddenly disappeared because of the 'divide and conquer' politics of the colonial era, later replaced by the creation of new national boundaries and through the establishment of Exclusive Economic Zones (EEZs). The Pacific Climate Warriors seek to reconnect Pacific communities through their campaigns and initiatives, overcoming the previous colonial borders of the Nation-State (also through the independence movements) and making mobility useful as a local resource instead of seeing it as a negative quality. This is visible in the structure of their organization: all the Warriors usually meet in one of the islands for their annual meeting, they lead regional campaigns working together on the same projects. In times of crisis, they raise funds for self-help and travel all over the world to give testimony in international fora about what they are experiencing because of climate change. The awareness of the situation and the problems it entails, not to mention the agency demonstrated in front of such issues, are what make the Pacific Climate Warriors protagonists of a new form of grassroots globalization. It is through the appropriation of Western discourse, implemented and enriched by their cultural heritage, that they can attract the public and media attention to spread their message and try to solve the climate crisis in the Pacific.

References

350 Pacific (2013) We are not drowning. We are fighting. https://350pacific.org/we-are-not-drowning-we-are-fighting/. Accessed 18 June 2020

350 Pacific (2014a) Pacific Climate Warriors. https://350pacific.org/pacific-climate-warriors/. Accessed 18 June 2020

350 Pacific (2014b) Become a Climate Warrior. https://350pacific.org/our-work/what-does-it-mean-to-be-a-warrior/. Accessed 18 June 2020

350 Pacific (2016a) #PrayforOurPacific. https://350pacific.org/prayforourpacific/. Accessed 20 August 2020

350 Pacific (2016b) Pacific Climate Warrior Europe Tour Sept/Oct 2015 by Koreti Tiumalu. https://350pacific.org/pacific-climate-warrior-europe-tour-septoct-2015-by-koreti-tiumalu/. Accessed 20 August 2020

350 Pacific (2018) Pacific Climate Warrior, Joe Moeono-Kolio speaking at the Closing Plenary of the Vatican Conference on Laudato Si. https:// www.facebook.com/350Pacific/videos/2046204442057395/. Accessed 18 June 2020

350 Pacific (2019) Matagi Mālohi. https://spark.adobe.com/page/rGA9VQOxnDtAn/. Accessed 20 August 2020

Anderson B (1983) Imagined communities. Verso, London

Appadurai A (1996) Modernity at large. University of Minnesota Press, Minneapolis

Appadurai A (2000) Grassroots globalization. Public Cult 12(1):1–19

Bonnemaison J (1985) The tree and the canoe: roots and mobility in Vanuatu societies. Pac Viewp 26(1):30–62

Boykoff M, Boykoff J (2007) Climate change and journalistic norms: a case-study of US mass-media coverage. Geoforum 38(6):1190–1204

Brunois F (2005) Ethnoécologie en Océanie. J Sociétés Océanistes. https://doi.org/10.4000/jso.299

Clifford J (2001) Indigenous articulations. Contemp Pac 13(2):468–490

Connell J (2016) Last days in the Carteret Islands? Climate change, livelihoods and migration on coral atolls. Asia Pac Viewp 57:3–15

Crate S (2011) Climate and culture: anthropology in the era of contemporary climate change. Annu Rev Anthropol 40:175–194

Crook T, Rudiak-Gould P (2018) Introduction: Pacific climate cultures. In: Crook T, Rudiak-Gould P (eds) Pacific climate cultures living climate change in Oceania. De Gruyter, Warsaw, pp 1–20

Cunsolo Willox A, Harper S, Edge V, 'My Word': Storytelling and Digital Media Lab, Rigolet Inuit Community Government (2012) Storytelling in a digital age: emerging narrative method for preserving and promoting indigenous oral wisdom. Qual Res 13(2):127–147

Demeritt D (2001) The construction of global warming and the politics of science. Ann Assoc Am Geogr 91(2):307–337

Demeritt D (2006) Science studies, climate change and the prospects for constructivist critique. Econ Soc 35(3):453–479

Dhillon J, Allooloo S (2016) Dismantling Columbus Day and the power of the present. https://truthout.org/articles/dismantling-columbus-and-the-power-of-the-present/. Accessed 24 June 2020.

Doyle J (2007) Picturing the clima(c)tic: Greenpeace and the representational politics of climate change communication. Sci Cult 16(2):129–150
Fair H (2015) Not drowning but fighting: Pacific islands activists. Forced Migr Rev 49:58–59
Fair H (2018) Three stories of Noah: navigating religious climate narratives in the Pacific Island region. Geo: Geogr Environ. https://doi.org/10.1002/geo2.68
Farbotko C (2005) Tuvalu and climate change: constructions of environmental displacement in the *Sydney Morning Herald*. Geogr Ann 87B(4):279–293
Farbotko C (2010a) Wishful sinking: disappearing islands, climate refugees and cosmopolitan experimentation. Asia Pac Viewp 51(1):47–60
Farbotko C (2010b) 'The global warming clock is ticking so see these places while you can': voyeuristic tourism and model environmental citizens on Tuvalu's disappearing islands. Singap J Trop Geogr 31:224–238
Farbotko C (2012) Skillful seafarers, oceanic drifters or climate refugees? Pacific people, news value and the climate refugee crisis. In: Moore K, Gross B, Threadgold T (eds) Migrations and the media. Peter Lang, New York, pp 119–142
Farbotko C, Lazrus H (2012) The first climate refugees? Contesting global narratives of climate change in Tuvalu. Glob Environ Chang 22(2):382–390
Farbotko C, Stratford E, Lazrus H (2015) Climate migrants and new identities? The geopolitics of embracing or rejecting mobility. Soc Cult Geogr 17(4):533–552
Farbotko C, McMichael C, Dun O, Ransan-Cooper H, McNamara K, Thornton F (2018) Transformative mobilities in the Pacific: promoting adaptation and development in a changing climate. Asia Pac Policy Stud 5:393–407
Favole A, Aria M (2010) Passeurs culturels, patrimonializzazione condivisa e creatività culturale nell'Oceania francofona. AM-Antropologia Museale 9(27):5–18
Ferdinand M (2018) Subnational climate justice for the French Outre-mer: postcolonial politics and geography of an epistemic shift. Isl Stud J 13(1):119–134
Gamson W (2004) Bystanders, public opinion, and the media. In: Snow E, Soule S, Kriesi H (eds) The Blackwell companion to social movements. Blackwell, Malden, pp 242–261
Gard R (2018) Talanoa for climate change. Paper presented at the 12th ESfO Conference, University of Cambridge, Cambridge, 7–10 December 2018
George N (2019) Climate change and 'architecture of entitlement': beyond gendered virtue and vulnerability in the Pacific Islands? In: Kinnvall C, Rydstrom H (eds) Climate hazards, disasters, and gender ramifications. Routledge, London, pp 101–121
Gnecchi Ruscone E (2012) Cristianesimi in Papua Nuova Guinea. In: Gnecchi Ruscone E, Paini A (eds) Antropologia dell'Oceania. Raffaello Cortina, Milano, pp 183–210
Hau'ofa E (1994) Our sea of islands. Contemp Pac 6(1):148–161
Hau'ofa E (1998) The ocean in us. Contemp Pac 10(2):392–410
Hulme M (2009) Why we disagree about climate change. Understanding controversy, inaction and opportunity. Cambridge University Press, Cambridge
IPCC (1992) Climate change: the 1990 and 1992 IPCC Assessments. https://www.ipcc.ch/report/climate-change-the-ipcc-1990-and-1992-assessments/. Accessed 2 April 2019
IPCC (2018) Annex I: glossary. In: Masson-Delmotte V et al. (eds) Global Warming of 1.5°C. An IPCC Special Report on the impacts of global warming of 1.5°C above pre-industrial levels and related global greenhouse gas emission pathways, in the context of strengthening the global response to the threat of climate change, sustainable development, and efforts to eradicate poverty. https://www.ipcc.ch/site/assets/uploads/sites/2/2019/06/SR15_AnnexI_Glossary.pdf. Accessed 4 April 2019
Jasanoff S (2010) A new climate for society. Theory, Cult Soc 27(2):233–253
Jetnil-Kijiner K (2017) Iep Jāltok. Poems from a Marshallese daughter. University of Arizona Press, Tucson
Jolly M (2018) Who are 'we' in the Anthropocene. Paper presented at the 12th ESfO Conference, University of Cambridge, Cambridge, 7–10 December 2018

Jolly M (2019) Engendering the anthropocene in Oceania: fatalism, resilience, resistance. Cult Stud Rev 25(2):172–195

Jumeau R (2013) Small island developing states, large ocean states. Paper presented at the ONU, Expert Group Meeting on Oceans, Seas and Sustainable Development: Implementation and follow-up to Rio+20, New York 18–19 April 2013

Kahn M (2011) Tahiti beyond the postcard. Power, place, and everyday life. University of Washington Press, Seattle

Kelman I (2010) Hearing local voices from Small Island Developing States for climate change. Local Environ 15(7):605–619

Kelman I, Stojanov R, Khan S, Gila O, Duzí B, Vikhrov D (2015) Viewpoint paper. Islander mobilities: Any change from climate change? Int J Glob Warm 8(4):584–602

Kempf W (2012) Climate change, migration and Christianity in Oceania. In: Hastrup K, Olwig KF (eds) Climate change and human mobility: challenges to the social sciences. Cambridge University Press, Cambridge, pp 235–257

Kempf W, Hermann E (2014) Epilogue. Uncertain futures of belonging. Consequences of climate change and sea-level rise in Oceania. In: Hermann E, Kempf W, van Meijl T (eds) Belonging in Oceania. Movement, place-making and multiple identification. Berghahn, New York, pp 189–213

Kim M (2020) Nesor Annim, Niteikapar (Good Morning, Cardinal Honeyeater). Contemp Pac 32(1):147–163

Lahsen M (2005) Seductive simulations? Uncertainty distribution around climate models. Soc Stud Sci 35(6):895–922

Lahsen M (2010) The social status of climate change knowledge: an editorial essay. WIREs Clim Chang 1:162–171

Latour B (2010) Cogitamus. Six lettres sur les humanités scientifiques. La Découverte, Paris

Latour B, Woolgar S (1979) Laboratory life: the construction of scientific facts. Sage, Beverly Hills

Lazrus H (2012) Sea change: island communities and climate change. Annu Rev Anthropol 41(1):285–301

Le Clézio J-M (2006) Raga. Approche du continent invisible. Seuil, Paris

Le Duff M, Dumas P, Cohen O, Allenbach M (2017) Coastal erosion monitoring on Ouvéa Island (New Caledonia): involving the local community in climate change adaptation. In: Leal Filho W (ed) Climate change adaptation in Pacific countries. Springer, Berlin, pp 255–268

Lévy J (2012) Globalizzazione della comunicazione. In: Saillant F, Kilani M, Graezer Bideau F (eds) Per un'antropologia non egemonica. Il Manifesto di Losanna. Elèuthera, Milano, pp 89–91

Luetz J, Havea P (2018) "We're not refugees, we'll stay here until we die!"—climate change adaptation and migration experiences gathered from the Tulun and Nissan Atolls of Bougainville, Papua New Guinea. In: Leal Filho W (ed) Climate change impacts and adaptation strategies for coastal communities. Springer, Berlin, pp 3–29

Maiava M (2018) Interactive conversation on advocacy in the Pacific. Lecture, Victoria University of Wellington, 22 February 2018

McMichael C, Katonivualiku M (2020) Thick temporalities of planned relocation in Fiji. Geoforum 108:286–294

McNamara K, Farbotko C (2017) Resisting a 'doomed' fate: an analysis of the Pacific Climate Warriors. Aust Geogr 48(1):17–26

Miller C (2004) Climate science and the making of a global political order. In: Jasanoff S (ed) States of knowledge: the co-production of science and social order. Routledge, London, pp 46–66

Miller D (2016) Academic studies of social media. In: Miller D (ed) How the world changed social media. UCL Press, London, pp 9–24

Mortreux C, Safra de Campos R, Adger N, Ghosh T, Das S, Adams H, Hazra S (2018) Political economy of planned relocation: a model of action and inaction in government responses. Glob Environ Chang 50:123–132

Nunn P (2007) Climate, environment and society in the Pacific during the last millennium. Elsevier, Amsterdam

Nunn P (2016) Fiji has already experienced climate change once. https://slate.com/technology/2016/06/fijis-experience-with-sea-level-rise-600-years-ago-shows-how-climate-change-can-inspire-violence.html. Accessed 4 April 2019
Nunn P (2017) Islands lost to the waves: how rising seas washed away part of Micronesia's 19th-century history. https://theconversation.com/islands-lost-to-the-waves-how-rising-seas-washed-away-part-of-micronesias-19th-century-history-82981. Accessed 24 June 2020
Nunn P, Kumar R (2019) Cashless adaptation to climate change in developing countries: unwelcome yet unavoidable? One Earth 1:31–34
Nunn P, Mulgrew K, Scott-Parker B, Hine D, Marks A, Mahar D, Maebuta J (2016) Spirituality and attitudes towards nature in the Pacific Islands: insights for enabling climate-change adaptation. Climatic Change 136:477–493
Nunn P, McLean R, Dean A, Fong T, Iese V, Katonivualiku M, Klöck C, Korovulavula I, Kumar R, Tabe T (2020) Adaptation to climate change: contemporary challenges and perspectives. In: Kumar L (ed) Climate change and impacts in the Pacific. Springer, New York, pp 499–524
Nurse L, McLean R (2014) Small islands. In: Barros V, Field C (eds) Climate change 2014: Impacts, adaptation, and vulnerability. Part B: regional aspects. Contribution of Working Group II to the Fifth Assessment Report of the Intergovernmental Panel on Climate Change. Cambridge University Press, Cambridge, pp 1613–1654
Orlove B, Lazrus H, Hovelsrud G, Giannini A (2014) Recognitions and responsibilities. On the origins and consequences of the uneven attention to climate change around the world. Curr Anthropol 55(3):249–275
Otto D (2017) Lived experience of climate change—a digital storytelling approach. Int J Glob Warm 12(3/4):331–346
Piggott-McKellar A, Pearson J, McNamara K, Nunn P (2019) A livelihood analysis of resettlement outcomes: lessons for climate-induced relocations. Ambio. https://doi.org/10.1007/s13280-019-01289-5
Roncoli C, Crane T, Orlove B (2009) Fielding climate change in cultural anthropology. In: Crate S, Nuttall M (eds) Anthropology and climate change: from encounters to actions. Left Coast Press, Walnut Creek, pp 87–115
Rudiak-Gould P (2012) Promiscuous corroboration and climate change translation: a case study from the Marshall Islands. Glob Environ Chang 22:46–54
Rudiak-Gould P (2013) "We have seen it with our own eyes": why we disagree about climate change visibility. Weather Clim Soc 5:120–132
Rudiak-Gould P (2014) The influence of science communication on indigenous climate change perception: theoretical and practical implications. Hum Ecol 42(1):75–86
Sabinot C, Juncker M (2018) Pêcher en milieu kanak. Est-il seulement question de rapporter du poisson? Paper presented at the Conférence au Centre Culturel Tjibaou, Nouméa, 29 March 2018
Sarewitz D (2000) Science and environmental policy: an excess of objectivity. In: Frodeman R (ed) The earth sciences, philosophy, and the claims of community. Pearson, New York, pp 79–98
Schlehe J (2010) Anthropology of religion: disasters and the representations of tradition and modernity. Religion 40:112–120
Scott-Parker B, Nunn P, Mulgrew Hine K, Marks A, Mahar D, Tiko L (2017) Pacific Islanders' understanding of climate change: where do they source information and to what extent do they trust it? Reg Environ Change 17:1005–1015
Steiner C (2015) A sea of warriors: performing an identity of resilience and empowerment in the face of climate change in the Pacific. Contemp Pac 27(1):147–180
Suliman S, Farbotko C, Ransan-Cooper H, McNamara K, Thornton F, McMichael C, Kitara C (2019) Indigenous (im)mobilities in the Anthropocene. Mobilities 14:298–318
Swain K (2017) Mass media roles in climate change mitigation. In: Chen W, Suzuki T, Lackner M (eds) Handbook of climate change mitigation and adaptation. Springer, Berlin, pp 167–219
Titifanue J, Kant R, Finau G, Tarai J (2017) Climate change advocacy in the Pacific. The role of information communication technologies. Pac Journal Rev 23:133–149

Tomlinson M, Tengan K (eds) (2016) New Mana. Transformations of a classic concept in Pacific languages and cultures. ANU Press, Acton
Trichet J, Leblic I (2008) Spécial environnement dans le Pacifique. J Société Océanistes. https://doi.org/10.4000/jso.1804
Weir T, Dovey L, Orcherton D (2017) Social and cultural issues raised by climate change in Pacific Island countries: an overview. Reg Environ Change 17:1017–1028
Worliczek E (2013) La vision de l'espace littoral sur l'île Wallis et l'atoll Rangiroa dans le contexte du changement climatique. Une analyse anthropologique de la perception des populations locales. Ph.D. dissertation, Université de la Nouvelle-Calédonie, Universität Wien
Worliczek E (2014) Le changement climatique à Rangiroa: le discours scientifique, le concept théorique, la responsabilité... Et ce qu'en pense la population de l'atoll. Bull Société Etudes Océaniennes 333:4–33

Claudia Ledderucci is a PhD candidate at the University of Florence and the University of Turin. She graduated in Cultural Anthropology at Sapienza—University of Rome presenting a dissertation on the social construction of climate change and indigenous narratives, based on her fieldwork research in New Caledonia.

Chapter 16
How Does Faith Influence Pro-Environmental Practice in Schools? Examples from the Pacific Islands

Thelma A. Raman and Johannes M. Luetz

Abstract Many environmental problems can be attributed to human behaviour, and education is seen as a means of shaping human behaviour that is pro-environmental. The role of the formal education system in supporting pro-environmental behaviour is considered pivotal and duly highlighted in key documents/plans on sustainable development. This chapter is based on empirical research conducted in Fiji, an island group situated in the South Pacific. The research investigated factors or circumstances that enable individuals to sustain or continue adopting pro-environmental behaviours after they have left the formal school environment. It explored how Environmental Education is provided in schools through the curriculum and pro-environmental practices. Twenty-seven schools participated in the research including faith-based schools (Christian, Hindu, Islamic, Sikh). The research highlighted the significance of religion in shaping how care for the environment is viewed within the schools. The research found differences and similarities between faith-based schools, both *within* Christian denominations and *between* religious schools (i.e., Christian, Islamic, Hindu, Sikh). The importance of religion in seeding and guiding their pro-environmental behaviour was also acknowledged by student research participants. This is unsurprising given the presence of various religions in Fiji and the calendar observance of key religious events. This chapter discusses the research findings and highlights the contribution that religion makes to pro-environmental behaviours and why it therefore deserves to be considered in discussions/forums about behaviour change.

T. A. Raman (✉)
Macquarie University, Sydney, NSW, Australia
e-mail: thelma.raman@mq.edu.au

J. M. Luetz
School of Social Sciences, Christian Heritage College (CHC), Brisbane, QLD, Australia

School of Law and Society, University of the Sunshine Coast (USC), Maroochydore, QLD, Australia

School of Social Sciences, University of New South Wales (UNSW), Sydney, NSW, Australia

J. M. Luetz
e-mail: jluetz@chc.edu.au

J. M. Luetz and P. D. Nunn (eds.), *Beyond Belief*, Climate Change Management,
https://doi.org/10.1007/978-3-030-67602-5_16

Introduction

Drawing connections between religion and the environment is not novel. Historian Lynn White (1967) claimed that the beginning of the notion of 'nature' as an exploitable source of resources is found in the first book of the Old Testament:

> Then God said: 'Let us make man in our image, in our likeness, and let them rule over the fish of the sea and the birds of the air, over the livestock, over all the earth, and over all creatures that move along the ground.' (Genesis 1: 26)

While White's view lends support to the understanding that religious beliefs and commitments have a potential to negatively impact environmental concerns, others contend that religion may generate a pro-environmental stewardship effect (Kearns 1997; Pepper and Leonard 2016; Woodrum and Wolkomir 1997).

Sherkat and Ellison's (2007) contention that there is a positive relationship between religious beliefs and environmental concern is supported by a UK study that asked 20,968 Year 9 and Year 10 students to provide information on their concern regarding environmental pollution, belief in God, church attendance, prayer, and their view on creationism. The study found that "belief in God, church attendance and personal prayer are all significant predictors of greater environmental concern" (Francis 1997, p. 7). It concludes that "contrary to the view that the Christian tradition promotes a cavalier attitude toward environmental issues, the data demonstrates a positive relationship between religiosity and environmental concern" (p. 7).

Hayes and Marangudakis (2000) were more equivocal in their findings. They examined the impact of religion on environmental attitudes and behaviour of adults within the USA, Canada, Great Britain and New Zealand. Their findings suggest that there is no significant difference between Christians and non-Christians regarding their concern for the environment, and "overall, religious identification is a relatively weak and inconsistent predictor of environmental attitudes and behavior across nation" (p. 159).

Pope Francis, leader of the Catholic Church, in his first papal encyclical dedicated to the environment, titled *Laudato Si'* (Praise Be), *On the Care of Our Common Home*, called for immediate action to stop environmental degradation. He advocates a change in lifestyle as he links environmental problems to "a throwaway culture which affects the excluded just as it quickly reduces things to rubbish" (Pope Francis 2015, p. 17). He also calls for "a conversation which includes everyone, since the environmental challenges we are undergoing, and its human roots, concern and affect us all" (p. 12). This encyclical has been shared with millions of Catholics all over the world and added to the body of the Church's social teachings (Nelson and Luetz 2019). Pope Francis also pleaded for action during his address at the United Nations in September 2015 when he likened harm done to the environment as harm done to humanity: "The ecological crisis, and the large-scale destruction of biodiversity, can threaten the very existence of the human species" (Goldenberg and Kirchgaessner 2015, p. 5).

Yencken et al. (2000) note that "in each of the three great Indian religious traditions, Hinduism, Jainism and Buddhism, humans are asked to show reverence and respect for nature, and conservation of nature is emphasized in their teachings" (p. 224). Acknowledging that Hinduism is a remarkably diverse religious and cultural phenomenon, Jain (2010) outlines what the diverse theologies of Hinduism suggest in their teaching on the environment. "The *earth* can be seen as a manifestation of the goddess and must be treated with respect, and *dharma* (duty) can be reinterpreted to include our responsibility to care for the earth" (p. 1).

Chan and Islam (2015) investigated the environmental viewpoints and behaviours of six key religions in Singapore—Buddhist, Catholic, Hindu, Islam, Protestant and Taoism—and concluded that "while an apparent gap between environmental beliefs and actual environmental action persists, there is a desire by religious members for more societal activism and support with regard to protecting the environment" (p. 177). They also found a striking contrast between Singapore and the level of environmental activism that religions exhibit in the world, which they attribute largely to 'environmental activism' being frowned upon by the Singapore Government. Yet they did find support for more societal activism: "all six religions felt that the environment can be better protected if religious groups work with each other, that they should do so, and do more about environmental matters" (p. 184).

Islam does not view nature as being exclusively for human use:

> The Quran requires humans to obey what God has ordered. Since all living things were created by God with different functions they need to be looked after and protected. Humans are warned against the abuse of natural resources and of consequent imbalances in nature. (Yencken et al. 2000, p. 224)

Islam also focuses on future generations as its teachings implore that "People should not abuse, misuse, or distort the natural resources as each generation is entitled to benefit from them but is not entitled to 'own' them in an absolute sense" (Oliver 2008, p. 3). Rice (2006) found support for the presence of an Islamic environmental ethic in Islamic religious teachings in Egypt. Practices such as recycling wastewater are in keeping with the spirit of Islamic teaching (Farooq and Ansari 1983). After conducting a review of recent research on a number of personal and social influences on pro-environmental behaviour, including religion, Gifford and Nilsson (2014) found that overall, only a few empirical studies have been conducted regarding the importance and potential of faith.

While environmental connections can be dated back to the establishment of a particular religion, the environmental crisis is acting to drive the momentum to acknowledge religion's influence in fostering behaviour change and transforming societies (Palmer and Finlay 2003; Luetz and Nunn 2020). If we accept the premise that the environmental crisis is ultimately a crisis of the mind and that "we see, do and are what we think, and what we think is shaped by our cultures, faith and beliefs" (Palmer and Finlay 2003, p. xv), we then need to consider how the mind is developed, engaged, and ultimately persuaded towards particular courses of action.

Environmental Education is broadly defined as "raising awareness, acquiring new perspectives, values, knowledge and skills, and formal and informal processes leading to changed behaviour in support of an ecologically sustainable environment" (Environment Australia 2000, p. 3). Luetz et al. (2020, p. 3) have analysed the literature on education for human behaviour change in the area of sustainable development and have distinguished two common approaches: The first conceives human behaviour as either "predicated on knowledge of scientific facts, hard evidence and robust peer reviewed information" (citing Boyes and Stanisstreet 2012) or as inherently "unpredictable and psychologically, socially, contextually and situationally constituted" (citing Sapolsky 2017; Kelly and Barker 2016).

The US Environmental Protection Agency (EPA) defines environmental education as:

> A process that allows individuals to explore environmental issues, engage in problem solving, and take action to improve the environment. As a result, individuals develop a deeper understanding of environmental issues and have the skills to make informed and responsible decisions. (EPA 2016b, p. 1)

Since the early 1970s, a number of international declarations and fora have highlighted the role which environmental education can play in grappling with environmental problems including climate change. Of particular significance to the role of education is the 1972 *Stockholm Declaration on the Human Environment* "which produced an immense sense of urgency about the environment and development issues" (Greenall and Womersley 1977, p. 17).

The 1975 *Belgrade Charter*, as well as the preceding 1972 *Stockholm Declaration*, and the subsequent 1977 *Tbilisi Declaration* (UNESCO 1978), are viewed as the early beginnings of formal international environmental education initiatives and stressed that "environmental education should be provided to people of all ages, all levels of academic aptitude, and must be delivered in both formal and non-formal environments" (Wright 2002, p. 204). The goals of environmental education included in the *Tbilisi Declaration* were:

1. To foster clear awareness of, and concern about, economic, social, political and ecological interdependence in urban and rural areas.
2. To provide every person with opportunities to acquire the knowledge, values, attitudes, commitment and skills needed to protect and improve the environment.
3. To create new patterns of behaviour of individuals, groups and society as a whole towards the environment (UNESCO 1978, pp. 26–27).

Agenda 21 (an outcome of the 1992 *United Nations Conference on Environment and Development*), in particular Chapter 36 (Promoting Education, Public Awareness and Training), also supported the need for formal and non-formal education and acknowledged the need for various countries to develop their own environmental education programs according to their own specific needs, policies and responsibilities (Wright 2002; Walid and Luetz 2018).

Pro-environmental behaviour may be referred to as "conservation behaviors, environmentally friendly behaviors, environmentally significant behaviors, environmentally sustainable behaviors, and responsible environmental behaviors" (Osbaldiston and Schott 2012, p. 258). Steg and Vlek (2009, p. 309) define pro-environmental behaviour as that which "harms the environment as little as possible or even benefits the environment."

Fiji

The Fiji group consists of some 330 islands of which about 100 are inhabited. The islands are situated in the South Pacific Ocean, north of New Zealand and East of Queensland in Australia. The two main islands are Viti Levu and Vanua Levu. About 75% of the population lives on Viti Levu and about 18% live on Vanua Levu (Fiji Bureau of Statistics 2008).

Fiji is a multicultural society with different ethnic and religious groups. According to the 2017 Census of Population, Fiji has a population of 884,887 (Fiji Bureau of Statistics 2018). The most recent ethnicity data was collected in the 2007 census and includes 475,739 Indigenous Fijians, 313,798 Indians and 47,734 Others. The 'Others' include European, Part European, Chinese and Rotuman and Other Pacific Islanders (Fiji Bureau of Statistics 2008).

Indigenous Fijians preserve and practice their culture through various traditions and rituals. The indigenous Fijians are predominantly of Melanesian and Polynesian descent and can trace their links to other nations in the Pacific including Solomon Islands, Tonga and Samoa (Tuimaleali'ifano 2011). The majority of Indians in Fiji are descendants of indentured labourers, originating mostly from Calcutta and Madras, who were brought to Fiji between 1879 and 1916 to work on sugar plantations. Free immigrants from Gujarat are also part of Fiji's Indian community (Ali 1997; Kelly 1992). Members of the Fiji Indian community continue to observe their heritage and culture through language, dress and various rituals.

A number of religions are practiced in Fiji. The breakdown of the population according to religion[1] includes Christians (64.4%), Hindus (27.9%), Islam (Muslims) (6.3%), Sikhs (0.3%) and Others (0.3%), with 0.8% people not identifying with a particular religion (Fiji Bureau of Statistics 2008). Key religious events are observed by public holidays: Christmas for Christians, Diwali for Hindus and Prophet Mohammed's Birthday for followers of Islam.

This research focused on secondary (high) schools and according to the Fiji Bureau of Statistics, in 2009 there were 172 secondary schools in Fiji with 67,072 students enrolled. Just 12 of these are government (secular) schools. The majority of schools in Fiji are managed by faith-based organisations (including Catholic, Anglican, Methodist, Mormons, Seventh Day Adventist, Assemblies of God, Sangam and Sabha (Hindu), and Muslim) and by local communities. These schools also receive financial assistance from the government. There is only a handful of private schools in Fiji.

Western education was brought to Fiji in the 1850s by missionaries who set up informal education systems focussing on reading and arithmetic. Senibulu (2000) states, "The goal of such an education system was basically to have a core group of students who could read and interpret the bible and to act as evangelists, spreading God's message to the locals" (p. 3). Formal education based on a European model was established in 1916 with the creation of the first Education Department to undertake the provision of all educational services in the country. Anglicans and Indian religious

[1]Based on 2007 census data with a total population of 837,271.

groups such as the Arya Samaj and Muslims had begun operating schools for Indians from 1916 (Tavola 1991).

While the Fiji Education Department registered all schools and centralised all educational services, from the curriculum to the recruitment of teachers, it did not centralise the ownership of schools but left this instead with the communities and other local-based authorities. This form of ownership has been acknowledged as a 'strength' of the education system in Fiji for "often in the face of economic hardships and adversities, communities devote significant human and physical resources to raise funds in the name of educating the young people of the nation" (Senibulu 2000, p. 3). An arrangement with the New Zealand Department of Education that lasted from the 1920s to the 1970s saw teachers working in Fiji enacting the same schooling system as in New Zealand which "enabled a European system to exist as an isolated enclave, with a totally New Zealand orientation: syllabus, textbooks, examinations and teachers" (Tavola 1991, p. 15).

In Fiji, environmental education became incorporated in formal education through its inclusion in the 2007 and 2013 *Fiji Islands National Curriculum Framework* (Ministry of Education 2007, 2013).

The Research

This research project aimed to identify the forms of environmental education provided in Fiji secondary schools that are most effective in creating sustained pro-environmental behaviour. It also looked at the role played by other factors such as influence from peers, parents, religion, culture and the media in students adopting and sustaining pro-environmental behaviours. These issues are also addressed in a recent study that looked at Fijian adolescents' understanding and evaluation of climate change (Scott-Parker and Kumar 2018). In order to respond to these questions, in 2014 and 2015 data were collected from current students attending a tertiary institution in Suva, Fiji, to determine how environmental education had been taught in their secondary schools, the forms of pro-environmental behaviours they adopted while they were in secondary school, and the forms of pro-environmental behaviours they maintained after leaving school. Data were also collected about how various 'influencers' influence their pro-environmental behaviours.

Data were collected from students via a questionnaire, followed by focus groups and semi-structured interviews. Non-probability sampling was used to reach the tertiary students as the tertiary institution was approached and purposively requested to gain access to students enrolled in both science and non-science subjects. A total of 208 students completed the questionnaire: 116 (56%) females and 92 (44%) males. 28 students participated in the focus group discussions and 13 students in semi-structured interviews.

The questionnaire started by requiring participants to provide 'demographic' information. This included their gender, age, where they attended secondary school (to determine whether they attended school in Fiji or overseas), final year in secondary school (to determine the length of time since they were in a school environment), country of birth (to determine whether they are Fiji citizens), and current programme of study. Participants were not required to provide any data on their religious beliefs.

The questionnaire was divided into two parts. Part A required participants to reflect on their time in secondary school. Part B required participants to focus on what they are presently doing. The questionnaire had a mixture of structured and open-ended questions. The first question focused on the forms of pro-environmental behaviours students participated in both when they were in secondary school and at present. Several behaviour options were provided, and participants could choose as many as they wished. The next question focused on the reasons why they participated in these forms of pro-environmental behaviours. Participants were presented with the following list of options: (1) My parents/guardians said I had to; (2) It was one of my secondary school rules; (3) My friends were doing it; (4) It is part of my religion; (5) It is part of my culture; (6) I wanted to (why? please explain); (7) Other (please explain). In Part B of this question, 'My workmates are doing it' was also presented as an option. What influenced respondents' attitude and behaviour towards the environment was also discussed in the focus groups and interviews. The influence of religion was specifically discussed only when it was identified as a factor by the participants.

Semi-structured interviews were held with principals and teachers in each school to ascertain how environmental education is delivered in secondary schools in Fiji, and also establish the principals' and teachers' views on how environmental education should be taught in schools. Data were collected from 27 secondary schools situated on the island of Viti Levu. Attention was given to ensure that the selection of schools was a 'fair representation' of secondary schools on Viti Levu. Factors for consideration included: school size, co-educational versus girls or boys-only schools, urban versus rural schools, religious, private, community and government schools.

Of the 27 participating schools, 18 (67%) were located in an urban or peri-urban area and 9 (33%) were located in a rural area. The schools represent the following types of secondary schools in Fiji: 6 (22%) were government schools; 11 (41%) were schools managed by community groups; 9 (33%) were faith-based (religious) schools and there was 1 (4%) private school; 21 (78%) schools were co-educational. With the faith-based schools, there were five Catholic schools, one Methodist school, one Muslim school, one Hindu school and one Sikh school. Focus groups and interviews were audio recorded and notes were taken. Thematic analysis was used to analyse the data generated through the open-ended questions in the questionnaire and the focus groups and interviews. Simple statistical analysis was conducted with the quantitative data collected.

Results and Discussion

The data collected through the questionnaires showed that the greatest influence on student participants' pro-environmental behaviour, both while in secondary school

Table 16.1 Influences on participant's pro-environmental behaviour

While in School (percentage of all respondents)	**Influences**	**Presently** (percentage of all respondents)
53	My parents/guardians said I had to	47
57	It was one of my secondary school rules	25
11	My friends were doing it	18
12	It is part of my religion	17
20	It is part of my culture	24
N/A	My workmates are doing it	6
49	I wanted to	51

and post school, were parents and guardians – see Table 16.1. Interestingly, religion was an influence for 12% of the respondents while in secondary school and increased to 17% at the time of completing the questionnaire (after they had left school). While the influence of parents/guardians remains consistently strong, it is worth noting that the number of participants identifying the influence of religion, culture and friends increased from the time they were in school to *presently* (at the time of completing the questionnaire).

Data collected from the focus groups and semi-structured interviews with students included the discussion of religion. Christian student comments included the following verbatim remarks:

> The environment is important in my religion.... we must look after the environment because the environment looks after us.
>
> In my school we were always reminded about how we must be stewards of creation.
>
> We have to keep our school clean like we keep our homes clean because it is showing respect to God.
>
> We used to go to our school grotto sometimes to say our prayers and we used to pray about what we were thankful for and I think because we were outside, many of us used to mention that we were thankful for the environment, the plants and the animals.

Comments made by a student who follows the Islamic faith showed awareness of links between religious beliefs and impact on the environment:

> In my religion we are not allowed to eat pork, and this is also good because of the methane gas and the damage to the environment.

A student who follows the Hindu faith showed awareness of links between religious beliefs and practices, and how the environment should be regarded:

> The environment must be kept clean like we keep our house and prayer places clean. We have to make sure that we clean everything and there is no rubbish, especially at Diwali time (*Hindu festival of lights*).

Another student discussed how religion was important in her family and her parents insisted that they did the right thing:

> We are not allowed to waste food or water because it is a sin, and my parents get angry when we waste things so even now, I try to remember that.

One student spoke about his involvement with the Church youth group and how they were getting involved with beach clean-ups and it was good to be doing something good for the environment. He also remarked that he didn't realise just how much rubbish was left on the beaches, especially plastic rubbish:

> At first we couldn't believe the amount of rubbish on the beach, especially the plastic rubbish and the empty bottles.

While religion has an influence on the pro-environmental behaviours of some student participants in this study, it is not as strong as the influence of culture. Previous studies conducted in the US and the UK have looked at the influence of religion and have suggested a positive relationship between religious beliefs and environmental concern (Francis 1997; Sherkat and Ellison 2007). Yet Hayes and Marangudakis (2000) found that overall "religious identification is a relatively weak and inconsistent predictor of environmental attitudes and behaviour across nation" (p. 159). In this study, participants who spoke of the influence that religion had on their concern for the environment had predominantly attended faith-based schools.

Contrary to Chan and Islam's (2015) finding that environmental activism is frowned upon by the Singapore Government, the recognition given to the practice of religion in Fiji with significant events in the religious calendar of the major religions observed with a public holiday, environmental activism by religious groups is unlikely to be frowned upon. For example, clean-up campaigns are organised by the church youth groups. The first author's personal experience of having participated in various clean-up activities organised by a church youth group in Fiji, lends support to the view that such groups have the potential to provide impetus for greater involvement of youth in environmental issues.

A number of religions are practiced in Fiji and public holidays are observed for significant events in the major religions. Therefore, given the significance of religion in Fiji, the influence of religion on the pro-environmental behaviours of some participants is not surprising. The challenge is to consider how religion can be used to influence the attitude and behaviour of Fiji citizens thereby increasing the adoption pro-environmental behaviours.

The principals of the faith-based secondary schools mentioned how the importance of caring for the environment is emphasized in their schools with links made to their particular religion. The data collected from semi-structured interviews with school principals of faith-based schools highlighted a commonality across all the religions represented: the focus on cleanliness, keeping the school environment clean as a sign of respect for God's creation, and linking it to the teachings of their particular religious beliefs. This drives the school practice of keeping the environment clean, including the school compound, classrooms and student desks.

Respect and care for others including plants and animals is another common element amongst the faith-based schools.

In the Catholic schools the environment is discussed in religious studies. Reference is made to the role one is expected to play as 'stewards of God's creation' which involves looking after the environment and all living creatures including plants and animals. A principal mentioned Pope Francis's encyclical *'Laudato si'* guiding conversations on the environment in religion classes:

> It is important for our students to see the level of concern being expressed by the Pope, so they know that religion is also about our current issues – it helps them to relate to what's happening around them.

There are also rituals that occur outside of the religion classes:

> We reflect on God's creation.... and to appreciate nature *and* have morning prayers and devotion and focus on God's creation and the environment around us. The time for reflection is very important because it allows us to think about our own actions.

Interview data from the principal of the Methodist school emphasised similar discussions about care and respect for the environment and living things in religion classes:

> We start our assembly with a prayer and give thanks to God for all that we have and remind students that we all have a job to protect our environment.

Some student projects and school events also focus on environmental topics such as climate change:

> We asked the students to decide the theme for the annual school concert and they chose climate change. Each form focused on something about climate change. They did poetry, songs, dances and storytelling – it was really good to see how interested they are about climate change.

The school that follows the Sikh religion has a 'Sikh Environment Day' every March:

> That is when we spend more time on God's gifts and we plant trees. Every form (*class*) plant a tree and they have to take care of it too. We also have an assembly that day and invite guest speakers to speak about the environment.

In addition, they have what the principal called 'religion/moral' classes. Each form has one class per week where they discuss the importance of the environment for survival:

> Because everything is connected and we must respect and look after each other, the environment and all other creatures. It is important that the students get this message.

The principal added that religious values must be reinforced in the home and parents need to work in partnership with teachers and teach good practices in the home.

The Islamic school has a 'values program' where the discussions are guided by Islamic teachings. Discussion on the environment is included in the program, with links made to Islam and general moral teachings:

> We discuss with students about walking the path of righteousness and showing respect and care for the earth because we will be held accountable for our actions.

The principal mentioned that despite the teachings about cleanliness and not throwing rubbish around the school, they still had students showing disregard for the environment.

> There is only so much we can do …. they should be taught about these things at home too. You can tell the students who are not being taught about these things in their homes by the way they treat things.

Discussion on environmental topics and their link to the Hindu religion is included within the 'moral and ethics' classes at the school. The focus is mainly on cleanliness and respect for things:

> We mention that cleanliness is godliness and students, especially the Hindu students understand this because we are always told this in our homes and by the pundits (priests).

The percentage of *iTaukei* (Indigenous Fijian) students at the Islamic, Sikh and Hindu school is 20%, 60% and 50% respectively. This suggests that some of the religious teachings may not be familiar to a number of students since the majority of *iTaukei* are Christians. In this study, while the Christian schools (Catholic and Methodist) have religion classes, schools based on the Islamic, Sikh and Hindu faith incorporate their teachings within values and morals programs.

The similarities between the Catholic and Methodist schools are based on their teachings about the environment from a Christian perspective. In addition, they both dedicate specified time to religious teaching and therefore have additional opportunities to reinforce the message of care for the environment, particularly from the perspective of 'being stewards of creation'. For example, the daily prayers and devotions providing occasions to connect with nature. On the other hand, the schools following the Islamic, Sikh and Hindu religion, while connecting their care for the environment to the tenets of their respective religions, embed their 'religious' teachings about care for the environment as part of moral and values programs and classes. The influence of religion on pro-environmental behaviours should not be overlooked: It drives the main environmental practices in these schools and has been identified as a factor influencing the pro-environmental behaviours of student participants.

Conclusion

The key findings of this study suggest that care for the environment, grounded in the tenets of religious teaching, provides a strong foundation for faith-based schools to focus their environmental practices. The emphasis on cleanliness, common across all the faith-based schools included in the study guides the pro-environmental behaviours that students practice while in school. This includes cleaning up the school environment (compound, classroom and desks), not throwing rubbish around the school, and planting trees. The findings of this study hold important implications for policy-makers and curriculum developers in terms of how pro-environmental practices can be integrated into values and moral programs within schools. As the findings show, not all faith-based schools have separate religion classes, yet still integrate the key message of care for the environment through the values and moral programs.

The research further suggests that school management should involve parents and guardians in the adoption and sustaining of pro-environmental behaviours. The need for parental support in instilling and reinforcing pro-environmental behaviour in the home has been highlighted by the school principals. Questionnaire data and student comments in focus groups and interview discussions also indicate that parents and guardians have the greatest influence on student participants' pro-environmental behaviour, both while in secondary school and post-school. School management, in their communications with parents and guardians, can encourage them to reinforce the message of care for the environment in the homes.

Lastly, the research findings have added to the overall body of research on the role of religion on environmental education in schools and the adoption of pro-environmental behaviours. This is particularly significant for Fiji since there is limited research in this area (Nunn et al. 2016; Scott-Parker and Kumar 2018).

The limitations of this study include the use of a single case study—Fiji. To some extent this limits the generalisability of the findings to other Pacific Islands. The Christian schools had greater representation. Only one school from the Hindu, Sikh and Muslim faith participated in the study which limited the data available for these faith-based schools.

References

Ali A (1997) Aspects of Fiji Indian history, 1879–1939: A society in transition. Economic and Political Weekly 12(42):1782–1790

Boyes E, Stanisstreet M (2012) Environmental education for behaviour change: which actions should be targeted? Int J Sci Educ 34(10):1591–1614. https://doi.org/10.1080/09500693.2011.584079

Chan A, Islam MD (2015) State, religion, and environmentalism: Fostering social cohesion and environmental protection in Singapore. Environmental Sociology 1(3):77–189

Environment Australia (2000) *Environmental education for a sustainable future: National Action Plan*. Canberra: Environment Australia

EPA United States Environmental Protection Agency (2016b) What is environmental education? Retrieved from https://www.epa.gov/education/what-environmental-education

Farooq S, Ansari ZI (1983) Wastewater reuse in Muslim countries: an Islamic perspective. Environ Manage 7(2):119–123

Fiji Bureau of Statistics (2008) Fiji Census Data 2007. Retrieved from http://www.statsfiji.gov.fj/index.php/chart-gallery/209-population-by-age-2007-census

Fiji Bureau of Statistics (2018) Fiji Census Data 2017. Retrieved from https://www.statsfiji.gov.fj/index.php/statistics/social-statistics/education-statistics43?highlight=WyJzY2hvb2xzIl0=

Francis L (1997) Christianity, personality and concern about environmental pollution among 13 to 15-year-olds. Journal of Beliefs & Values 18:7–16

Gifford R, Nilsson A (2014) Personal and social factors that influence pro-environmental concern and behaviour: A review. International Journal of Psychology 49(3):141–157

Goldenberg S, Kirchgaessner S (2015, September 26) Pope Francis demands UN respect rights of environment over thirst for power. The Guardian. Retrieved from http://www.theguardian.com/world/2015/sep/25/pope-francis-asserts-right-environment-un

Greenall AE, Womersley JC (eds) (1977) Development of environmental education in Australia-Key issues. The Curriculum Development Centre, Canberra, Australia

Hayes BC, Marangudakis M (2000) Religion and environmental issues within Anglo-American democracies. Review of Religions Research 42(2):159–174

Jain P (2010) Ten key Hindu environmental teachings. Retrieved from http://www.greenfaith.org/media/press-clips/ten-key-hindu-environmental-teachings-1

Kearns L (1997) Noah's ark goes to Washington: A profile of evangelical environmentalism. Social Compass 44:349–366

Kelly JD (1992) Fiji Indians and 'commoditization of labor'. American Ethnologist 19(1):97–120

Kelly MP, Barker M (2016) Why is changing health- related behaviour so difficult? Public Health 136:109–116. https://doi.org/10.1016/j.puhe.2016.03.030

Luetz JM, Nunn PD (2020) Climate Change Adaptation in the Pacific Islands: A Review of Faith-Engaged Approaches and Opportunities. In: Leal Filho W (ed) *Managing Climate Change Adaptation in the Pacific Region* (Climate Change Management ed., pp. 293–311). Springer Nature, Switzerland. https://doi.org/10.1007/978-3-030-40552-6_15

Luetz JM, Margus R, Prickett B (2020) Human Behavior Change for Sustainable Development: Perspectives Informed by Psychology and Neuroscience, In: Leal Filho W, Marisa Azul A, Brandli L, Gökcin Özuyar P, Wall T (eds) *Quality Education, Encyclopedia of the UN Sustainable Development Goals*. Cham, Springer. https://doi.org/10.1007/978-3-319-69902-8_12-1

Ministry of Education (2007) *The Fiji Islands National Curriculum Framework—Education for a better future*. Republic of the Fiji Islands. Retrieved from http://www.paddle.usp.ac.fj/collect/paddle/index/assoc/fj33.dir/doc.pdf

Ministry of Education (2013) *The Fiji National Curriculum Framework—Quality Education for Change, Peace and Progress*. (2nd ed). Republic of the Fiji Islands

Nelson W, Luetz JM (2019) What can we learn from pope Francis about change management for environmental sustainability? A case study on success factors for leading change in change-resistant institutional envi- ronments. In: Leal FW, Consorte MCA (eds) Sustainability and the humanities. Springer, Cham. https://doi.org/10.1007/978-3-319-95336-6_29

Nunn P, Mulgrew K, Scott-Parker B, Hine DW, Marks ADG, Mahar D, Maebuta J (2016) Spirituality and attitudes towards nature in the Pacific Islands: insights for enabling climate-change adaptation. Clim Change 136:477–493

Oliver R (2008, January 28) All about: Religion and the environment. *CNN*. Retrieved from http://edition.cnn.com/2008/WORLD/asiapcf/01/27/eco.about.religion/?iref=nextin

Osbaldiston R, Schott JP (2012) Environmental sustainability and behavioral science: Meta-Analysis of pro-environmental behavior experiments. Environment and Behaviour 44(2):257–299

Palmer M, Finlay V (2003) Faith in conservation: new approaches to religions and the environment. The World Bank, Washington, DC

Pepper M, Leonard R (2016) How Eco-theological Beliefs Vary Among Australian Churchgoers and Consequences for Environmental Attitudes and Behaviours. Review of Religious Research 58(1):101–124

Pope Francis (2015) Laudato Si. Encyclical letter of the Holy Father on the care of our common home. Retrieved from https://laudatosi.com/watch

Rice G (2006) Pro-environmental behavior in Egypt: Is there a role for Islamic environmental ethics? J Bus Ethics 65:373–390

Sapolsky R (2017) Behave: the biology of humans at our best and worst. Penguin Random House, London, UK

Scott-Parker B, Kumar R (2018) Fijian adolescents' understanding and evaluation of climate change: implications for enabling effective future adaptation. Asia Pacific Viewpoint 59:47–59

Senibulu I (2000) *Education for All: The Year 2000 Assessment. Country Report by Fiji*, National Assessment Coordinator, Ministry of Education, Marela Houses, Suva, Fiji. Retrieved from http://www.accu.or.jp/litdbase

Sherkat DE, Ellison CG (2007) Structuring the Religion—Environment Connection: Identifying Religious Influences on Environmental Concern and Activism. Journal for Scientific Study of Religion 46(1):71–85

Steg L, Vlek C (2009) Encouraging pro-environmental behavior: An integrative review and research agenda. Journal of Environmental Psychology 29:309–317

Tavola H (1991) Secondary education in Fiji: A key to the future. Institute of Pacific Studies, University of the South Pacific, Suva, Fiji

Tuimaleali'ifano M (2011) *Origin of Fiji's People.* Unit 1 Year 11 History Textbook. Fiji Ministry of education, Heritage and Arts. Retrieved from http://www.education.gov.fj/images/TextbookResources/ECopies/YEAR%2011/HISTORY%20CULTURAL%20INTERACTION/Unit%201%20final.pdf

UNESCO (1978) Intergovernmental Conference on Environmental Education: Tbilisi (USSR), 14–26 October 1977, Paris: UNESCO

Walid M, Luetz J M (2018) From education for sustainable development to education for environmental sustainability: reconnecting the disconnected SDGs. In: Leal Filho W (eds) Handbook of sustainability science and research. Springer Nature, Switzerland, pp 803-826. https://doi.org/10.1007/978-3-319-63007-6_49

White L (1967) The Historical Roots of Our Ecologic Crisis. *Science, 155*(3767), 10 March, pp. 1203–1207

Woodrum E, Wolkomir MJ (1997) Religious effects on environmentalism. Sociological Spectrum 17:223–234

Wright TSA (2002) Definitions and frameworks for environmental sustainability in higher education. Journal of Sustainability in Higher Education 3(3):203–220

Yencken D, Fien J, Sykes H (eds) (2000) Environment, education and society in the Asia-Pacific: Local traditions and global discourses. Routledge, London

Part VI
Faith for Survival

Chapter 17
Wonders in the Heavens Above, Signs on the Earth Below: Pacific Islands Pentecostalism, Climate Change and Acts 2

John D. Griffiths

Abstract Pentecostalism's understanding of creation is of growing importance as Pentecostalism continues to grow in areas most effected by climate change. As such, there is a need to identify, reflect upon and develop a Pentecostal ecology in order to establish an effective and lasting response to the present ecological crises in areas like the Pacific Islands. Pentecostalism's enchanted understanding of creation, where the material is imbued with the spiritual, leads Pentecostals to propose spiritual solutions (repentance, exorcism) to material problems (ecological crises). This chapter develops this enchanted understanding of creation by incorporating an ecological reading of Joel 1–2 and Acts 2: 17–21, which brings to the fore the ecological themes of interconnectedness and voice. This chapter finishes with a concluding synthesis, which reflects upon the ways that Pentecostalism's ecology can be further advanced through these ecological themes, and ways these insights can be applied to the modern context of the Pacific Islands.

Introduction

Pentecostalism has firmly established itself as a significant force in Christianity over the past 120 years, and in the South Pacific Islands, Pentecostalism "exceeds church growth rates anywhere else in the world" (Gore 2019, p. 297; Brison 2007; Anderson 2013). A significant factor in Pentecostalism's growth in the Pacific Islands is attributed to the similar cosmological worldview that Pentecostalism shares with the traditional ancestral worship in the Pacific Islands (Smith 2010; Eriksen 2017; Rio et al. 2017; Gallagher and Gallagher 2019; Macdonald 2019). For example, Gallagher and Gallagher have argued that Pentecostals and Papua New Guineans have a similar worldview based on the same "cosmological assumptions", which leads to a shared understanding concerning spirituality and the "spirit world" (Gallagher and Gallagher 2019, p. 328). This similar cosmological worldview, sometimes described

J. D. Griffiths (✉)
Alphacrucis College, Parramatta, NSW, Australia
e-mail: john.griffiths@ac.edu.au

J. M. Luetz and P. D. Nunn (eds.), *Beyond Belief*, Climate Change Management,
https://doi.org/10.1007/978-3-030-67602-5_17

as enchanted or non-dualistic, leads to a unique, spiritually charged, understanding of creation.

This enchanted understanding of creation has come to the fore as the devastating effects of the climate crisis increase in the Pacific Islands. Indeed, as others have pointed out, the Pacific Islands are particularly vulnerable to the effects of climate change (Keener et al. 2012; Cox et al. 2018). The warming of the Earth's ocean is leading to rising sea levels, which is leading to "coastal inundation and the submergence of low-lying islands" (Cox et al. 2018, p. 382). Moreover, the rise in temperature in the seas surrounding the Pacific Islands has led to an increase in the location and severity of cyclones (Keener et al. 2012). These two effects, coupled with the limited economic resources available, leads to Pacific Islands being on the "front lines" of the effects of climate change (Cox et al. 2018). However, as the editors of this book have clearly elucidated, foreign intervention from Western technocratic nations in the Pacific Islands has not been effective in gaining the support of local Pacific Island communities (Luetz and Nunn 2020).

In this chapter, I seek to explain the Pentecostal conceptualisation of creation, and by extension, its understanding of the current climate crisis. Understanding the "enchanted" aspect of Pentecostalism's worldview of creation is pivotal, as Pentecostalism is a major expression of Christianity in the Pacific Islands. Therefore, climate change mitigation strategies in the Pacific Islands will not be wholly effective unless Pentecostalism is addressed. In the first section of this chapter, I draw upon the work of James K. A. Smith, who has convincingly argued that Pentecostalism has an enchanted view of creation. This enchanted understanding of creation leads to an inclination towards spiritual solutions (repentance, exorcism or faith) for material problems (drought, rising sea levels). I then engage with two examples of this enchanted view of creation, noting how spiritual solutions are often proposed by Pacific Island Pentecostals in order to address the effects of climate change.

From here, in the following four sections, I suggest a way for Pentecostal spirituality to be more ecologically sensitive, by addressing the images of creation used in the quotation of Joel 2: 28–32 at Pentecost (Acts 2: 17–21). As I outline in the third section of this chapter, Acts 2 is the most obvious starting place for engaging with Pentecostal spirituality, as Luke's portrayal of Pentecost is the foundational text for Pentecostalism and its spirituality. In the fourth section, I set out two ecological lenses, interconnectedness and voice, that I will apply to Joel 1–2 and Acts 2: 17–21. I then engage with the intertextuality of the Acts 2: 17–21, noting the ecological lens of interconnectedness and voice that creation has in Joel 1–2. Finally, I follow this theme of interconnectedness and voice into Acts 2: 17–21, from which I conclude with some applications for engaging with Pacific Island Pentecostalism and climate change.[1]

[1] In this chapter, I focus on the Pentecostal understanding of creation, and how the ecological lens of interconnected and voice help inform this Pentecostal ecology. As this Pentecostal understanding of creation is quite different from a technocratic understanding of nature, and due to space constraints, I do not significantly address the practical applications of this research. Further research could be done in the area of the practical application of this research.

Pentecostalism's Enchanted Worldview

Pentecostalism has a unique worldview compared to other expressions of Christianity, a worldview that is often in conflict with materialistic or technocratic worldviews (Yong 2005, 2009; Smith and Yong 2010). Smith notes "that embedded in the embodied practices and spirituality of pentecostalism are the elements of a latent but distinctive *understanding* of the world, an affective 'take' on the world that constitutes more of a social imagery than a cognitive framework" (Smith 2010, p. 30). While different expressions of traditional Christianity may differentiate themselves based on doctrines, it is their epistemology, which derives from its unique worldview, that most clearly differentiates Pentecostals. In broad strokes, if traditional expressions of Christianity tend to adopt a modern worldview, then Pentecostalism adopts a distinctively counter-modern worldview (Smith 2010).

Smith argues persuasively that Pentecostalism has five elements to its worldview: "a radical openness to God…, an 'enchanted' theology of creation and culture…, a nondualistic affirmation of embodiment of materiality…, an affective, narrative epistemology…, [and] an eschatological orientation to mission and justice" (Smith 2010, p. 32). Each one of these elements of Pentecostal worldview could be explored in relation to views on creation and more pressingly, climate change.[2] However, I see the second element, the enchanted theology of creation, as the most pivotal element for engaging with Pacific Island Pentecostalism's understanding of the current climate crisis.

Smith defines the 'enchanted' nature of Pentecostalism's view of creation as "the implicit affirmation of the dynamic, active presence of the Spirit not only in the church, but also in creation" (Smith 2010, p. 39).[3] For Pentecostals, the material world is imbued with spiritual significance because the Holy Spirit is seen as imminently present in creation. As Smith further explains, "nature, in a sense, is 'suspended' in the Spirit of creation; or we might say that creation is 'charged' with the Spirit's presence. Nature, then, is always more than 'the natural.' It is suffused with something more; there is always more than meets the naturalizing eye" (Smith 2010, p. 40). Creation is not simply material, but rather the material creation is deeply charged with spiritual significance.

However, this enchantment of creation with the Holy Spirit does have an inverse belief, namely, a possible mis-enchantment of creation that is associated with the other (often demonic) spiritual forces. Smith notes that Pentecostalism is also "deeply attentive to what we might describe as the mis-enchantment of the world by other

[2]Pentecostal ecologies often start with Smith's first element, a radical openness to God (Clifton 2009; Kärkkäinen 2014), or the affirmation of embodiment, e.g. healing (Ede 2014; Tallman 2014), or they wrestle with the eschatological orientation of Pentecostalism (Waddell 2012). This chapter takes a different approach by focussing on the enchanted aspect of Pentecostalism's worldview.

[3]It is worth emphasising that not all forms of Pentecostalism have this clear enchanted worldview. Smith (2010, p. 41) notes, "While North American [and I'd add Australian] pentecostalism is increasingly 'naturalized' on this score, all commentators agree that the implicit cosmology assumed by spiritual warfare is one of the primary factors in the explosion of Christianity in the majority world, particularly where indigenous or 'primal' religions emphasize a similarly enchanted cosmology.".

spirits" (Smith 2010, p. 41). This mis-enchantment of creation is often found in situations where creation is having harmful effects on humans (e.g. sickness, famine, rising sea levels). This enchanted worldview, which combines the material and the spiritual, then opens the door to a mis-enchantment of creation, which attributes the effects of the degradation of creation to other spiritual forces.[4]

When Pentecostal theologians engage with this tendency towards the mis-enchantment of creation, it is usually in the context of healing (Warrington 2000, 2005). Once the Pentecostal worldview of an enchanted creation is accepted, their theology of healing can easily become dualistic; the material (the sickness) is associated with the spiritual (an unrepented sin, a demonic force, a lack of belief) and the cure for the material is therefore spiritual. This obviously presents numerous pastoral and theological issues, which Pentecostal theologians have been addressing in recent history (Warrington 2000, 2005).

This enchanted view of creation, as embodying both the Holy Spirit and other spirits, has implications for understanding how Pentecostals view ecological crises. When exploring geographically how Pentecostals understand creation and more specifically climate change, Pentecostal scholars have focussed on North America (Chan 2014), Africa (Tallman 2014), Europe (Ede 2014), and Latin America (Waldrop 2014). However, less attention has been directed towards the Pacific Islands and how this enchanted understanding of creation understands ecological crises.

This Enchanted Worldview in the Pacific Islands

In the last decade, there has been increasing research into Pentecostalism in the Pacific Islands from anthropologists (Brison 2007, 2017; Ryle 2011, 2012; Eriksen 2017). However, little of this work specifically engages with Pentecostalism's understanding of climate change (Cox et al. 2018). Interestingly, Eriksen's anthropological insights from researching Pentecostalism in Vanuatu, while being quite independent of Pentecostal scholarship articulating their worldview, has articulated this enchanted view of creation (Eriksen 2017). Eriksen (2017) notes that for Pentecostals in Melanesia any place can be,

> a 'place' where the immediate experience of the Holy Spirit is a defining feature of everyday life. … In this world, it is not only the believers who can see and feel the Spirit. Rather, the Spirit is already there: it is a taken-for-granted part of this 'place,' and some engage with it and others do not. Thus, the 'distance' between the materially 'real' and the spiritually 'real' is negated analytically. (p. 166)

The worldview that Eriksen is describing here, is an enchanted worldview, where the 'place' (creation) is imbued with spiritual significance. There is no distance

[4]When building a theology of creation, Pentecostal theologians rarely address this possibility of a negative enchantment of creation, possibly due to the desire to be prescriptive rather than descriptive. That is, a Pentecostal theology of creation must account for the degradation of our environment and Pentecostalism's tendency to understand this degradation through the lens of a negative enchantment.

between the material and the spiritual. However, Eriksen does not extend her discussion to a Pentecostal understanding of creation, rather keeping her focus on the social realm. In this section, and building upon the first section, I will give two examples, both from Fiji,[5] for this enchanted understanding of creation and how it engages with the effects of climate change.

First, Ryle addresses the Pentecostal Fijian cosmological understanding of drought and unproductive crops, when addressing the traditional reconciliation ceremony on Vitilevu, Fiji's main island, in November 2003 (Ryle 2012). This reconciliation ceremony was conducted by the ACCF (the Association of Christian Churches in Fiji), which is a combination of Methodist and Pentecostal churches with a "particular focus… [on] a Pentecostal healing of the land and exorcism of the ancestral spiritual power inherent in the land" (Ryle 2012, p. 93).

This reconciliation ceremony was held in light of the murder of Thomas Baker and his associates in 1867 at Nubutautau (in central Vitilevu), with the murder being seen as cursing the land, resulting in poor crops and a lack of rain (Ryle 2012). As Ryle (2012) recounts,

> Brokenness between clans needs to be restored through rite of reconciliation that will heal social relations, relations with the spirit world and the land, i.e. reconciliation will bring nourishment and fertility to the social barrenness of human relations and the material barrenness of the land—and will transform the powers of the spirit world from 'inimical' to 'supportive'. (p. 96)

The material state of creation is attributed to the spiritual (and by extension moral) state of the people of Fiji.

Importantly, while Pentecostal churches associated with the ACCF participated in the reconciliation ceremony, they attributed the barrenness of the land, not to the murders of Thomas Baker as his associates, but to evil spirits that needed to be exorcised through the power of the Holy Spirit (Ryle 2012). The fertility or barrenness of the land was linked directly to the presence of ancestral spirits (which Pentecostals saw as evil, see for example Newland 2004) and the moral state of the people. To heal the land, the people of Fiji, according to the Pentecostals, needed to repent, fast and exorcise the evil spirits through the power of the Holy Spirit. As Ryle (2012) notes,

> While the 2003 Baker reconciliation in outer form was a traditional Fijian reconciliation ceremony, prior to the event ACCF pastors had been engaged in preparing the people of this area beforehand, through Pentecostal prayer and ritual to confess their sins and reconcile with God, which would lead to healing the land and social relations. (p. 93)

This enchanted worldview – seeing the material as imbued with the spiritual – leads to spiritual solutions (repentance, fasting, exorcism) to material problems (lack of rain, poor harvests).

Second, Cox et al. have explored the Fijian understanding of the effects of climate change, as they elucidated the Christian interpretation of recent tropical cyclones

[5]My reliance on Fiji for examples are solely due to the fact that the majority of scholarship on Pentecostalism in the Pacific Islands has focussed on Fiji.

in Fiji (Cox et al. 2018). They show that Cyclone Winston (February 2016) was commonly understood by Fijian Christians as "an act of divine judgement and punishment" or as a consequence of "original sin" (Cox et al. 2018, p. 381). When grappling with ecological disasters, like flooding or cyclones, Pentecostal Fijians use "key Old Testament texts, such as 2 Chronicles 7: 14, [which] are thought to set out the paradigm of post-disaster humility and repentance" (Cox et al. 2018, p. 392). When faced with ecological disasters, Fijian Pentecostals tend to interpret these disasters as a divine judgement for the sin and lack of repentance among the people.

Cox et al. (2018, p. 389) note that "interpreting natural disasters as an act of divine judgment is an ancient practice in Christianity," however, technocratic worldviews seek scientific answers to these *natural* disasters. This leads to a clash between the enchanted and technocratic worldviews as,

> Secular state and NGO disaster management practices derive from this shift in perspective [from enchanted to technocratic] and rarely acknowledge local perceptions of divine agency. Nevertheless, … religious interpretations of "natural" disasters are highly important for understanding local and national responses to these disasters in many contexts outside the West, including the Pacific. (Cox et al. 2018, p. 389)

The technocratic worldview dismisses any divine agency in ecological disasters, meaning that disaster management efforts presuming this technocratic worldview rarely connect with localised responses to ecological disasters. This is shown in the religious response to the secular hashtag#StrongerthanWinston, which Cox et al. (2018) note that:

> Despite its broad appeal, "Stronger than Winston" did not sit comfortably with some evangelical [which would include Pentecostal] Christians who regarded the cyclone as an act of chastisement from God. For these groups, the claim that the community was "Stronger than Winston" was a hubristic and anthropocentric claim that ignored divine agency and therefore diverted attention from the urgent need to demonstrate a humble disposition toward God through public acts of repentance. (p. 389)

The clash of enchanted and technocratic worldviews leads to foreign technocratic responses being seen as hubristic by the local community, leading to a lack of engagement by local inhabitants and a low longevity to foreign technocratic responses (Luetz and Nunn 2020).

These two examples show this enchanted worldview in action, as ecological disasters, whether they are severe like cyclones or flooding, or long-term like drought or failing crops, are understood as indicating a spiritual problem. In this enchanted worldview, ecological disasters require a spiritual solution, whether that be repentance of particular sins or the exorcism of ancestral spirits. Having elucidated the enchanted nature of Pentecostal worldview that is present in the Pacific Islands, I will now engage with this worldview, through an ecological reading of Acts 2: 17–21, which is one of the key biblical foundations of Pentecostal worldview.

Acts 2 and Pentecostal Worldview

In the next four sections, I examine the imagery of creation found in Acts 2: 17–21, which is a quote from Joel 2: 28–32.[6] This quote begins with the description of the Spirit of God being poured out on "all flesh" in the last days (Acts 2: 17a). From here, the effect on humans is described, as both males and females prophesy, dream and see visions (2: 17b–18). From this human perspective, the orientation of the text turns cosmological, describing wonders in the heavens and signs on the Earth (2: 19a). The imagery of blood, fire, smoke and darkness is used as a prelude to the coming of the day of the Lord (2: 19b–20). The passage ends with an exhortation for all to call upon the name of the Lord for salvation (2: 21).

I have chosen this passage due to its significance to Pentecostal worldview. Acts 2 stands at the very foundation of Pentecostalism's spirituality, as Pentecostals see their spirituality as mimicking the spiritual experience of the early believers at Pentecost described in Acts 2 (Kay 2011). Experiencing a renewed spiritual outpouring of the Holy Spirit at the beginning of the twentieth century, the early Pentecostal movement legitimised their spiritual manifestations and outlook by identifying it as analogous with Acts 2. This identification remains strong into the present form of Pentecostalism. Indeed, so pivotal is Luke's account of Pentecost to Pentecostals, that they describe themselves in light of this narrative. To engage with Acts 2, is to engage with the central biblical foundation of Pentecostalism.

Pentecostal (and charismatic) interpretations of this passage either have a solely anthropological focus (Keener 2012; Cho and Park 2019) or have seen verses 19 and 20 as describing the decay, destruction, or decimation of the Earth. For example, after quoting Acts 2: 19–20, Yong (2005, p. 178–179) states "this is metaphorical language functioning, in both Acts and its original context (the prophecy of Joel), as apocalyptic discourse that calls attention to the cataclysmic events attending the arrival of the Day of the Lord and the salvation and vindication of the people of God." Yong sees Acts 2: 19–20 as describing a cataclysmic event, which is the prelude for the salvation of the *people* of God.

While there has been a tendency to understand Acts 2: 17–21 in the light of other texts like 2 Peter 3:10, which can be more easily read as describing a destruction of the Earth (Waddell 2012), I will offer a more ecologically sensitive interpretation of Acts 2: 17–21. I trust this will guide Pentecostals, globally and in the Pacific Islands, to a more collegial and conscientious relationship with creation. I will approach Acts 2: 17–21 intertextually, noting the central role that creation plays in Joel 1–2,[7]

[6]There are seven variations that Luke makes to the quotation from Joel 2: 28–32 (Keener 2012, p. 874–877).

[7]Wall (2000, p. 542) describes intertextuality "as a broad reference to the various ways by which biblical writers presume the continuing authority of their Scripture that is cited or 'echoed'... when it is exegeted to amplify the meaning of this sacred tradition (*traditium*) as the word of God (*traditio*) for new readers or auditors." The intertextuality of Acts 2: 17–21 with the wider work of Joel 1–2 is well established (Wenk 2000).

while also drawing upon the ecological lens' of interconnectedness and voice to draw ecological insights from Joel 1–2 and Acts 2: 17–21.

The Ecological Principles of Interconnectedness and Voice

In the following two sections, I will read Joel 1–2 and Acts 2: 17–21 from an ecological perspective. I will focus on two ecological lenses developed from ecological readings of the Bible, that is the lens of interconnectedness and voice (Habel 2000; Horrell 2014).[8] The interconnectedness between humans and the rest of creation is a major ecological lens for reading the Bible (Habel 2000; Horrell 2014). Horrell (2014) notes that the ecological lens of interconnectedness speaks of the,

> inextricable connection between the fate of humans, animals, plants and land. Whether this be in a negative way, as when human disobedience stands at the commencement of a story of general degradation and decay…, or in a positive way, as when prophets dream of a time of fertility and flourishing in a future time of righteousness…, the essential connections remain the same. (p. 131)

Rather than seeing humans as independent creatures that are detached from the fate of creation, this principle of interconnectedness sees humans as dependent on and deeply connected to creation.

Importantly for this chapter, the focus on interconnectedness leads to a broader understanding of repentance. As Horrell (2014, p. 132) notes, this principle "offers both challenge and contribution to a theology of sin that needs to move away from a classical focus on interpersonal relationships to a broader compass, to the whole network of (ecological) relationships in which the human animal is inextricably embedded." The ecological lens of interconnectedness helps to point us to the ecological effect of human sin and the need for repentance that is ecologically sensitive.

The ecological lens of interconnectedness then leads to a second ecological lens, namely creation having a voice (Habel 2000; Horrell 2014). As both humans and all of creation are bound in covenant relationship to God (such as the Noahic Covenant), "the whole of creation, and not just humanity, is bound in covenant relationship with God and called into being to express praise of God" (Horrell 2014, p. 133). While Horrell focusses mainly on creation's ability to express praise of God, the Earth Bible Team note that creation can also express anguish and lament in the face of injustice (Habel 2000; see also Hayes 2002). That is, creation has a voice, either to express praise to God or to express anguish and lament at the ecological ramifications of human sin. In the next section, I explore the ecological themes of interconnectedness

[8] The ecological readings in the following two sections will rely on the ecological readings Braaten (2008) for Joel 1–2 and Trainor (2020) for Acts 2: 17–21. Further research in the ecological themes of these passages is needed, with could also explore the role that the priesthood and Temple play in Joel 1–2 and Acts 2: 17–21.

and voice in Joel 1–2, which will establish the intertextuality context for Acts 2: 19–20.

Joel 2: 28–32 and the Imagery of Creation

Joel 1 is set in a scene of the interconnectedness between humans and the rest of creation, where human overconsumption has led to the judgement of God and to creation raising its voice in lament. Joel begins with a call for all the inhabitants of the land, including non-human inhabitants (e.g. Hosea 4: 1–3), to listen to God's word and remember a great locust plague that ravaged the land (Joel 1: 2–4). This coming judgement of God is for the present sin among the people, yet this sin is not explicitly stated in Joel 1–2. Concerning this unknown sin, Braaten (2008) notes that,

> the only possible hint concerning this human sin might be given in the odd beginning, where 'drunkards' and 'wine drinkers' are the first groups called to lament. The prophets sometimes connect excessive wine drinking with conspicuous consumption that accompanies apathy for the poor, and oppression of the weak. (p. 69)

As is found in the wider prophetic literature, drunkenness is associated with excessive consumption (Amos 6: 4–7; Isa 5: 20–23) and is even found in the context of land theft (Isa 5: 8–13). Moreover, this suggestion is further supported by the fact that the suffering of creation is given significant space in Joel 1–2, indicating that the sin that the prophet calls the people to repent of is in some way a sin that has an ecological effect. Therefore, Joel 1 could be set in a situation where excessive human consumption has led to an ecological judgement from God.

In this possible context of excessive consumption, God calls the drunkards (Joel 1: 5), the priests (1: 9), and an unnamed female subject (1: 8) to lament. This unnamed female subject is most probably the Earth, as the surrounding verses describe the lament of creation (Braaten 2008). Hayes notes the various verbs that describe the damage to the Earth in Joel 1: 7 and 10 are used elsewhere for humans who are lamenting and mourning (Hayes 2002). It is then creation—the fields, the ground and the trees—that first respond to this call from God to lament.

Yet, humans seemingly remain ignorant in Joel 1 as there is no indication that humans repent or lament. As Braaten (2008, p. 68) notes, "the hazy condition of the drunkard in Joel 1: 5 is [then] symptomatic of the entire community; perhaps they care about nothing except their self-gratifying consumption which insulates them from feeling the pain of others." While the Earth raises its voice in response to God's call and the injustice in its midst, humans in Joel 1 remain unmoved.

This lamenting called upon by God and fulfilled by creation brings implications for the worship of God, as "grain offering and drink [wine] offerings are cut off from the house of the Lord" (1: 9, 13). In this context a holy fast is called (1: 14), which could be out of necessity, as "food has been cut off before our very eyes" (1: 16), "the grain has dried up" (1: 17) and "the streams of water have dried up" (1: 20).

This overconsumption and ecological neglect then has ramifications for the religious and the spiritual. Braaten (2008) observes that,

> while it is common understanding that the proper functioning of the cult [the Temple and related priesthood] maintains the order of creation, or Earth, here the opposite is stated: it is the proper functioning of Earth that maintains the order of the cult! The divine-human interaction in the cult is incomplete without a third partner, Earth. (p. 66)

These images of creation in Joel 1 then communicate an interconnectedness of morality, spirituality and creation. Human overconsumption leads to the judgement of God and the lament of creation.

The second chapter of Joel echoes the first, as Zion faces another imminent threat (Joel 2: 1). Joel proclaims that the day of the Lord is coming (2: 1b), which brings darkness, gloom and blackness (2: 2). This coming army will bring God's judgement upon Zion, which is likened to turning the Garden of Eden into a wilderness (2: 3). That is, the judgement of God brings calamity not only on humanity but the whole of creation, as "the earth shakes [and] the heavens tremble" (2: 10a), and the sun, moon and stars lose their light (2: 10b).

Joel urges the people to repent and fast (Joel 2: 12–17), however, it is not clear whether the human audience actually repents or fasts (Braaten 2008). It is God who ultimately relents (2: 18). The primary reason God relents is because God is jealous for God's land (2: 18a), while also having compassion on God's people (2: 18b). This lifting of judgement on the people brings an ecological flourishing. Wild animals are told not be afraid because the wilderness pastures will be lush again (2: 22a). The trees will again bare plentiful harvests (2: 22b) and there will be abundant rainfall (2: 23). This restoration of Israel brings a reinstatement of the grain offerings and drink offerings (2: 14), which is then returned two-fold in the restoration from God (2: 19, 24). As with chapter one, Joel 2 again speaks of an interconnectedness of morality, spirituality and creation, as immorality begins a severe eco-cosmic judgement, while the relentment of God brings a reinstatement of spirituality and ecological flourishing.

These ecological themes of interconnectedness and voice then flow into Joel 2: 28–32, where the Spirit of God will be poured out on "all flesh" (Joel 2: 28). In the wider context of Joel 1–2, where creation is more attentive to the commands of God than humans, this "all flesh" that Joel refers to surely includes the previously lamenting cattle and sheep (2: 18) and wild animals (2: 20), as well as, the fields, streams, ground and trees. Perhaps paralleling the Pentecostal understanding of an enchanted creation, Joel here sees the promise of God's spirit as imminently present in creation. Rather than remaining inattentive and unmoved, humans will join creation in responding to the leading of God, as the spirit of God will be poured out on all flesh, both human and non-human.

In this context, the "wonders in the heavens and on the Earth" (Joel 2: 30), along with the cosmic imagery of blood, fire, smoke and darkness, echo the looming judgement that humans and creation have already been confronted with in Joel's prophetic opening chapters (1: 19, 2: 2–3, 5, 10–11). Joel indicates that one can be saved from this terrible cosmic judgment, by calling on the name of the God (2: 32). Importantly, Joel 1–2 has consistently described the lament of creation and its call for salvation from judgement. The difference with this third coming of the day of God described in Joel 2: 31, is that humans, with the spirit of God, will join with creation in crying out for this salvation. As Horrell (2014, p. 135) states, the principle of voice indicates "that the hope and vocation of salvation are not only a matter for humans but for the whole created order."

To summarise, the ecological principles of interconnectedness and voice run throughout Joel 1–2. Humans and creation are deeply interconnected, as human sin (possibly overconsumption) leads to ecological judgement, while God's relentment of judgement on this human sin brings an ecological flourishing. Moreover, while humans in Joel 1–2 remain unresponsive to God's call to lament, it is creation that fulfils this call and laments. These principles are then found in Joel 2: 28–32, as the spirit of God is poured out on all flesh, human and non-human (indicating an interconnectedness), and humans join with creation in a call for salvation (indicating the theme of voice).

Implications for Acts 2: 17–21

There are two indications that these ecological themes of interconnectedness and voice are taken up by Luke's use of Joel 2: 28–32 in Acts 2: 17–21. First, on the theme of interconnectedness, Luke has used the LXX's version of Joel 2: 28, which uses the Greek word *sarx* for "flesh" at this point. While most Pentecostal (and charismatic) commentators see Luke as indicating "all people" in his use of *sarx* (Keener 2012; Cho and Park 2019), *sarx* can have a broader ecological context. Trainor notes that the Jewish historian Josephus (37–100 CE) understands *sarx* as encompassing non-human flesh, which is also a common understanding of the broader Greek literature (Trainor 2020). This understanding of *sarx* does have implications for Acts 2:17, as Trainor (2020) notes that Luke's use of *sarx*,

> implies an added layer of relationship that is more than anthropocentric. *Sarx* emphasizes humanity in its relationship to the wider environment and world in which human being live. The expression underscores a person's environmental network or relationships that makes one human. … [*Sarx*] is not exclusively anthropocentric; it also implies the non-human world. (p. 38)

Luke's adoption of *sarx* in Acts 2: 17 then indicates the interconnectedness of humans and creation, as God pours out the Spirit upon all flesh, both human and non-human. While the first half of the quote from Joel 2: 28–32 goes on to describe the human effects of this outpouring (Acts 2: 17–18), the second half addresses the effects of this outpouring on creation (2: 19–20), which leads to the ecological theme of voice.

Second, on the ecological theme of voice, Luke has modified Joel 2: 30 to emphasise the voice of creation. Luke slightly changes the "wonders in the heavens and on the earth" (Joel 2: 30) to "wonders in the heavens above and signs on the earth below" (Acts 2: 19), with the addition of "signs" connecting it to the work of Jesus

(Act 2: 22) and the apostles (Acts 2: 43). It is God who will show these signs (2: 19), which could lead to us as seeing creation as a passive partner in the prelude to the day of the Lord. However, two things should be noted. First, God also works signs and wonders through both Jesus and the apostles, hence, creation is not simply a passive partner in these signs and wonders. Second, the intertextuality of Joel 2: 30–32 and Acts 2: 19–21 reminds us that creation has been an active and vocal partner in Joel 1–2, often raising its voice in lament and calling for salvation (Braaten 2008).

Indeed, this intertextuality could lead us to see Acts 2: 19–20 as describing a prophesying creation. Just as the Spirit of God has particular effects on humans (Acts 2: 17–18), so a broader cosmological and ecological effect of the outpouring of the Spirit is then developed (2: 19–20). As Trainor (2020) notes:

> Creation [in Romans 8] participates in God's redemptive act. For Luke, a similar agency is potentially at work in the non-human world ... which becomes a prophetic voice of God's presence and action through the presence of the Spirit. In this case, the Spirit is not about vivifying human beings exclusively, but every organism that constitutes Earth. (p. 38)

Luke's understanding of the outpouring of the Spirit at Pentecost is not solely focussed on the Spirit's prophetic effect on humans, but also encompasses the ecological effect of the outpouring of the Spirit.

Concluding Synthesis

What does this ecological reading of Joel 1–2 and Acts 2: 17–21 mean for a Pentecostal ecology and spirituality? It both supports and directs Pentecostalism's enchanted understanding of creation and its understanding of the effect of climate change, for the Pacific Islands and globally.

First, it is clear that the ecological themes of the interconnectedness and voice can co-exist with an enchanted understanding of creation. Rather than the naturalistic and materialistic view of creation as purely a passive object, creation is an active partner in Pentecostalism's spirituality. God's Spirit is poured out on all flesh, both human and non-human, and so humans are not only interconnected with the rest of creation materially, but Pentecostals would add spiritually. The enchanted worldview implies a spiritual as well as material interconnectedness between humans and creation.

Indeed, technocratic nations[9] should not be quick to judge this enchanted worldview as backward, seeking to "naturalise" it, in order to implement climate change mitigation strategies. As research into deep ecology reveals, having a deep spiritual understanding of and connection to creation is pivotal for lasting ecological change. For example, Capra and Luisi (2014, p. 13) note, "ultimately, deep ecological awareness is spiritual awareness." That is, while the Pentecostal understanding

[9] In this discourse 'technocratic nations' are understood as countries where the majority of people can be characterised as adhering to a predominantly technocratic worldview orientation.

of an enchanted creation does have its difficulties, such as the possibility of mis-enchantment, it is no less a valid starting point than a technocratic understanding of creation. Indeed, Capra and Luisi note that it may be more scientific, as:

> The extensive explorations of the relationships between science and spirituality over the past three decades have made it evident that the sense of oneness, which is the key characteristic of spiritual experience, is fully confirmed by the understanding of reality in contemporary science. ... Connectedness, relationship, and interdependence are fundamental concepts of ecology... and connectedness, relationship, and belonging are also the essence of spiritual experience. (p. 290–291)

An enchanted understanding of creation, which sees the material imbued with the spiritual, can be a productive starting place for an effective Pentecostal ecology. Moreover, this enchanted understanding of creation may be a more effective and productive starting point for developing a deep ecology than a technocratic understanding of creation.

However, this ecological reading of Acts 2: 17–21 does also direct Pentecostalism's understanding of an enchanted creation, specifically in the area of mis-enchantment of creation. This ecological reading highlights the need for a biblically guided discernment of the voice of creation, in particular in two ways. First, Pentecostalism should be open to understanding ecological disasters as a lament of creation. Joel 1–2 clearly establishes that creation can lament in the face of injustice and possibly even over-consumption. Second, the tendency for Pentecostals to understand ecological crises as a call to repentance can be guided to include repentance for ecological sins. This has already been suggested by other Pentecostal scholars as an important aspect of a Pentecostal ecology (Clifton 2009; Tallman 2014). For example, Tallman (2014, p. 199) states, "the confession of our own sinful complicity in harming and destroying God's creation is a good starting point for Pentecostal Christian worship and environmental praxis."

However, when considering the specific context for climate change mitigation strategies in the Pacific Islands, perhaps technocratic nations should begin by being self-reflective, as the technocratic worldview can easily distance human culpability from ecological disaster. When reflecting upon the ecological disasters occurring on Fiji, Cox et al. (2018) note:

> In the face of Australian government recalcitrance on climate change, disaster "management," "preparedness," or "risk reduction" have become the preferred language by which donors and recipient governments discuss responses to the effects of climate change in the region. Rendering events such as Tropical Cyclone Pam or Tropical Cyclone Winston as singular "disasters" and the region as "vulnerable to natural disasters" allows for official responses that do not consider climate change as a cause or amplifier of these disasters and do not address the implications of future intensifications of storms and cyclones or rising sea levels. Rather, international disaster discourse eschews blame, frames catastrophic events as natural phenomena, and allows government and nongovernment organizations' relief efforts to be incorporated into moral political narratives that construct Australia as a generous donor. (p. 383)

It is hard to call Pacific Islanders to repentance for ecological sins when, in fact, it is the technocratic nations—who portray themselves as generous in their relief efforts—which are the main contributors to climate change.[10] Indeed, there are many similarities between technocratic nations and the drunkard at the beginning of Joel, both are overconsuming and are unwilling to change (repent).[11] Perhaps technocratic nations like Australia could possibly reframe their "mitigation strategies" as acts of change, or to use Pentecostal language, an act of repentance for ecological sins.

On the theme of voice, it is also important to note the role that human prophetic voices play in Joel 1–2 and Acts 2: 17–21. While creation responds to the call of God to lament in Joel 1–2, it is the prophetic human voice—the author of Joel—who communicates this call of God and this lament of creation to the people of Judah. That is, a key element to an effective Pentecostal ecology should be an openness to hearing the prophetic voices of those that are describing the lament of creation and the need for ecological justice and repentance (Luetz et al. 2018; Timon et al. 2019).

The enchanted understanding of creation present among Pentecostal Pacific Islanders is a productive starting place for a deep ecology. When informed by the ecological themes of interconnectedness and voice, this may provide an effective Pentecostal ecology for the Pacific Islands, Oceania and more broadly global Pentecostalism.

References

Anderson A (2013) An introduction to pentecostalism: Global charismatic christianity. Cambridge University Press, Cambridge

Braaten LJ (2008) Earth community in Joel: a call to identify with the rest of creation. In: Habel N (ed) Exploring Ecological Hermeneutics. SBL, Atlanta, pp 63–74

Brison KJ (2007) The empire strikes back: pentecostalism in Fiji. Ethnology 46(1):21–39

Brison KJ (2017) The power of submission: self and community in Fijian pentecostal discourse. American Ethnologist 44(4):657–669

Capra F, Luisi PL (2014) The systems view of life: a unifying vision. Cambridge University Press, Cambridge

Chan MJ (2014) Sins of the ancestors: generational sin, pentecostalism, and the ecological crisis. In: Swoboda AJ (ed) Blood cries out: pentecostals, ecology, and the groans of creation. Pickwick, Eugene, pp 169–184

Cho P, Park HD (2019) Acts: part one: introduction and chapters 1–12. NCCS, Cascade, Eugene

Clifton S (2009) Preaching the "Full Gospel" in the Context of Global Environmental Crises. In: Yong A (2009) (eds) The Spirit Renews the Face of the Earth: Pentecostal Forays in Science and Theology of Creation. Pickwick, Eugene, pp 117–134

Cox J, Finau G, Kant R, Tarai J, Titifanue J (2018) Disaster, Divine Judgement, and Original Sin: Christian Interpretations of Tropical Cyclone Winston and Climate Change in Fiji. The Contemporary Pacific 30(2):380–411

Ede P (2014) River from the temple: the spirit, city earthkeeping and healing urban land. In: Swoboda AJ (ed) Blood cries out: pentecostals, ecology, and the groans of creation. Pickwick, Eugene, pp 205–224

Eriksen A (2017) Going to 'pentecost': how to study pentecostalism—in Melanesia, for example. JRAI 24:164–180

[10] For example, Australians are one of the largest per capita contributors of carbon-dioxide on Earth.

[11] Repentance in its most basic means to turn from something, to change.

Gallagher SD, Gallagher LJ (2019) Pentecostalism in Papua New Guinea. In: Austin DA, Grey J, Lewis PW (eds) Asia Pacific Pentecostalism. Brill, Leiden, pp 325–346
Gore K (2019) The pentecostal movement in the South Pacific Islands. In: Austin DA, Grey J, Lewis PW (eds) Asia Pacific Pentecostalism. Brill, Leiden, pp 297–324
Habel NC (2000) (eds) Readings from the perspective of Earth. The Earth Bible Volume One. Sheffield Academic Press, Sheffield
Hayes KM (2002) The Earth mourns: prophetic metaphor and oral aesthetic. Brill, Leiden
Horrell DG (2014) The Bible and the environment: towards a critical ecological biblical theology. Routledge, London
Kärkkäinen VM (2014) The greening of the spirit: towards a pneumatological theology of the flourishing of nature. In: Swoboda AJ (ed) Blood cries out: pentecostals, ecology, and the groans of creation. Pickwick, Eugene, pp 84–97
Kay WK (2011) Pentecostalism: a very short introduction. Oxford University Press, Oxford
Keener CS (2012) Acts: an exegetical commentary: introduction and 1: 1–2:47. Baker Academic, Grand Rapids
Keener VW, Marra JJ, Finucane ML, Spooner D, Smith MH (2012) (eds) Climate change and Pacific Islands: indicators and impacts. Report for the 2012 Pacific Islands Regional Climate Assessment (PIRCA). Island Press, Washington, DC
Luetz JM, Buxton G, Bangert K (2018) Christian theological, hermeneutical and eschatological perspectives on environmental sustainability and creation care-the role of holistic education. In: Luetz JM et al (eds) Reimagining Christian education. Springer Nature, Singapore, pp 51–73
Luetz JM, Nunn PD (2020) Climate change adaption in the Pacific Islands: A review of faith-engaged approaches and opportunities. In: Leal Filho W (ed) Managing climate change adaption in the Pacific Region. Springer, Switzerland, pp 293–311
Macdonald F (2019) 'God was here first': value, hierarchy, and conversion in a Melanesian Christianity. Ethnos 84(3):525–541
Newland L (2004) Turning the spirits into witchcraft: pentecostalism in Fijian villages. Oceania 75(1):1–18
Rio K, MacCarthy M, Blanes R (2017) (eds) Pentecostalism and witchcraft: spiritual warfare in Africa and Melanesia. Contemporary anthropology of religion. Palgrave Macmillan, London
Ryle J (2011) Laying our sins and sorrows on the altar: ritualizing catholic charismatic reconciliation and healing in Fiji. In: Lindhardt M (ed) Practicing the faith: the ritual life of pentecostal-charismatic christians. Berghahn, Oxford, USA
Ryle J (2012) Burying the past-healing the land: ritualising reconciliation in Fiji. Archives de sciences sociales des religions 157:89–111
Smith JAK (2010) Thinking in tongues: pentecostal contributions to christian philosophy. Eerdmans, Grand Rapids
Smith JAK, Yong A (2010) (eds) Science and the spirit: a pentecostal engagement with the sciences. Indiana University Press, Bloomington
Tallman M (2014) Healing for a sick World: models of pentecostal environmentalism in Africa. In: Swoboda AJ (ed) Blood Cries Out: Pentecostals, Ecology, and the Groans of Creation. Pickwick, Eugene, pp 187–204
Timon T, Kaunda CJ, Hewitt RR (2019) Re-Envisioning tangintebu theological college in the context of climate change: an emerging model of coconut theological education and ministerial formation. HTS Teologiese Studies/ Theological Studies 75(1):1–8
Trainor M (2020) Acts: an Earth Bible commentary: about Earth's children: an ecological listening to the acts of the apostles. Bloomsbury, London
Waddell R (2012) Apocalyptic sustainability: the future of pentecostal ecology. In: Waddell R (ed) Perspectives in Pentecostal Eschatologies: World Without End. James Clarke & Company, Cambridge, UK, pp 95–110
Waldrop RE (2014) Spirit of creation, spirit of pentecost: reflections on ecotheology and mission in latin American pentecostalism. In: Swoboda AJ (ed) Blood Cries Out: Pentecostals, Ecology, and the Groans of Creation. Pickwick, Eugene, pp 225–235

Wall RW (2000) Intertextuality, Biblical. In: Evans CA, Porter SE (eds) Dictionary of new testament background. Inter-Varsity, Westmont, pp 541–551

Warrington K (2000) Jesus the healer: paradigm or unique phenomenon?. Paternoster, Milton Keynes

Warrington K (2005) Healing and suffering: Biblical and pastoral reflections. Paternoster, Milton Keynes

Wenk M (2000) Community-forming power: the socio-ethical role of the spirit in Luke-Acts. T&T Clark, London

Yong A (2005) The spirit poured out on all flesh: pentecostalism and the possibility of global theology. Baker Academic, Grand Rapids

Yong A (2009) (eds) The spirit renews the face of the Earth: Pentecostal forays in science and theology of creation. Pickwick, Eugene

Chapter 18
Christianity, Creation, and the Climate Crisis: Ecotheological Paradigms and Perspectives

Johannes M. Luetz and Richard G. Leo

Abstract Ecotheology is an area of growing interest to faith communities both denominationally and ecumenically. This includes Pacific Island nations, which are characterised by a comparatively high degree of Christianisation. Ecotheological ideas commonly explore the interrelationships between religion and nature in the light of contemporary environmental concerns. More specifically, ecotheology typically ponders questions of how spirituality impinges on sustainability. A review of the literature reflects nuanced and contested perspectives that suggest that religion may either enable or disable environmental conservation in different contexts. This chapter conceptualises religion as an effective cultural adaptation resource whose potential in the Anthropocene remains grossly underappreciated. It presents epistemologies and 'green prisms' that explain Christian environmental action and inaction, pointing the way to effective and sustainable climate-change adaptation in the future.

Introduction: The Three 'Rs' of Christian Environmental Thought

Pentecostal theologian Amos Yong (2005) once suggested that, if we neglect to include the sciences and the natural world in our theological worldview, we will be severely handicapped in trying to understand many of the pressing issues of our time (p. 276). Over recent years there has been a growing interest among faith

J. M. Luetz (✉)
School of Social Sciences, Christian Heritage College (CHC), Brisbane, QLD, Australia
e-mail: jluetz@chc.edu.au

School of Law and Society, University of the Sunshine Coast (USC), Maroochydore, QLD, Australia

School of Social Sciences, University of New South Wales (UNSW), Sydney, NSW, Australia

R. G. Leo
School of Education, Humanities and Business, Christian Heritage College (CHC), Brisbane, QLD, Australia
e-mail: rleo@chc.edu.au

J. M. Luetz and P. D. Nunn (eds.), *Beyond Belief*, Climate Change Management,
https://doi.org/10.1007/978-3-030-67602-5_18

communities in the sustainability of the planet (Hore-Lacy 2006; Spencer and White 2007; Brunner et al. 2014; Nelson and Luetz 2019). This includes Pacific Island nations, which are characterised by a high degree of Christianisation (Ernst 2012; Gibson et al. 2013; Nunn et al. 2016; Rubow and Bird 2016; Fair 2018; Havea et al. 2018). Scholars have used a range of terminological and conceptual approaches to describe different expressions of this faith-informed concern (or faith-informed lack of concern) for environmentally sympathetic practices (Hessel and Radford Ruether 2000; Orr 2005; Zaleha and Szasz 2015; Kempf 2020).

It was Lyn White Jnr's seminal (1967) article which described Christianity, at least in its Western form, as "the most anthropocentric religion the world has ever seen" (p. 1205). This publication effectively birthed the field of contemporary ecotheology, yet sparked numerous responses, many of which took issue with his key criticism, namely that Christianity was both the source and propagator of our current environmental crisis (Björnberg et al. 2017; Braterman 2020). More recently, ecotheology has been discussed under headings such as creation care, theology of earth care, faith-based ecosystem management, and biblically-based stewardship, among others (Ayre 2014; Brunner 2014). Relevant ecotheological ideas commonly explore the interrelationships between religion and nature in the light of contemporary environmental concerns, thus forcing the conversation to become more theistically centred (Moo and Moo 2018). More specifically, ecotheology typically grapples with questions of how spirituality impinges on sustainability (Hodson and Hodson 2008; Spencer et al. 2009). In the literature, we identify three broad strategies of biblical interpretation (or hermeneutic[1]): recovery, resistance or revision (Horrell 2010; Moo and Moo 2018).

Recovery approaches believe that Christians have simply failed to see the ecological wisdom which already exists in the bible. Adherents of this perspective argue that green perspectives have just been hidden or obscured by the way that people have previously interpreted the scriptures. Practitioners of this approach are trying to 'recover' these missing perspectives, principally by trying to rescue Christian thought from the accusation that Christianity is an anthropocentric religion that is underpinned by the dominion (or domination) theory (more on this in Sect. 5). Examples of this approach would include the *Green Bible* where green print was used for those passages deemed ecologically relevant by the editors (Maudlin and Baer 1989) or the Exeter Project which used history to understand what the bible might have to say about ecology and theology (University of Exeter n.d.).

Others accept that White had a valid case. Those who hold this view are happy to acknowledge that the bible, at least in parts,[2] is an anthropocentric text. Authors who hold to this perspective are often strongly committed to ecological values already and

[1]In the world of ecotheology it is important to note that there is a subtle difference between 'environmental hermeneutics' and 'ecological hermeneutics', whereby the latter is a sub-branch of the former. This difference may seem subtle, but it is the difference between talking about the scientific field of 'ecology' versus 'the environment' (Nilsen and Solevåg 2016, p. 672). Much general Christian discussion does not seem to care about this difference. Whilst acknowledging this subtlety of difference, to avoid any confusion, this chapter will use the term 'ecological hermeneutic' to also refer to an 'environmental hermeneutic'.

[2]This perspective acknowledges the diverse authorship of the bible over many centuries.

approach the bible through a hermeneutic of *resistance*[3] (Horrell 2010, pp. 13–15). Environmental interpretations of resistance argue that historic Christianity must learn how to read the scriptures in light of our contemporary environmental circumstances. An assumption is made that the bible primarily reflects the interests of humans and that we should learn to read the scriptures "from the perspective of Earth" (Habel 2000, p. 34). Examples of this approach include the team of the *Earth Bible* (Habel 2008, 2012), Hilary Marlow's 'ecological triangle' (Marlow 2009), the eco-feminism of Elaine Wainwright (2006) or the post-colonial interdisciplinary work of Scandinavian scholars Tina Nilsen and Anna Solevåg (2016) that was inspired by the Earth Charter.

A third option is the *revisionist* approach. Adherents accept that one cannot avoid reading the bible through the lens of one's personal experiences. It attempts to find a middle ground that is faithful with historic Christianity whilst taking insights from 'resistant' scholars and integrate them with progressive perspectives gleaned from 'recovery' scholars (Santmire 1985; Horrell 2010, pp. 11–32).

Accepting that all ecotheological positions can be broadly framed within one of these three perspectives helps contextualise the field of ecotheology. This then becomes the starting position of this chapter which extends previous research by building on these three frameworks. This chapter attempts to analyse and synthesise connections between political ideologies and diverse ecotheological interpretations that are essential for effectively engaging different publics towards more environmentally sympathetic practices.

This chapter is written with a worldview orientation in mind that accepts the bible as a key revelatory document that complements climate scientific and sociocultural perspectives (more on this in Sect. 4). The multiple and diverse authorship of the bible across many centuries offers rich perspectives on the interrelationships between humanity, the natural world and the realm of the divine. Advocating for a tripartite model (see Fig. 18.1) that integrates perspectives from the natural or Earth sciences, social sciences and theological traditions, we argue that all three realms have a vital contribution to make for effective and sustainable climate change adaptation in the Pacific Island region and beyond. With this objective in mind, the following section argues from the perspective that metaphysics and science are inseparably intertwined and that they conceptually complement, rather than displace or eclipse each other.

[3]Inspired by feminist theologians, scholars adopt three principles of *suspicion*, *identification* and *retrieval* (Habel 2012, p. 97) that shape 'ecojustice'. Accordingly, eco-justice theologians *suspect* that the Earth is likely to be treated as an object (anthropocentric perspective), *identify* that an 'Earth perspective' means seeing humanity as interlinked with, and not separate from, the exploited cosmos, and, *retrieve* the voice of the Earth that has been hidden in the gaps and silences of the bible. There is also an alternative version of the *resistant* approach, one which resists the resistants. It tends to be found in popular Christian literature and usually claims an overt biblicism and refuses to bow to any perceived cultural pressure applied by an environmental agenda. Concerns are often raised about New Age spiritualities coming in through the back door of environmentalism or that natural disasters and signs of earthly decay should be celebrated as signals that the 'End Times' are near. These types of approaches were particularly evident in popular books such as Hal Lindsey's *The Late Great Planet Earth* or the *Left Behind* novels of Tim LaHaye and Jerry Jenkins (Horrell 2010, pp. 14–16).

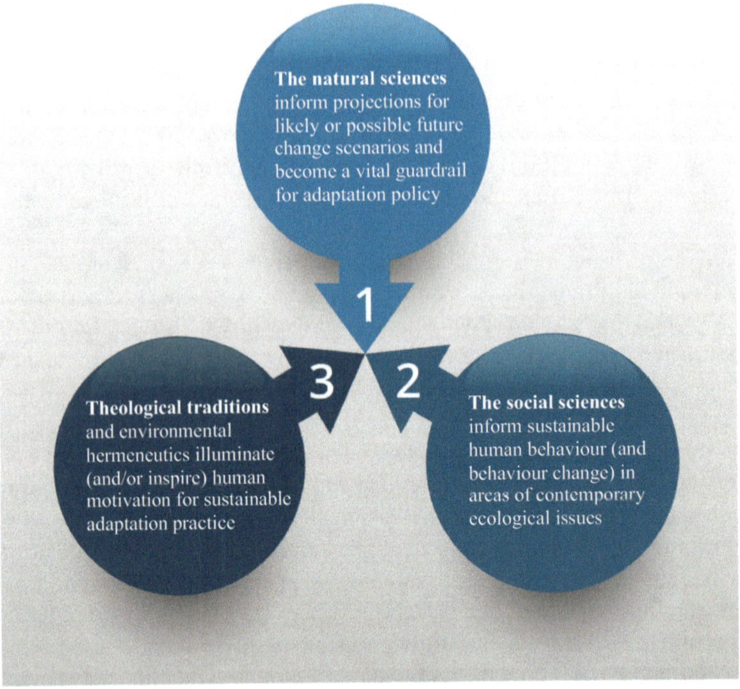

Fig. 18.1 The natural sciences, social sciences and theological traditions comprise three vital dimensions for effective and sustainable climate change adaptation practice. Representational figure by Authors. Based on Horrell (2010, pp. 124–127) and Luetz and Nunn (2020, p. 305)

"[T]he almost total Christianization of the Pacific Islands within the past century" (Nunn et al. 2016, p. 2) makes it appropriate to broach these interrelated topics from a worldview of Christian theism (Sire 2009; Ernst 2012; Ernst and Anisi 2016; Havea et al. 2018). As authors we acknowledge that this paradigm may differ significantly from other religions and worldview orientations.

This chapter is organised as follows: Sect. 2 introduces contemporary climate change as a gargantuan challenge that is best understood within the context of the Great Acceleration and Anthropocene. Section 3 argues that climate change (as a symptom of the Great Acceleration) cannot be meaningfully understood or tackled from a science-only perspective, wherefore this section explores the role of religion as a critical and underappreciated resource. Given that 'metaphysical beliefs' strongly impinge on human behaviour, Sect. 4 offers analysis of truth, truth-seeking and knowledge creation in the context of evangelical Christian faith traditions. Section 5 builds on this foundation and presents five bible-informed 'green prisms'[4] that help to explain faith-informed motivations behind differences in environmentally sustainable practices. Thereafter, Sect. 6 offers analysis and synthesis on impediments that have stymied environmental action within some faith communities (Orr 2005; Zaleha and Szasz 2015). Finally, Sect. 7 concludes the discussion with a call for new conceptual capital for a new era. The epistemologies, prisms and paradigms which are presented in this chapter help explain Christian worldview-informed environmental action and inaction and point the way to effective and sustainable climate-change adaptation in the future.

The Great Acceleration and Anthropocene

In seeking to comprehend contemporary Christian environmental thought, it is helpful to contextually ground the discussion in the period since the end of World War II. In this era, known as the 'Great Acceleration', rapid global industrialisation was marked by accelerating changes in population, conflict, resource usage, globalisation, consumerism and increasing economic inequities. Environmental historian John McNeil argues that, if we take a long-range view of the past, we see that, without intending to, humanity has "undertaken a gigantic uncontrolled experiment" in changing the global climate that has fundamentally altered our relationship with the environment. It may turn out, he says, to be more important than any of the other political, social or economic changes that occurred in the twentieth century (McNeil 2000, p. 4). These changes have been depicted in a series of graphs[5] produced by the International Geosphere-Biosphere Programme in 2004 (Steffen et al. 2005), showing a range of factors, including carbon and methane pollution, world population, consumerism, motor vehicle usage and so forth, that all spike after 1950 in a way unprecedented in the Earth's history. A study that updated the 2000–2010 data found that, whilst there were some encouraging small corrections, the overwhelming trend remains the same. The authors of the updated study conclude that "planetary stewardship has yet to emerge" (Steffen et al. 2015, p. 94).

What is concerning about the Great Acceleration is that human-driven changes are operating with unprecedented rapidity, magnitude, and at the same time. Often their impacts are exceeding the range of natural variability observed in scientific records for the past several millennia such that various Earth systems are starting to become unstable in ways that have no precedent in the geological record (MEA 2005; IPBES 2019; UN 2019). We just do not know what will happen when these systems become unstable enough to cross a planetary ecological threshold. Climatologists now can no

[4]Peter Hay first used the idea of a 'green prism' to describe when people reframed how they understood diverse political or economic issues within the context of environmental thought as early as 1987 (Doyle 2000, p. 61).

[5]A version of the 2004 graphs were published by *NewScientist* magazine in 2008 and available at: https://www.newscientist.com/article/dn14950-special-report-the-facts-about-overconsumption/.

longer rule out irreversible and abrupt planetary-wide changes (Steffen et al. 2005; Hinkel et al. 2018). The historian Paul Kennedy (1993) has described the situation as follows:

> The environmental crisis we confront [today] is quantitatively and qualitatively different from anything before, simply because so many people have been inflicting damage on the world's ecosystem during the past century that the system as a whole—not simply its various parts—may be in danger (p. 96)

Given the planetary changes caused by the Great Acceleration, since the turn of the twenty-first century, the construct of the Anthropocene has come to hold increasing importance in discussions about planetary and climate changes (Hamilton 2015, 2016). The term 'Anthropocene' is generally regarded as having originated with Dutch climatologist Paul Crutzen who, whilst attending a conference in 2000, was becoming more and more frustrated at claims that humans were still living in the geological age of the Holocene. The story goes that his frustration boiled over to the point that he blurted, "We're no longer in the Holocene. We are in the Anthropocene!" (Christian 2008, p. 101). In 1993, environmental historian Robert Gottlieb also observed that, if we focused only on 'Nature' and how it was managed or developed, then we would end up missing a crucial link. For him, it was imperative that we learn to link the changes in material life after World War II—the rise of petrochemicals, the dawning of the nuclear age, the tendencies towards over-production and over-consumption—to the rise of new social movements focused on issues of quality of life (Gottlieb 1993, pp. 3–4). Or in the words of the philosopher-historian Dipesh Chakrabarty,

> … the wall between human and natural history has been breached. We may not see ourselves as a geological agent, but we appear to have become one at the level of the species. And without that knowledge that defies historical understanding there is no making sense of the current crisis that affects us all (Chakrabarty 2009, p. 221)

Scientists claim that the human imprint on the global environment has become so "large and active that [we] now rival some of the great forces of Nature in [our] impact on the functioning of the Earth system" (Steffen et al. 2011, p. 843). These are bold claims, that human activities and not natural processes are now the dominant driver of change across the Earth, yet it is the crux of the argument for the Anthropocene.

In May 2019, the Anthropocene Working Group (AWG) recommended that a new geologic age starting in the post-war boom of the late 1940s and early 1950s be accepted as the Anthropocene's start date.[6] The group had spent several years debating and researching an answer to a simple but key question: "When did human

[6]Readers need to be aware that at the time of preparing this chapter the idea of 'Anthropocene' is simultaneously a cultural and contested geologic reality, with neither the International Commission on Stratigraphy (ICS) or the International Union of Geological Sciences (IUGS) officially endorsing the term. At the 35th International Geological Congress held in Cape Town, South Africa, August 2016, the AWG recommended that the term be adopted as a scientific reality, beginning in 1950. In May 2019, the AWG voted to submit a formal proposal that the term be adopted as a geologic era to the ICS by 2021.

influence over the environment reach a tipping point?" To define a new epoch, scientists need to find a 'signal' in the geologic record around the world, recently settling on the nuclear bomb tests from the late 1940s and early 1950s. Prior to the first nuclear explosion, radionuclides such as long-lived plutonium-239, do not appear naturally in the geologic record. It is not until these tests occurred that we can identify undeniable human influence on the global environment. The chair of the AWG, Jan Zalasiewicz, said it is very clear in the data that "the most significant event in humanity's life on the planet is the Great Acceleration, the period of rapid global industrialization that followed the Second World War" (Meyer 2019, para. 7).

In relation to climate change, scientists have computed the absorption of energy by the Earth brought on by the human release of heat-trapping greenhouse gasses to be "equivalent to exploding 400,000 Hiroshima atomic bombs per day 365 days per year" (Hansen 2012; cf. Braasch 2013; Cook 2013). Annualising this daily energy release reinforces the point that change is occurring on an order of magnitude not previously witnessed; "exploding 146 million Hiroshima atomic bombs annually makes it quite clear that anthropogenic climate change is likely to have severe long-term consequences that may well have … 'apocalyptic' end results" (Luetz et al. 2018, p. 64).

Taken together, changes now underway progressively threaten to destabilise vital Earth systems, and alarmed scientists have repeatedly and solemnly warned of their disequilibrium, chaos or even collapse (Kendall 1997; Diamond 2005; Ripple et al. 2017). As human-induced global warming continues unchecked, so do increases in global temperatures, extreme weather events, droughts, floods, heat stress, bushfire risk, human health problems, species extinctions and rises in the impacts of disasters, among others (MEA 2005; Hanson 2009; Deryng et al. 2014; Matthews et al. 2017; Kinney 2018; Ashbolt 2019; IPBES 2019; Lou et al. 2019; Luetz 2020, pp. 158–163). Such changes, attributable to the interplay of the Great Acceleration and the Anthropocene, have created a multi-dimensional and complex historical situation that we now confront as a species. In this era, climate change is no longer a 'one-dimensional challenge' of achieving piecemeal reductions in greenhouse gas emissions in the coming decades while continuing along the same unsustainable perpetual 'global growth' economic paradigm (Luetz and Walid 2019). Climate change is already a 'wicked problem', one of 'a family of interlocking problems' that have combined to create the 'phenomenon of ecological overshoot' by humanity. Our challenge now is to intelligently manage both the Earth and our resources (food, water, land, population) to solve the problems that will inevitably emerge as we move forward (Chakrabarty 2016). This challenge is so formidable and unprecedented that it requires a new strategy and a new ally, to which we will now turn our attention.

Conceptualising Religion as an Effectual Cultural Adaptation Resource

Those who work in the environmental humanities have begun to increasingly sense that the thoughts, institutional frameworks and practices of religion have important contributions to make in how humanity responds to the challenge of the Anthropocene and climate change (Gard and Veitayaki 2017). For example, with reference to the Pacific Islands region, Haluza-Delay (2014) notes that whilst theologians have been working to indigenise the faith, developing an 'oceanic' theology appropriate for a 'liquid continent', the broader belief structures of Christianity still hold significant influence across the Pacific. Around 90% of the population of the Solomon Islands identifies as Christian and other Pacific Island nations have similar rates (Nunn et al. 2016), with "echoes of pre-Christian spiritual beliefs pervad[ing] many environmental interactions" (p. 2). As such, a discussion of the wider perspectives in the field of ecological hermeneutics holds particular relevance in the Pacific region. Indeed, it is well recognised that religious faith is often highly motivational, wherefore such "groups and their members, scholars of the Abrahamic and other world religions can help by researching their traditions, reflecting carefully on them for their fruitfulness in addressing climate change and other ecological problems" (Schaefer 2014, p. 42). In the case of Christianity, careful reflection is becoming quite urgent given that a commitment to ecology is not yet central to Christian self-understanding (Eugene Bianchi 1998; Edwards 2006, p. 3). Christians are still not used to thinking about God's work of "redeeming the nonhuman world" (Bianchi 1998, p. 68).

For some time, a popular view has been that scientific and religious perspectives are by their nature in conflict (Barbour 1997, 2000; Orr 2005; Weldon 2017). Richard Blackwell has suggested we can adopt one of two approaches towards the status of religion and its relationship with scientific perspectives: strategies of separation or integration. In his view, the former is a defensive approach, seeking to identify differences, whereas the latter is a more "confident state of mind" that perceives that the two can be useful to each other (Blackwell 1998, p. 14). In contrast, Harrison (2006a, 2015) counters that the territories of science and religion should be more appropriately seen as being in conversation with each other rather than in conflict, echoing ideas about Non-Overlapping Magisteria (NOMA) introduced by geologist Stephen Jay Gould (1999). While conflict models continue to "exercise a tenacious hold on the popular imagination" (Harrison 2015, p. 24; see also Barbour 1997, 2000), Kjaergaard (2016) has described such conceptual perspectives as regrettably simplistic and no longer useful in explaining the relationship between scientific and religious positions. His views agree with the 'complexity thesis' advanced by Brooke (1991), who proposed avoiding simplistic representations of conflict and harmony and instead, as Lindberg and Numbers (2003) suggest, revel in a "thoroughly entangled relationship" (p. 3) that both motivates and regulates ongoing conversation.

Much recent work shows how religion can be helpful in motivating both individuals and communities to enact positive adaptive change within a construct that combines religious perspectives and contemporary science. Mike Hulme has conceptualised climate change as not just a physical process but also a scientific narrative that has now expanded well beyond the sciences. Climate change and the Anthropocene, in his view, are as much an objective physical reality as a sociocultural construct (Hulme 2009, p. 329). When applying his views to a physical and social narrative such as the Anthropocene, he argues that climate change is change that modifies both the physical world (scientifically understood) and social worlds (as understood in the humanities). Lynch and Veland (2019) reinforce this notion when they argue that our response to such a crisis challenges human values, government policy and systems of sovereignty and law. In their work, they encourage us to perceive the interplay of the physical and social narratives through the theme of 'myth'. For them, myth is neither false nor factual but an "underlying narrative by which we make sense of the world" (Lynch and Veland 2019, p. x), a sociocultural construct that is informed, but not constrained, by the objective physical reality of the natural and physical sciences. In addition, Patrick Nunn (2020a) observes that "many ancient stories, previously regarded only as legends or myths, may be based on observations of natural phenomena by preliterate peoples" (p. 118).[7]

While secular assumptions are deeply embedded in Western-influenced secularised governance frameworks, Erin Wilson (2012) suggests that religious frameworks can bring new perspectives to bear on public policy questions that are informed by the social sciences in areas such as human rights, aid and development, conflict resolution, and climate change. She warns us that by foregrounding only secular approaches, activists and social justice theorists are missing helpful insights originating in the interplay of neo-liberal economic regimes and climate change through the period of the Great Acceleration.

In the Pacific Islands region and similar contexts elsewhere, the tradition of storytelling as an embodied and embedded religious practice in dealing with wicked problems like climate change is gaining prominence as a promising adaptation strategy (Buxton et al. 2021). Lisa Stenmark (2015) perceives storytelling as indispensable when scientific data and discourses have clearly proved insufficient to sway opinion. Moreover, Alex Evans argues that storytelling increases the possibilities and alternatives available through offering stories which help people to take action in the midst of uncertainty (Evans 2016). In his study of how three separate Pacific communities attempted to harmonise the Noah story with scientific discourse, Wolfgang Kempf (2017) found that in both praxis and politics, it was necessary to take religious (Christian) perspectives seriously within the context of climate change, sea-level rise and migration.

Hannah Fair (2018) came to similar conclusions, recommending that communities embrace the "ambiguity, complexity and partiality of story-telling" in her Vanuatu based study of how the Noah story shapes what people are saying and doing about

[7]Nunn (2020b) adds a provocative corollary when he states "science could have learned more and sooner had it treated the 'stories' of ancient Australians (and people of other non-western cultures) more seriously as potential sources of information about the past, had it not pejoratively dismissed such stories as entertainments rather than expository, characterized the storytellers as 'literati' rather than true scientists communicating their wisdom along unfamiliar pathways" (p. 29).

climate change in three separate island communities. She says tapping into so-called "'thick' moral accounts rather than thin evocations of global values seems particularly pressing in the Pacific" (Fair 2018, p. 4). Importantly, she explains how spiritually-shaped local worldviews may be engaged as "a motivational force not mirrored by economics or science" (p. 4). There is therefore an argument that, rather than constituting a barrier, religion might be leveraged as a "cultural resource" (Hulme 2017, p. 15) because "the most influential messages are those that engage with people's spiritual beliefs" (Nunn 2017, para 14; cf. Nunn et al. 2016). This observation agrees with Knight (2006) who posits that "[m]en and women at their deepest level are motivated by metaphysical beliefs" (p. 19). This has particular relevance and application in the Pacific region (Rubow and Bird 2016; Havea et al. 2018).

Amitav Ghosh (2016) likewise recognises the power of storytelling, arguing that eschewing religious responses to the challenges of climate change will inevitably prove ineffectual. These examples highlight for us that there are religious cultural resources that can be brought to bear in motivating individuals and communities towards taking appropriate adaptive action on climate change in ways that science alone is unable to achieve. Hence, in line with Hulme (2017), we argue that religious perspectives have a crucial role to play in sociocultural constructs that are shaped and informed by the physical realities of climate change.

Building on the work of Horrell (2010), we advocate for a tripartite model (see Fig. 18.1) that integrates perspectives from the natural or Earth sciences, social sciences and theological traditions. We argue that each of these three realms has a vital contribution to make to effective and sustainable climate change adaptation in the Pacific Island region and beyond. Expressed simply, the tripod of adaptation responses in the Pacific needs three firm legs to stand solidly. First, the natural sciences are indispensable for generating projections for likely or possible future change scenarios; second, the social sciences are needed to inform sustainable human behaviour (and behaviour change) in areas of contemporary ecological issues; third, theological traditions, exegesis and hermeneutics can inspire human motivation for action. The key is that all three dimensions are conjointly engaged (Horrell 2010, pp. 124–127; Luetz et al. 2020).

In summary, in view of the many pressing problems facing Pacific Island countries and communities, there is a case that climate change adaptation and mitigation will stand a better chance of being effective and sustainable if the science-informed rationale for action may additionally leverage faith-informed motivations (Nunn et al. 2016; Luetz and Nunn 2020).

Nature, Creation, Knowledge and Truth-Seeking: Leveraging Epistemologies

This section will intentionally accommodate, quote and interpret the biblical scriptures. In doing so, the discourse will aim to comprehend how 'reality' is constituted or constructed by members of the Christian faith and how conceptions of knowledge and truth inform Christian environmental action or inaction. Unless otherwise indicated, all scripture will be quoted from the New International Version (NIV) of the bible. Moreover, references to 'God' with the pronoun 'his' (as in 'his creation') reflect the customary gender usage in the Pacific Island region.[8]

Truth and truth-seeking occupy an eminent place in the bible and in Christianity. Invariably a quest for the truth cannot be divorced from a quest for knowledge, which brings the realm of epistemology and the study of knowledge into focus, namely what is known and what can be known. In Christian literature 'sources of knowledge' or 'ways of knowing' are presented as multiplex and complementary. According to Knight (2006), there are five sources of knowledge (the senses, revelation, authority, reason, intuition), and according to Wesley, there are four sources of knowledge (experience, tradition, reason, and the bible) (Maddox 1994; Hey 2018). This section introduces these multiple sources of knowledge and explores how truth-seeking and knowledge generation can play an important epistemological role in supporting more environmentally friendly practices among Christian communities of faith.

According to Knight (2006), "[t]here is no one source of knowledge that supplies people with all knowledge" (p. 25). In Knight's epistemological conceptualisation there are rather five sources of knowledge that are "in a complementary relationship rather than one of antagonism" (p. 25). They are as follows: (1) the senses; (2) revelation; (3) authority; (4) reason; and (5) intuition (see Fig. 18.2). A short overview of these five sources of knowledge (and their complementarity) is helpful. First, according to Knight (2006), *empirical knowledge* is gathered through the senses and has the advantage that "many sensory experiences and experiments are open to both replication and public examination" (p. 23). Secondly, *revealed knowledge* presupposes an omniscient and transcendent source, which people of faith typically hold as "absolute and uncontaminated" (p. 23). Notwithstanding, for Knight (2006) the need for both faith and "human interpretation" (p. 23) make revealed knowledge unreliable as an only way of knowing. Third, *authoritative knowledge* defers to the expertise of other authorities and thereby "saves time and enhances social and scientific progress" (p. 23). Fourth, *reason* or *rationalism* organise the raw data into "meaningful system[s]" (p. 24), thereby generating new insights and coherent knowledge. Finally, *intuition knowledge* may occasionally overcome human limitations by sparking "sudden flash[es] of insight" (p. 24). Knight cautions that each way of knowing is highly unreliable on its own and stresses the need to draw on multiple knowledge sources concurrently. Thus 'multiple knowledges' become a crucial safeguard against the limitations inherent in each one.

According to Fair (2018) it is possible to "hold various knowledges (Christian, *kastom* and scientific) in balance, exploring their convergence, connections and tensions" (p. 7). During her research in Vanuatu she used the Bislama term

[8]If it is true that "God created man in His own image; [as] male *and* female" (Gen 1: 27, RSV; emphasis added), then God evidently comprises both masculine *and* feminine traits. See Johnson (2002) for a richer discussion on God and gendered perspectives and portrayals.

Fig. 18.2 Knight (2006) proposes drawing on five sources of knowledge as the basis for truth and knowledge. Representational figure by Authors

'*tufala save*' (lit. "double knowledge") to illustrate how Ni-Vanuatu respondents hold religious and scientific knowledge in tandem:

> Oh well that's like kind of climate change, 'cause you know the rainbow came out, that's a promise, but and then scientists say it's 'cause of the water or something … creating a spectrum. It's good to have two beliefs. Both of them are right. (Fair 2018, p. 7)

Her example demonstrates how revealed knowledge and empirical knowledge can be integrated in daily life and illustrates how humans tacitly draw on multiple sources of knowledge conjointly. Even so, most people tend to privilege one knowledge source over others, which then becomes the "background against which other means of obtaining knowledge are evaluated" (Knight 2006, p. 25). In the world of science, "empirical knowledge generally is seen as the most basic source [and] biblical Christianity sees revelation as providing the basic framework" (p. 25). This holds an important application for human motivation. Fair (2018) found that spiritually-shaped local worldviews are "a motivational force not mirrored by economics or science" (p. 4). This accords with the view of Hulme (2017) who advocates that "climate policies need to tap into intrinsic, deeply-held values and motives if cultural innovation and

Fig. 18.3 John Wesley's 'Quadrilateral' framework proposes four sources of knowledge as the basis for truth and knowledge. Representational figure by Authors. Based on Maddox (1994)

change are to be lasting and effective" (p. 15). There is therefore a compelling argument that 'metaphysical motivation' is auspicious because it exceeds the efficacy of other motivational forces, including reliance on rational economic or scientific factuality (Boyes and Stanisstreet 2012; Luetz et al. 2020). As noted above, "[m]en and women at their deepest level are motivated by metaphysical beliefs" (Knight 2006, p. 19). In short, drawing on multiple knowledges in concert produces holistic perspectives and thus becomes a solid motivational foundation for roundly informed human adaptation practice.

The influential theologian, evangelist and founder of Methodism, John Wesley (1703–1791), whose influence in the Pacific Island region endures to this day (Gibson et al. 2013),[9] similarly engaged multiple knowledge sources. His framework has since become well-known as the 'Wesleyan Quadrilateral' (Maddox 1994; Hey 2018). A short overview of Wesley's four sources of knowledge (and their complementarity) is helpful (see Fig. 18.3). First, Wesley held *experience* in high esteem as a kind

[9]For example, according to Plant (2013), the majority of ethnic Fiji islanders today are Methodist (p. 359).

of "wisdom acquired through living" (Maddox 1994, p. 46;). Second, he esteemed *tradition* and Christian Antiquity "as a 'history of saints' to be emulated" (p. 45). Third, he regarded *reason* as "common sense" (p. 41) and as an important "processor (*organon*) of knowledge" (p. 40) that can organise and conceptualise human experiences. Fourth and finally, Wesley considered the *bible* "as the most basic authority" (Maddox 1994, p. 36). Moreover, being "explicitly aware that Scripture did not definitively address every possible issue" (p. 39), Wesley "interpreted the Protestant *sola Scriptura* … to mean that Scripture is the *primary*, rather than exclusive, Christian authority" (p. 37; emphasis original). In short, Wesley advocated using multiple knowledge sources for meaning making and often engaged "two or three of them jointly" (Maddox 1994, p. 36).

Wesley's belief in complementary knowledges also encapsulated the natural world or creation as an eminent source of 'revelation' about God (Maddox 1994, p. 29). He claimed that "God's basic attributes of omnipresence, omnipotence, and wisdom can be deduced *from creation*" (p. 30; emphasis added). In other words, Wesley argued that if people wish to learn about God, they could look to creation.

This idea is by no means new. The French theologian and protestant reformer John Calvin (1509–1564) postulated a similar view two centuries before Wesley's time. He noted that while God was "by nature incomprehensible and hidden from human understanding, He engraved in each one of His works certain signs of His majesty by which He offered Himself to be known by us according to our small capacity" (Calvin 1541/2009, p. 30). This principle has been variously noted since, with theologians giving "the term 'special revelation' to the scriptures [and] view[ing] nature as a 'general revelation'. Rightly understood, both special and general revelation give the *same message*, because they have the *same author*" (Knight 2006, p. 180; emphasis added). This perspective is essentially that of Brunner et al. (2014) who conceive of creation as God's second 'inspired' bible, which complements the traditional sacred writings found in scripture:

> Reformation emphasis on *sola Scriptura* (Scripture alone) has desensitized us to God's other book, one that has spoken truthfully for millennia … [Although the] Christian community has long held that the Scriptures of the Old and New Testament are 'inspired' (*theopneustos*, 2 Tim. 3: 16) by God and the basis for faithful reflection in the church … Christian writers have maintained that Creation can be read as a text about God and that God has presented us with 'two books'—Scripture *and* the natural world. (Brunner et al. 2014, pp. 23–24)

Brunner et al. (2014) conclude that "[i]n the same way that the Bible offers four distinctive accounts of the gospel story, so Scripture and nature offer two accounts of God's story" (p. 24). Merritt (2010) picks up this idea of the 'two bibles' when he likens the sanctity of scripture to the sanctity of creation, whose defilement in both instances is similarly sacrilegious:

> "There are two forms of divine revelation: the special revelation in Scripture that is able to lead us to salvation and the general revelation we receive through nature. Both are from God. So when we destroy creation, which is God's revelation, it's similar to tearing a page out of the Bible." (Merritt 2010, p. 2; citing Professor of Theology John Hammett).

If we consider Nature as being thus inviolable and imbued with divine revelatory or sacrosanct qualities, today's dramatic environmental destruction certainly gives pause for thought. For if it is true that destroying creation is akin to tearing a page out of the bible, what might it be called when—as the United Nations recently warned—"1,000,000 species [are now] threatened with extinction"? (UN 2019; IPBES 2019, pp. 3, 13) This situation has been described by McKibben (2005) as "tantamount to … 'decreation' of the natural order we found on this earth" (p. 8).

Several scriptures hold important hermeneutical implications for nature conservation efforts,[10] and especially for those Christians who accept that on the authority of the bible the Earth was formed by God and declared 'very good': "God saw all that he had made, and it was very good" (Genesis 1: 31). Consequently, if it is true that nature reflects something of who God is, then its conservation becomes a strategic epistemological concern. In other words,

> if God can be known through creation, then creation has inherent worth in and of itself—apart from utilitarian value to humans. And importantly, if God can be 'known' through what He has made, then 'creation care' is an epistemological priority of the highest order. Following this argument, epistemologically speaking, God becomes progressively less 'knowable' from what He has created as the defacing and careless destruction of His creation continues … Or inversely, the knowledge of God can be preserved through the conservation of His creation. (Luetz et al. 2018, pp. 59, 69)

In short, conserving creation conserves 'revelatory knowledge' about God. This then translates into a clear and unambiguous mandate for caring for the Earth. Vociferous Christians may also stand relieved of self-imposed efforts to verbally evangelise about God, knowing that by simply safeguarding his creation they can support him while he himself continues to disseminate his own message to all who are prepared to receive it.

Having looked at the 'book of creation', we now turn to consider the 'book of the bible' itself. Many scriptures attest to the preeminent place of 'the truth' in the biblical canon. First and foremost, Jesus is quoted in the Gospel of John as embodying the truth: "Jesus said: 'I am the way, the truth, and the life. No one comes to the Father except through me" (John 14:6). Stated differently, with Christ laying claim to 'the truth', the bible seemingly appropriates 'the truth' as belonging to God. This reading of scripture is reminiscent of the widely quoted truism attributed to St. Augustine, one of the Latin Fathers of the Church, who famously posited at the end of the fourth century: "All truth is God's truth" (Book Two, XVIII, 28; in Green 1997, p. 47). Accepting this precept holds an important implication, for if all truth is God's truth, then it might be discovered in (and mined from) a range of sources, including (as Augustine reminds his readers), "even in pagan literature".[11]

A modern-day appropriation of this principle can be found in climate change, which science has unequivocally established as a fact: "Warming of the climate

[10] "The heavens declare the glory of God; the skies proclaim the work of his hands" (Psalm 19: 1). "For since the creation of the world God's invisible qualities—his eternal power and divine nature—have been clearly seen, being understood from what has been made, so that people are without excuse" (Romans 1: 20).

[11] In the original rendition St. Augustine's quote is recorded and translated by Green (1997) as follows: "A person who is a good and true Christian should realize that truth belongs to his Lord, wherever it is found, gathering and acknowledging it even in pagan literature…" (p. 47).

system is unequivocal, and … unprecedented over decades to millennia" (IPCC 2014, p. 2). Incidentally, a study of 11,994 peer reviewed scientific articles concluded that the scientific consensus surrounding climate change is robust: "97.1% endorsed the consensus position that humans are causing global warming" (Cook et al. 2013, p. 1). If the best peer-reviewed published science available today overwhelmingly concludes that humans are primarily responsible for the growing climate crisis, in other words, if it is actually and factually 'true', then those who claim to be followers of a God of truth must also stand on the side of 'truth' and not be counted among the liars or deniers (Orr 2005; Oreskes and Conway 2011; Björnberg et al. 2017; Maslin 2019). In other words, this truth about the world can then simply be accepted, esteemed and believed, rather than feared, suppressed, resisted or denied. Importantly, only where the truth is known and accepted can its liberating properties be expected to operate effectively as posited in scripture: "Then you will know the truth, and the truth will set you free" (John 8: 32). Accepting the scientific 'truth' about anthropogenic climate change thereby becomes the rationale for addressing its adverse consequences through robust action on mitigation and adaptation.

Furthermore, an understanding of the truth as belonging to God also implies that consilience is possible between the domains of science and faith, for the faithful may lean comfortably and confidently on the realm of science for the generation of knowledge, facts and truth. According to the bible, such truth-informed living seemingly pleases God who reportedly desires 'truth' as the basis for honest worship: "God is spirit, and his worshippers must worship in the Spirit and *in truth*" (John 4: 24). Furthermore, according to the biblical account, God is committed to propagating 'the truth' by means of the Holy Spirit, whom scripture refers to as the 'Spirit *of truth*': "But when he, the Spirit of truth, comes, he will guide you into all the truth" (John 16: 13). There is therefore a strong case to view God as the primordial propagator, promoter and protector of truth, who according to scripture stands opposed by the primeval enemy of truth, whom the bible simply refers to as

> the devil [who] was a murderer from the beginning, not holding to the truth, for there is no truth in him. When he lies, he speaks his native language, for he is a liar and the father of lies (John 8: 44).

It is therefore incumbent upon bible-abiding Christians to face the truth because it comes from God, the fountain of truth. Furthermore, standing on the 'side of truth'[12] also implies a tacit duty to counter misinformation, lies, deceit and untruth, including in the area of climate science denialism (Björnberg et al. 2017; Cook et al. 2018; Oreskes and Conway 2011; Maslin 2019). If all truth is indeed God's truth, then it can (and should be) stridently pursued in multiple arenas. Knight (2006) comments,

[12]This understanding of being on the 'side of truth' is reminiscent of scripture where the Apostle John reports the following conversation about 'truth' between Jesus Christ and the Roman governor Pontius Pilate who presided over his trial and ultimately ordered his crucifixion: '"You are a king, then!" said Pilate.—Jesus answered, "You say that I am a king. In fact, the reason I was born and came into the world is to testify to the truth. Everyone on the side of truth listens to me."—"What is truth?" retorted Pilate. With this he went out again to the Jews gathered there and said, "I find no basis for a charge against him."' (John 18: 37–38).

> God is the Creator of everything. Therefore, truth in all fields stems from Him. Failing to see this point clearly has led many to develop a false dichotomy between the secular and the religious. That dichotomy implies that the religious has to do with God, while the secular is divorced from Him. From this point of view, the study of science, history and mathematics [or climate science] is seen as basically secular, while the study of religion, church history, and ethics is viewed as religious. That is not the biblical perspective. In the Bible, God is seen as the Creator of the objects and patterns of science and math … In essence, there is no such thing as 'secular' [truth] (Knight 2006, p. 225)

In summary, striving after truth and knowledge provides the Christian community (and other faith communities) a strong incentive—and a robust multiplex foundation—to care for and safeguard the Earth. To this end, "the various sources of knowledge available to the Christian are complementary" (Knight 2006, p. 183). This involves conserving creation as a palpable source of revelatory knowledge about its Creator. Given this rationale for metaphysically-informed environmentalism, the following section will now explore why Christian conservation efforts have faced sometimes vociferous opposition from within.

Contextualising Green Prisms

The Australian ecotheologian Clive Ayre (2014, pp. 91–116) provides us with a useful summary of five different 'green prisms' to outline a scriptural basis for creation care. For our purposes, these prisms can be broadly described as the Dominion, Stewardship, Partnership, Covenant and Sacrament Prisms (see Fig. 18.4). These have evolved over the course of the twentieth century as Christian thinkers and theologians have increasingly wrestled with the ecological challenges of our time

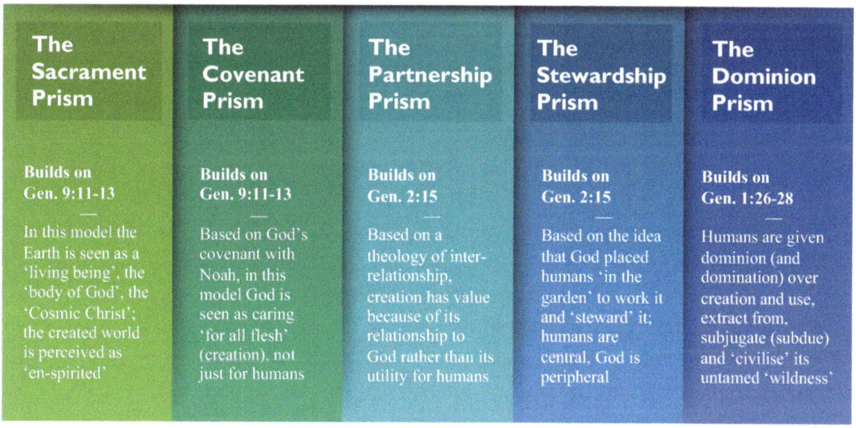

Fig. 18.4 Five 'green prisms' explain five possible perspectives or relationships that Christians may have with the Earth. These five prisms are discussed in Ayre (2014, pp. 91–116). Representational figure by Authors

whilst working within their respective theological traditions. Perspectives shown in Fig. 18.4 are presented along a right-left continuum: The Dominion Prism is presented first, whereafter discussion of the remaining prisms will then move from 'right' to 'left'.

1. The Dominion Prism

The dominion prism is the one that has the longest history in the wider public consciousness, probably due to Lyn White's (1967) famous article (see above). The dominion prism is based on the passage of scripture found in Genesis 1: 26–28. Interpretations of this passage hinge on how we understand the divine command in verse 28 to "fill the Earth and subdue it". It is the use of the word 'subdue' in most bible translations that gives the dominion prism its name. In recent years, there has been some attempt to reform the dominion prism. The historian Peter Harrison has found that in the early years of the industrial era people saw themselves as restorers of the natural world rather than exploiters, believing they were restoring a 'fallen world' and 'cultivating' or 'civilising' the 'wildness' of nature caused by 'the Fall' as recounted in Genesis 3. To them, "[v]irgin territory was … a constant reminder of past human sin and present human idleness" (Harrison 2006b, p. 29). Others have tried to 'soften the language' and place the Genesis 1 command within the wider socio-historical context of ancient Mesopotamia. Theologian John Rogerson's argument that the command to 'subdue' the Earth is actually a "reconciliation with nature" strategy whereby subjugation becomes a process of bringing "an ongoing and dynamic process" under a managed control rather than enforced domination (Rogerson 2010, p. 28). Nevertheless, the concept of subjugation has primarily been associated with acts of (dis)possession, ecological destruction, conquest and colonialism.

2. The Stewardship Prism

Based on Genesis 2: 15 in which God places humans in the garden "to work it and take care of it", this verse is the hinge on which the stewardship prism rests. The garden is seen as a metaphor for the Earth as a whole and humanity has a divine injunction to 'steward' or care for it. The Earth is also seen as a gift for human use due to the command to 'work it'. Yet the instruction to 'care' for the Earth whilst working it is usually seen as a moderating command that limits and prevents inflicting wanton destruction and extractive overuse. This prism is the dominant perspective in evangelical churches today (Goldberg et al. 2019). It is worth noting, that as a prism, it still tends to separate humanity from the natural world as it encourages a way of thinking that positions humanity as the 'source' of the solution. According to Genesis 2: 15, humanity is placed *into* the garden as an 'addendum' to an already operating and self-sustaining ecosystem. Theologians Himes and Himes (1990) view the anthropocentric tendencies of Christian stewardship to be quite dangerous as it can encourage a propensity to focus on human agency and problem solving (by 'stewarding' the environment) whilst silently removing God from the scene. As a result, human beings tacitly assume oversight of the Earth and move to centre stage in the creation narrative rather than having the Creator as the focus in conversations about sustainability. Contemporary scientific principles of sustainability and ecology challenge the largely anthropocentric perspective of Christian 'stewardship' as not in keeping with the ecological "web of life" perspective from the Earth sciences (Macy and Brown 2014, p. 4).

3. The Partnership Prism

This green prism explains that creation has value because of its relationship to God rather than its utilitarian function. Built on the redemptive teachings of Jesus about the Kingdom of God, his resurrection is taken as not just renewing divine-human relationships, but also renewing the divine-nature relationship. In this prism perspective, creation has value to God, not because humans are doing anything to it (or 'stewarding it') but simply because it exists. Whilst still based on a reading of Genesis 2: 15 and the principle of 'care and work' (gardening), now the resurrection of Jesus is used to signal a renewing of the relationships between human and non-human creations. To only think about the Earth as a source of resources that can be accessed through care and work overlooks the knowledge that the creation itself is a source of divine delight. Animals and plants have value "in and for themselves" as well as "for God", quite apart from their usefulness to humans (McDaniel 1990, p. 165). Using New Testament teaching in this way forces a reappraisal of stewardship and establishes the partnership prism on a theology of interrelationship (Romans 8: 19–23). This prism is still largely human-centred given that, in practice, partnership often means that humans rely on technology to improve or solve ecological problems rather than on behavioural change as the solution to today's environmental crisis.

4. The Covenant Prism

This prism is the first from among these five perspectives to move away from the Genesis 1–2 creation accounts. The covenant God made with Noah following the flood (Genesis 9: 11–13) shows us for the first time that God cares for all life in his creation and not just for human life. This perspective favours the view that the "cosmic covenant" acts to restore the "primal harmony between human beings and the animal and physical worlds" (Barton 2010, p. 51). The covenant reinstates a direct link between human conduct and wider, more universal forms of human flourishing. In this prism, the "cosmos flourishes when … humans act harmoniously with one another" and it becomes a disastrous "world turned upside down when bad behaviour brings about the divine destabilizing of cosmic order" (Stavrakopoulou 2010, p. 17). The Covenant prism also draws on the Psalms or wisdom literature of the Old Testament which is often sharply critical of anthropocentrism. Texts such as Psalm 104 portray a world that does not involve humans at all and suggests the possibility that humans can learn something from the non-human creation about the interactions and relationships between the created and the Creator (Dell 2010, p. 57). Theologians who argue for or from this viewpoint see that the complexity of the natural world requires an acknowledgment of the richness and diversity of all non-human life that is both created and sustained by its Creator (Dell 1994).

5. The Sacrament Prism

This fifth and final prism draws on traditions and perspectives, particularly teachings from Eastern Orthodoxy about the 'cosmic Christ', where the "fabric of the material world [is] viewed as spiritually" endowed (Balabanski 2010, p. 104). In Orthodox teaching, the universe is a 'universal seed' and the Spirit, hovering over the waters, is seen as a bird nesting over its eggs preparing the water to produce living beings. The world and waters, through this sacramental view of the creation, are pregnant with the possibilities of all living species originating from a single source or 'divine seed' implanted by God in the 'beginning'. Seeing creation as a 'divine seed' has a rich tradition in Orthodox teaching. This cosmic vision sees creation not as a one-time event but rather a continuing and evolving process. The ongoing creative dance of the cosmos has Christ at the centre, reconciling and creating the world around us. Such language reflects Orthodox teaching that encourages us to become a "compassionate friend in and of the world" (Chryssavgis 2019, pp. 138–146). For Western Christians who are more familiar with declaring Jesus as 'Lord', 'Saviour', or even 'friend', the idea of a 'cosmic Christ' who is the reconciler of a broken and exploited cosmos, one through whom and from whom the world is created, can be quite foreign (Balabanski 2010, pp. 104–105). Western Christians may be more willing to accept such a view of the world through Pentecostal or charismatic teaching. Pentecostal philosopher James K. A. Smith argues that Pentecostals see the world through an "enchanted worldview" (Smith 2010, p. 87). Pressing this theme further, fellow Pentecostal theologian Amos Yong argues strongly for a pneumatological theology[13] of the environment where we encounter God sacramentally within creation (Yong 2005, p. 299). In Yong's understanding, the spiritual and material worlds are intertwined and as such the Earth becomes our neighbour, whom we are obliged to love. These acts do not simply shape creation for today but also anticipate the new heavens and new Earth. Thus, the "Holy Spirit's activity" is both life-giving and life-sustaining (Yong 2011, pp. 23–24). Such an 'en-spirited' sacramental prism sees "the resurrection of Christ, and thus of all those saved through him [in the] transfiguration of the cosmos, such that all of nature is taken up and made into the New Creation" (Russell 2008, p. 275). According to Jürgen Moltmann (1985), this perspective presupposes the presence of the Spirit and that the "Spirit is poured out on everything that exists, and that the Spirit preserves it, makes it live and renews it" (p. 10).

Towards Synthesis: A Spectrum of Responses

Some years back, James Lovelock, the NASA scientist who developed the Gaia Hypothesis, expressed his concern about how environmental ethics had become just like a new 'religion'. He was "dismayed by the iconic significance of a giant

[13] A form of theology that privileges the role and action of the Holy Spirit in our world. It is common in Charismatic and Pentecostal faith traditions. The term is derived from the Greek word *pneuma*, meaning breath, blow or air in motion.

Ecocentric Paradigm: The natural world exists for its own sake and benefit. Humans are one of many companion species, ecologically embedded within the wider web of life. The intrinsic value of organic life and inorganic matter is incalculable. Nature is inviolable and imbued with sanctity.

Anthropocentric Paradigm: The natural world exists for the sake and benefit of humans. Natural resources should be managed for maximum long-term human gain. The extrinsic value of natural resources is calculable in terms of their economic benefits to humans. Nature is an asset in the utilitarian service of humans.

Fig. 18.5 Two alternative paradigms; ecocentric and anthropocentric attitudes expressed along a left-right continuum. Representational figure by Authors

wind turbine on a green hill [that seemed] to mock the Cross" (Lovelock 2009, p. 225). He believed it was hypocritical to see fossil fuel energy companies advertising their 'green' credentials with wind farms whilst simultaneously making the lion's share of their profits from petrochemicals. Through this complaint, Lovelock was expressing his concern about two distinct environmentalisms. The first is anthropocentric, regards the needs of humans as central, and views the resources offered by the natural world in light of these needs (Hoffman and Sandelands 2005). The second, an ecocentric approach, is one that places the needs of the natural world first and sees humanity as an ecologically embedded companion species within the organic and non-organic natural world (Cooper 2003; Siegel et al. 2018). These alternative paradigms are depicted in Fig. 18.5.[14]

Presenting these two environmentalisms in such a fashion can help us to conceptualise how environmental attitudes shape—and are shaped by—political ideologies. There is some suggestion in the literature that an inverse relationship between Christian beliefs and environmental concern may actually be more from the influence of conservative political ideology rather than any specific pre-determined theological analysis (Clements et al. 2013). Indeed, the authors of that study advise that future analysis of the nexus between religious beliefs and environmental concern should account for the effect of political ideology. That said, whilst individuals may find their opinions falling along a spectrum of possibilities on these matters, research has found that climate science denial is strongly correlated with anthropocentrism, neo-liberal capitalist ideologies, conservative values, individualism and religious evangelical beliefs (Orr 2005; Björnberg et al. 2017; Braterman 2020). Polls across the Anglosphere (Australia, UK, North America) frequently show that evangelical Christians have higher rates of climate change denial (Haluza-Delay 2014) and those with a conservative voting intention and low environmental values tended to be the

[14]In explaining these concepts, Hoepper et al. (2017) envisage six categories along a spectrum, however, for our purposes we are representing these two categories along a left-right/ecocentric-anthropocentric continuum (see Fig. 18.5).

most sceptical about the reality and severity of climate change (Poortinga et al. 2011, p. 1017).

Human values and spirituality play an important role in human behaviour (Luetz et al. 2020; Luetz and Nunn 2020). Given what the body of literature suggests about the interplay between political ideologies, pro-environmental behaviour and Christian thought, we propose taking our framework of the various green prisms proposed earlier (see Fig. 18.4) and ordering these positions along a left-right continuum which mirrors progressive (ecocentric) to conservative (anthropocentric) perspectives (see Fig. 18.6). In proposing this model, we recognise that any environmental interpretation of the bible "demands a multivalent and creative approach" (Stavrakopoulou 2010, p. 10) that locates palatable climate adaptation solutions within communities. In this way, spirituality can contribute to 'thick' understandings of Christian worldview-informed solutions to climate change in the Anthropocene.

As we read these varying descriptions, it is quite possible that several of these green prisms resonate with people of Christian faith to some degree. Clive Ayre (2014), from whose work these prisms are drawn, acknowledges this is most likely

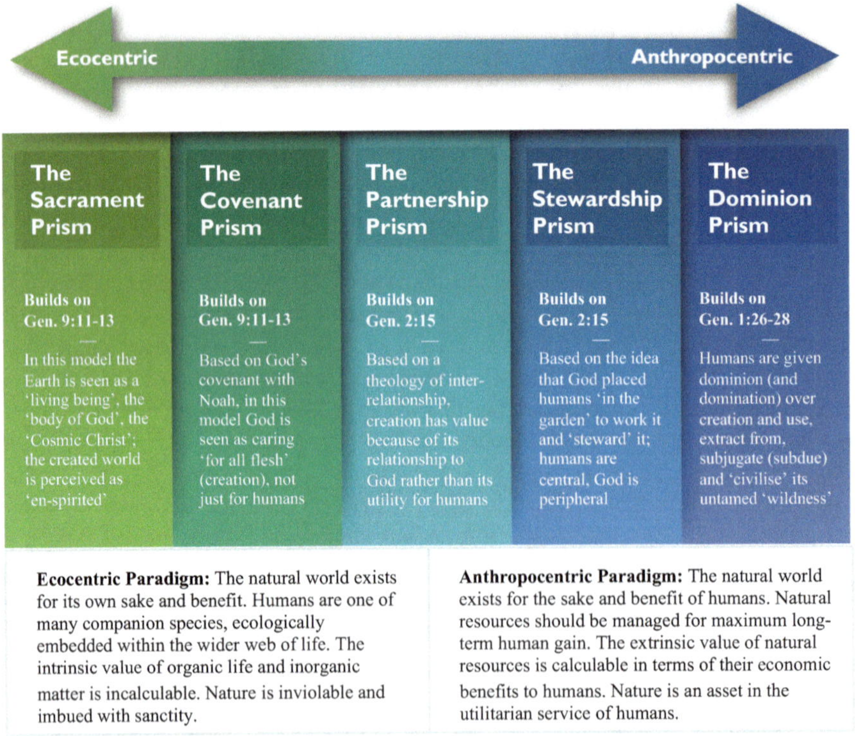

Fig. 18.6 'Five Green Prisms' (based on Ayre 2014) of Christian thought overlaid with 'Two Paradigms' (based on Hoffman and Sandelands 2005) showing a relationship between ecotheological and politico-ideological positions. Representational figure by Authors

the case. Our analysis suggests that interpretations that may be common in one part of the Christian world (e.g., stewardship among North American evangelicals) may not hold similar relevancy in other parts of the world (e.g., the Pacific) that are more visibly bearing the impacts of climate change. Haluza-Delay (2014), for example, notes how North American Christians or Buddhists enact their religion in different ways to South Pacific Christians or Bhutanese Buddhists whilst still adhering to those larger common global belief systems (p. 273). In the aftermath of the Aitape tsunami of 1998 in Papua New Guinea, pastoral teachings explaining the disaster did "not indicate that the pastors were *not* evangelical, but rather that they had a *Melanesian* evangelical theology [emphasis original]" (Fountain et al. 2004, p. 342). This shows the importance of local contexts in shaping ecotheological perspectives.

Likewise, Hagevi (2014) found that at a societal level, public opinion in countries dominated by Catholic or Orthodox culture tended to show a higher level of environmental concern than those societies shaped by a Protestant ethic. Certainly, in their study of how Southern Baptists responded to ideas of 'stewardship', Zaleha and Szasz (2015) found anti-environmentalism within North American conservative Christianity to be connected to fears that 'stewardship' of creation is drifting towards neo-pagan nature worship (Zaleha and Szasz 2015). Figure 18.6 can provide some guidance towards understanding why this fear of neo-pagan nature worship may dampen environmental sympathies within more theologically conservative faith communities.

Haluza-Delay (2014) argues that one advantage of listening to religious environmental thought is that it provides "alternative ways of conceptualizing the human-environment relationship" (p. 262). While 'stewardship' may be the most popular 'flagship'[15] word (Ayre 2014, p. 102) and the most widespread concept in evangelical circles today (Goldberg et al. 2019), our analysis shows that the model still rests largely on an anthropocentric paradigm. In consequence, faith communities which are directly facing the challenges posed by our environmental crisis today may resort to other conceptual models of environmental concern, both to inform and motivate their adaptation practices.

Conclusion: New Conceptual Capital for a New Era

Religion is as much practiced as it is a set of beliefs; practices are contextualised within the communities in which beliefs are acquired and taught. Earlier in this chapter we identified that it is possible to hold various knowledges concurrently and these multiple 'ways of knowing' contribute to creating 'thick' cultural meanings. Such cultural meanings require critical attention.

[15]Clive Ayre stresses that his use of the term comes with several qualifications whereby it should be used in a 'composite sense' and not in the way that it is typically used by North American evangelicals. In his work (Ayre 2014, p. 101) he lists six qualifying points that do not frame the term in an anthropocentric manner (pers. comm. 23 Nov 2020).

Calvin DeWitt (1998) has alluded to the perceived vulnerability of spirituality as a cultural resource for adaptation: "Human and cultural degradation are threatening and eliminating long-standing knowledge held by native and some Christian communities on living sustainably and cooperatively with creation … Among the most serious degradation is the extinguishing of longstanding cultures that have learned to live sustainably on earth" (p. 19). In addition, dialogue between scientific and theological perspectives can be a productive process that creates common ground (Kasak and Kull 2018). In line with Knight (2006), who argues that most people choose one knowledge source as primary or basic, knowing that there are diverse environmental theo-political concepts can encourage individuals and faith communities to shift their environmental readings of the bible towards more ecocentric positions.

While faith communities continue to wrestle with the metaphysical challenges of our era, it is obvious that appropriate language and conceptual constructs constitute an important part of the required adaptation portfolio. In the era of the Anthropocene and climate change, we need to learn to wrestle with and seek spiritual explanations for concepts such as 'ecocide' (the killing of the planet's life systems) or 'geocide' (the killing of the planet). Thomas Berry challenges us by querying why we have difficulty using these two terms when we are very accustomed to speaking about suicide, homicide and genocide (Collins 2009). Whereas the term "Genocide" was coined by Raphael Lemkin (1944/2008) towards the end of World War II to create "a new term and new conception for destruction of nations … or of an ethnic group" (p. 79), the Anthropocene is now ushering in conditions that put the health of our planet in serious jeopardy and therefore require new conceptual language yet again (MEA 2005; IPBES 2019; UN 2019). Importantly, the new world of the Anthropocene creates opportunities for religions to mobilise and leverage theo-cultural capital as part of the required mix of solutions to today's growing climate crisis. Communities on the frontlines of climate change, like those in the Pacific Islands (Nunn and Kumar 2018), which are already experiencing its impacts have spiritual capital that can assist them in directly facing these challenges (Havea et al. 2018). 'Thin' anthropocentric evocations that are grounded in merely 'stewarding' our seriously injured ecosystems will not suffice. More holistic and 'thicker' ecocentric perspectives are urgently required at the table of climate change policy discourse.

Acknowledgements The authors thank Clive Ayre, Ben Myers and Patrick Nunn for their excellent comments on draft versions of this manuscript.

References

Ashbolt NJ (2019) Flood and infectious disease risk assessment: health in ecological perspectives in the anthropocene. In: Watanabe T, Watanabe C (eds) Health in ecological perspectives in the anthropocene, pp. 145–159, Springer, https://doi.org/10.1007/978-981-13-2526-7_12

Ayre C (2014) Earth, faith and mission: the theology and practice of earthcare. Morning Star Publishing, Northcote, Vic

Balabanski VS (2010) Hellenistic cosmology and the letter to the colossians: towards an ecological hermeneutic. In: Horrell DG, Cherryl H, Southgate C, Stavrakopoulou F (eds) Ecological hermeneutics: Biblical, historical and theological perspectives. T&T Clark, London, pp 94–107

Barbour IG (1997) Religion and science: historical and contemporary issues. HarperCollins, New York

Barbour IG (2000) When science meets religion: enemies, strangers or partners?. HarperCollins, New York

Barton J (2010) Reading the prophets from an environmental perspective. In: Horrell DG, Cherryl H, Southgate C, Stavrakopoulou F (eds) Ecological hermeneutics: Biblical, historical and theological perspectives. T&T Clark, London, pp 46–55

Bianchi E (1998) The ecological challenge to Christianity. New Theology Review, 11(1)

Björnberg KE, Karlsson M, Gilek M, Hansson SO (2017) Climate and environmental science denial: a review of the scientific literature published in 1990–2015. J Clean Prod 167:229–241

Blackwell RJ (1998) The aquinas lecture, 1998: science, religion and authority: lessons from the Galileo affair. Marquette University Press, Milwaukeeo

Boyes E, Stanisstreet M (2012) Environmental education for behaviour change: which actions should be targeted? International Journal of Science Education 34(10):1591–1614. https://doi.org/10.1080/09500693.2011.584079

Braasch G (2013) Climate change: is seeing believing? Bulletin of the Atomic Scientists 69(6):33–41. https://doi.org/10.1177/0096340213508628

Braterman P (2020) 'God intended it as a disposable planet': meet the US pastor preaching climate change denial, The conversation, 13 October, https://theconversation.com/god-intended-it-as-a-disposable-planet-meet-the-us-pastor-preaching-climate-change-denial-147712

Brooke JH (1991) Science and religion: some historical perspectives. Cambridge University Press, Cambridge, UK

Brunner DL, Butler JL, Swoboda AJ (2014) Introducing evangelical ecotheology: foundations in scripture, theology, history, and praxis. Baker Academic

Buxton G, Luetz JM, Shaw S (2021) Towards an embodied pedagogy in educating for creation care. In Luetz JM, Green B (eds), Innovating Christian Education Research: Multidisciplinary Perspectives (pp. 349–375). Springer Nature, Singapore. https://doi.org/10.1007/978-981-15-8856-3_20

Calvin J (2009) Institutes of the Christian religion: 1541 French edition (trans: McKee EA). William B. Eerdmans, Cambridge, UK

Chakrabarty D (2009) The climate of history. Critical Inquiry 35(2):197–222

Chakrabarty D (2016) Humanities in the anthropocene: the crisis of an enduring kantian fable. New Literary History 47:377–397

Christian D (2008) This fleeting World: a short history of humanity. Berkshire Publishing Group, Great Barrington, Mass

Chryssavgis J (2019) Creation as sacrament: reflections on ecology and spirituality. T&T Clark, London

Clements JM, McCright AM, Xiao C (2013) Green Christians? an empirical examination of environmental concern within the US general public. Organization and Environment 27(1):85–102

Collins P (2009, June 11) Life of a 'geologian'. Eureka Street, 11 June, http://www.eurekastreet.com.au/article.aspx?aeid=14319

Cook J (2013) 4 Hiroshima bombs worth of heat per second. Skeptical Science. Retrieved Jan 31, 2018, from https://www.skepticalscience.com/4-Hiroshima-bombs-worth-of-heat-per-second.html

Cook J, Nuccitelli D, Green SA, Richardson M, Winkler B, Painting R et al. (2013) Quantifying the consensus on anthropogenic global warming in the scientific literature. Environmental Research Letters, 8(2), 024024, 1–7. https://doi.org/10.1088/1748-9326/8/2/024024/pdf

Cook J, Bedford D, Mandia S (2018) Raising climate literacy through addressing misinformation: case studies in agnotology-based learning. J Geosci Educ 62(3):296–306. https://doi.org/10.5408/13-071.1

Cooper S (2003) Review: on extending the half-life of green politics. Arena Magazine, June-July, pp 39–42
Dell KJ (1994) Green ideas in the wisdom tradition. Scottish Journal of Theology 47:423–451
Dell KJ (2010) The significance of the wisdom tradition in the ecological debate. In: Horrell DG, Cherryl H, Southgate C, Stavrakopoulou F (eds) Ecological hermeneutics: Biblical, historical and theological perspectives. T&T Clark, London, pp 56–69
Deryng D, Conway D, Ramankutty N, Price J, Warren R (2014) Global crop yield response to extreme heat stress under multiple climate change futures. Environmental Research Letters 9(3):034011. https://doi.org/10.1088/1748-9326/9/3/034011
DeWitt CB (1998) Caring for creation: responsible stewardship of God's handiwork. Baker Books Center for Public Justice, Grand Rapids, MI
Diamond JM (2005) Collapse: How societies choose to fail or survive, Viking
Doyle T (2000) Green power: the environment movement in Australia. UNSW Press, Sydney
Edwards D (2006) Ecology at the heart of faith: the change of heart that leads to a new way of living on Earth. Orbis, New York
Ernst M (2012) Changing Christianity in Oceania: a regional overview. Archives de sciences sociales des religions. http://doi.org/10.4000/assr.23613
Ernst M, Anisi A (2016) The historical development of Christianity in Oceania. In Sanneh, L. & McClymond, M.J. (Eds.), The Wiley Blackwell Companion to World Christianity, 588–604, Wiley, https://doi.org/10.1002/9781118556115.ch44
Evans A (2016) The myth gap: what happens when evidence and arguments aren't enough. Transworld Publishers, London
Fair H (2018) Three stories of Noah: navigating religious climate change narratives in the Pacific Island region. Geo Geogr Environ 5(2). https://doi.org/10.1002/geo2.68
Fountain PM, Kindon SL, Murray WE (2004, Fall) Christianity, calamity, and culture: the involvement of Christian churches in the 1998 Aitape tsunami disaster relief. The Contemporary Pacific, 16(2), 321–355
Gard AR, Veitayaki J (2017) In the wake of Winston—climate change, mobility and resiliency in Fiji. International Journal of Safety and Security Engineering 7(2):157–168
Ghosh A (2016) The great derangement: climate change and the unthinkable. University of Chicago Press, Chicago
Gibson W, Forsaith P, Wellings M (2013) The ashgate research companion to World methodism. Routledge Handbooks. https://doi.org/10.4324/9781315613789
Goldberg MH, Gustafon A, Ballew MT, Rosenthal SA, Leiserowitz A (2019) A social identity approach to engaging christians in the issue of climate change. Science communication, 1–22
Gottlieb R (1993) Reconstructing environmentalism: complex movements, diverse roots. Environmental History Review 17(4):1–19
Gould SJ (1999) Rock of ages: science and religion in the fullness of life. Vintage, London, UK
Green RPH (1997) Saint Augustine on Christian teaching. Oxford University Press
Habel NC (2000) Readings from the perspective of the Earth. Sheffield Academic Press
Habel NC (2008) Introducing ecological hermeneutics. In: Habel NC, Trudinger P (eds) Exploring ecological hermeneutics. Society of Biblical Literature, Atlanta, GA, pp 1–8
Habel NC (2012) Introducing ecological hermeneutics. Lutheran Theological Journal, 46(2), 97–105
Hagevi M (2014) Religion and the environmental opinion in 22 countries: a comparative study. Revue Internationale de Sociologie 24(1):91–109
Haluza-Delay R (2014) Religion and climate change: varieties in viewpoints and practices. WIREs Climate Change, 5(March/April), 261–279
Hamilton C (2015) Getting the Anthropocene so wrong. The Anthropocene Review 2(2):1–6. https://doi.org/10.1177/2053019615584974
Hamilton C (2016) The Anthropocene belongs to Earth system science. The conversation, 18 August, https://theconversation.com/the-anthropocene-belongs-to-earth-system-science-64105

Hansen J (2009) Storms of my grandchildren: the truth about the coming climate catastrophe and our last chance to save humanity. Bloomsbury

Hansen J (2012) James Hansen: why I must speak out about climate change. TED Talk. Video file. Retrieved Feb 2, 2018, from www.ted.com/talks/james_hansen_why_i_must_speak_out_about_climate_change.html

Harrison P (2006a) 'Science' and 'Religion': constructing the boundaries. The Journal of Religion 86(1):81–106

Harrison P (2006b) Having dominion: Genesis and the mastery of nature. In: Berry RJ (ed) Environmental stewardship: critical perspectives, past and present (pp 17–31). London: T&T Clark. Retrieved from Academia: https://uq.academia.edu/PeterHarrison

Harrison P (2015) The territories of science and religion. The University of Chicago Press, Chicago

Havea PH, Hemstock SL, Des Combes HJ, Luetz JM (2018) "God and Tonga are my inheritance!"—climate change impact on perceived spirituality, adaptation and lessons learnt from Kanokupolu, 'Ahau, Tukutonga, Popua' and Manuka in Tongatapu, Tonga. In W. Leal Filho (eds) Climate change impacts and adaptation strategies for coastal communities. Springer Nature, Switzerland, pp 167–186. https://doi.org/10.1007/978-3-319-70703-7_9

Hessel DT, Radford Ruether R (2000) (eds) Christianity and ecology: seeking the well-being of Earth and humans. Harvard University Press

Hey S (2018) The Wesleyan quadrilateral as an aid to formation in tertiary education. In: Luetz JM, Dowden T, Norsworthy B (eds), Reimagining Christian education—cultivating transformative approaches (pp 145–161). Springer Nature, Singapore. https://doi.org/10.1007/978-981-13-0851-2_11

Himes MJ, Himes KR (1990) The sacrament of creation: toward an environmental theology. Commonweal Magazine, 26 January, https://www.commonwealmagazine.org/sacrament-creation-toward-environmental-theology

Hinkel J, Aerts JCJH, Brown S, Jiménez JA, Lincke D, Nicholls RJ, Scussolini P, Sanchez-Arcilla A, Vafeidis A, Addo KA (2018) The ability of societies to adapt to twenty-first-century sea-level rise. Nature Climate Change 8:570–578

Hodson MJ, Hodson MR (2008) Cherishing the Earth: how to care for God's creation. Monarch Books

Hoepper B, Cutter-Mackenzie A, Gilbert R (2017) Sustainability. In: Humanities Teaching, Sciences Social (eds) R Gilbert & B Hoepper. Cengage Learning, South Melbourne, Victoria, pp 428–457

Hoffman AJ, Sandelands LE (2005) Getting right with nature: anthropocentrism, ecocentrism, and theocentrism. Organization & Environment 18(2):141–162. https://doi.org/10.1177/1086026605276197

Hore-Lacy I (2006) Responsible dominion: a christian approach to sustainable development. Regent College Publishing

Horrell DG (2010) The Bible and the environment. Equinox, London

Hulme M (2009) Why we disagree about climate change: understanding controversy, inacation and oppurtunity. Cambridge University Press, Cambirdge

Hulme M (2017) Climate change and the significance of religion. Economic and Political Weekly 52(28):14–17

IPBES - Intergovernmental Science-Policy Platform on Biodiversity and Ecosystem Services (2019) Summary for policymakers. https://www.ipbes.net/sites/default/files/downloads/spm_unedited_advance_for_posting_htn.pdf

IPCC - Intergovernmental Panel on Climate Change (2014) Climate change 2014: Synthesis report summary for policymakers. https://www.ipcc.ch/site/assets/uploads/2018/02/AR5_SYR_FINAL_SPM.pdf

Johnson E (2002) She who is: the mystery of God in feminist theological discourse. Crossroad, New York

Kasak E, Kull A (2018) On a productive dialogue between religion and science. Scientia et Fides 6(1):129–153

Kempf W (2017) Climate change, christian religion and songs: revisiting the Noah story in the Central Pacific. In: Dürr E, Pascht A (eds) nvironmental Transformations and Cultural Responses. Palgrave Macmillan, New York, pp 19–48
Kempf W (2020) Introduction: climate change and Pacific Christianities. Anthropological Forum 30(3):215–232. https://doi.org/10.1080/00664677.2020.1812052
Kendall H (1997) World scientists' warning to humanity. Union of concerned scientists: Cambridge, MA., https://www.ucsusa.org/sites/default/files/attach/2017/11/World%20Scientists%27%20Warning%20to%20Humanity%201992.pdf
Kennedy P (1993) Preparing for the twenty-first century. Random House, New York
Kinney PL (2018) Interactions of climate change, air pollution, and human health. Current Environmental Health Reports 5(1):179–186. https://doi.org/10.1007/s40572-018-0188-x
Kjaergaard PC (2016) Peter Harrison's 'Territories of science and religion': a symposium. Zygon 51(3):684–697
Knight GR (2006) Philosophy & education: an introduction in christian perspective (4th ed), Andrews University Press
Lemkin R (2008) Axis rule in occupied Europe: laws of occupation, analysis of government, proposals for redress. Lawbook Exchange, Clark, NJ. ISBN 978-1-58477-901-8
Lindberg DC, Numbers RL (2003) Introduction. In: Lindberg DC, Numbers RL (eds.) When science and christianity meet, 1–5. Chicago: University of Chicago Press
Lovelock J (2009) The vanishing face of Gaia: a final warning. Basic Books, New York
Lou J, Wu Y, Liu P, Kota SH, Huang L (2019) Health effects of climate change through temperature and air pollution. Current Pollution Reports 5(3):144–158. https://doi.org/10.1007/s40726-019-00112-9
Luetz JM (2020) Disaster-resistant schools for disaster-resilient education. In: Leal Filho W, Marisa Azul A, Brandli L, Gökcin Özuyar P, Wall T (eds) Quality education, encyclopedia of the UN sustainable development goals. Cham, Springer, pp 158–174. https://doi.org/10.1007/978-3-319-69902-8_15-1
Luetz JM, Nunn PD (2020) Climate change adaptation in the Pacific Islands: a review of faith-engaged approaches and opportunities. In: Leal Filho W (ed) Managing climate change adaptation in the Pacific region. Springer Nature, Cham, pp 293–311. https://doi.org/10.1007/978-3-030-40552-6_15
Luetz JM, Walid M (2019) Social responsibility versus sustainable development in United Nations policy documents: a meta-analytical review of key terms in human development reports. In: Leal Filho W (ed) Social responsibility and sustainability. Springer Nature, Cham, pp 301–334. https://doi.org/10.1007/978-3-030-03562-4_16
Luetz JM, Buxton G, Bangert K (2018) Christian theological, hermeneutical and eschatological perspectives on environmental sustainability and creation care—The role of holistic education. In: Luetz JM, Dowden T, Norsworthy B (eds) Reimagining christian education—cultivating transformative approaches (pp 51–73). Springer Nature, Singapore. https://doi.org/10.1007/978-981-13-0851-2_4
Luetz JM, Margus P, Prickett B (2020) Human behavior change for sustainable development: perspectives informed by psychology and neuroscience. In: Leal Filho W, Marisa Azul A, Brandli L, Gökcin Özuyar P, Wall T (eds) Quality Education: Encyclopedia of the UN Sustainable Development Goals (pp 1–16). Springer Nature, Switzerland. https://doi.org/10.1007/978-3-319-69902-8_12-1
Lynch AH, Veland S (2019) Urgency in the Anthropocene. The MIT Press, Cambridge, Mass
Macy J, Brown M (2014) Active hope: how to face the mess we're in without going crazy. New Society Publishers, Gabriola Island, Canada
Maddox RL (1994) John Wesley's practical theology, kingswood books. Abingdon Press, Nashville, TN
Marlow H (2009) Biblical prophets and contemporary environmental ethics. Oxford University Press, Oxford

Maslin M (2019) The five corrupt pillars of climate change denial, The conversation, 29 Nov, https://theconversation.com/the-five-corrupt-pillars-of-climate-change-denial-122893

Matthews TKR, Wilby RL, Murphy C (2017) Communicating the deadly consequences of global warming for human heat stress. Proceedings of the National Academy of Sciences - PNAS 114(15):3861–3866. https://doi.org/10.1073/pnas.1617526114

Maudlin MG, Baer M (eds) (1989) NRSV: The Green Bible. HarperOne, San Francisco

McDaniel J (1990) Where is the Holy Spirit anyway? Response to a sceptic environmentalist. The Ecumenical Review 42(2):162–174

McKibben B (2005) The comforting whirlwind: God, Job and the scale of creation. Cowley Pub, Cambridge, MA

McNeil J (2000) Something new under the sun: an environmental history of the twentieth-century world. WW Norton, New York

MEA - Millennium Ecosystem Assessment (2005) Living beyond our means: Natural assets and human well-being. https://www.wri.org/publication/millennium-ecosystem-assessment-living-beyond-our-means

Merrit J (2010) Green like God: unlocking the divine plan for our planet. Hachette, New York, NY

Meyer R (2019) The cataclysmic break that (maybe) occurred in 1950. Retrieved from The Atlantic, 16 April, https://www.theatlantic.com/science/archive/2019/04/great-debate-over-when-anthropocene-started/587194/

Moltmann J (1985) God in creation: an ecological doctrine of creation, the Gifford lectures 1984–1985 (trans: Kohl M) London: SCM

Moo DJ, Moo JA (2018) Creation care: A Biblical theology of the natural world. Zondervan, Grand Rapids, MI

Nelson W, Luetz JM (2019) What can we learn from Pope Francis about change management for environmental sustainability? A case study on success factors for leading change in change-resistant institutional environments. In Leal Filho W, Consorte McCrea A (eds) Sustainability and the humanities. Springer Nature, Switzerland, pp 503–524. https://doi.org/10.1007/978-3-319-95336-6_29

Nilsen TD, Solevåg AR (2016) Expanding ecological hermeneutics: the case for ecolonialism. Journal of Biblical Literature 135(4):665–683

Nunn PD, Mulgrew K, Scott-Parker B, Hine DW, Marks ADG, Mahar D, Maebuta J (2016) Spirituality and attitudes towards nature in the Pacific Islands: insights for enabling climate-change adaptation. Clim Change 136(3–4):477–493. https://doi.org/10.1007/s10584-016-1646-9

Nunn PD (2017) Sidelining God: why secular climate projects in the Pacific Islands are failing. The conversation, 17 May. https://theconversation.com/sidelining-god-why-secular-climate-projects-in-the-pacific-islands-are-failing-77623

Nunn P, Kumar R (2018) Understanding climate-human interactions in Small Island Developing States (SIDS): Implications for future livelihood sustainability. International Journal of Climate Change Strategies and Management 10(2):245–271

Nunn PD (2020a) In anticipation of extirpation: how ancient peoples rationalized and responded to postglacial sea level rise. Environmental Humanities 12(1):113–131. https://doi.org/10.1215/22011919-8142231

Nunn PD (2020b) Endings. *Chicago Quarterly Review* 30: 20–32)

Oreskes N Conway EM (2011) Merchants of doubt: how a handful of scientists obscured the truth on issues from tobacco smoke to climate change. Bloomsbury

Orr DW (2005) Armageddon versus extinction. Conserv Biol 19(2):290–292. https://doi.org/10.1111/j.1523-1739.2005.s04_1.x

Plant SJ (2013) Methodism and politics: mapping the political on the methodist Genome, In: Gibson W, Forsaith P, Wellings M (eds) The Ashgate Research Companion to World Methodism, Routledge Handbooks, https://doi.org/10.4324/9781315613789.ch20

Poortinga W, Spence A, Whitmarsh L, Acapstick S, Pidgeon N (2011) Uncertain climate: an investigation into public scepticism about anthropogenic climate change. Glob Environ Change 21:1015–1024

Ripple WJ, Wolf C, Newsome TM, Galetti M, Alamgir M, Crist E et al (2017) World's scientists' warning to humanity: a second notice. Bioscience 67(12):1026–1028. https://doi.org/10.1093/biosci/bix125

Rogerson JW (2010) The creation stories: their ecological potential and problems. In: Horrell DG, Cherryl H, Southgate C, Stavrakopoulou F (eds) Ecological Hermeneutics: Biblical, historical and theological perspectives. T&T Clark, London, pp 21–31

Rubow C, Bird C (2016) Eco-theological responses to climate change in Oceania. Worldviews 20:150–168

Russell RJ (2008) Cosmology: from Alpha to Omega. Fortress Press, Minneapolis

Santmire P (1985) The travail of nature: the ambiguous ecolgical prmise of Christian theology. Fortress, Philadelphia

Schaefer J (2014) Religious motivation for mitigating human-forced climate change: scientifically informed, politically astute, and collaborative. International Journal of Climate Change Strategies and Management 6(1):34–46

Siegel L, Cutter-Mackenzie-Knowles A, Bellert A (2018) Still 'Minding the Gap' sixteen years later: (re)storying pro-environmental behaviour. Australian Journal of Environmental Education 34(2):189–203

Sire JW (2009) The Universe next door: a basic Worldview catalogue, Inter-Varsity Press (IVP) Academic

Smith J (2010) Thinking in tongues: pentecostal contributions to Christian philosophy. Wm B Eerdmans Publishing Company, Grand Rapids, MI

Spencer N, White R (2007) Christianity, climate change and sustainable living. Society for Promoting Christian Knowledge, London

Spencer N, White R, Vroblesky (2009 Christianity, climate change, and sustainable living. Hendrickson Publishers

Stavrakopoulou F (2010) Introduction to Part I. In: Horrell DG, Cherryl H, Southgate C, Stavrakopoulou F (eds) Ecological hermeneutics: Biblical, historical and theological perspectives. T&T Clark, London, pp 15–20

Steffen W, Sanderson A, Tyson P, Jäger J, Matson P, Moore B, Wasson R (2005) Global Change and the Earth system: a planet under pressure. SpringerVerlag, Heidelberg

Steffen W, Grinevald J, Crutzen P, McNeill J (2011) The Anthropocene: conceptual and historical perspectives. Philosophical Transactions of the Royal Society 369:842–867

Steffe W, Broadgate W, Deutsch L, Gaffney O, Ludwig C (2015, January 16) The trajectory of the Anthropocene: the great acceleration. The Anthropocene Review, 2(1), 81–98

Stenmark L (2015) Storytelling and wicked problems: myths of the absolute and climate change. Zygon 50(4):922–936

University of Exeter (n.d.) Uses of the Bible in environmental ethics. Retrieved July 31, 2019, https://web.archive.org/web/20160630133108/, http://humanities.exeter.ac.uk/theology/research/projects/uses/projectoutline/

UN - United Nations (2019) UN report: nature's dangerous decline 'unprecedented'; species extinction rates 'accelerating'. https://www.un.org/sustainabledevelopment/blog/2019/05/nature-decline-unprecedented-report/

Wainwright E (2006) Women healing/healing women: the generization of healing in early Christianity. Equinox, London

Weldon SP (2017) Science and Religion. In: Ferngren GB (ed) Science and religion: a historical introduction. John Hopkins University Press, Baltimore, pp 3–19

White L (1967) The historical roots of our ecologic crisis. science, 155(3767), 10 March, pp 1203–1207

Wilson E (2012) After secularism: rethinking religion in Global politics. Palgrave, Basingstoke

Yong A (2005) The spirit poured out on all flesh: pentecostalism and the possibility of global theology. Baker Academic, Grand Rapids, MI

Yong A (2011) The spirit of creation: modern science and divine action in the Pentecostal-charismatic imagination. Wm B Eerdmans, Grand Rapids, MI

Zaleha BD, Szasz A (2015) Why conservative Christians don't believe in climate change. Bulletin of the Atomic Scientists 71(5):19–30. https://doi.org/10.1177/0096340215599789

Chapter 19
Faith and a Sustainable Pacific

Clive W. Ayre

Abstract The serious threats generated by Climate Change in the Pacific region highlight a much wider global threat which inevitably raises questions about our approach and response as Christians. This paper is based on a Christian ecotheological frame of reference and posits that our response must not merely be pragmatic, for Christian mission is inevitably based on the nature and theology of the Church. Ecotheology is a recognition that the care of creation is an essential part of our calling and implies an ecomissiology. Our approach to God's creation is also expressed in the theme of the covenant between God and all life. We are not required to act alone or simply in our own denomination, but in partnership with others who have a concern for the environment; that includes climate scientists, political leaders, and followers of other religions. In order to relate the theoretical and the practical, it is necessary to gain some understanding of the ecological and sustainability threats faced by Pacific Islanders. A brief summary of the history of Christian mission in the pacific will also provide some context. Thus, the theological, cultural, and ecological aspects provide the essential context for faith-based opportunities in mission. Importantly, the concerns and actions of the Pacific Churches need to be noted. The paper will conclude with a summary of faith-based opportunities on behalf of a sustainable Pacific.

Introduction

In recent years, and in different parts of the world, there have been unprecedented and well-publicised disasters such as fires and floods, or even the melting of polar ice. Many of those events have been beyond the coping capacity of the people immediately involved; the cost in financial terms, in the loss of life and of wildlife, has been enormous. The response of both humans and animals during these crises suggests that while it is true that different species tend to prey on one another, most of the evidence points to the inter-connectedness of life.

C. W. Ayre (✉)
University of Queensland, Brisbane, QLD, Australia

J. M. Luetz and P. D. Nunn (eds.), *Beyond Belief*, Climate Change Management,
https://doi.org/10.1007/978-3-030-67602-5_19

But there is another developing crisis that has to some extent escaped the attention of many people, and that is the vulnerability of Pacific Island nations in seeking to deal with the impacts of climate change. The key question in this emerging situation is, "In the context of a changing climate, what does a sustainable future look like for Pacific Islands people"? Part of that involves the role faith can play if it is to be part of the answer (Ayre 2013). The purpose of this chapter is to explore those questions.

It is important to note that life, in whatever form it takes, does not exist in a vacuum. Similarly, nations themselves cannot ultimately cocoon themselves against the impacts of global climatic change. In the Pacific region as elsewhere, it follows that there must be a balance between global and local factors; neither side can achieve the outcomes needed for survival without the other. Where, then, is this leading us?

First, I will outline some of the theological underpinning of Christian mission as it relates to the care of the earth and show that our actions do not arise out of pure pragmatism. Second, the need to form partnerships with other groups or philosophies is essential, and an interfaith dimension of Earthcare is an important part of that. Third, it will be helpful to develop an understanding of the current environmental situation in the Pacific region; this will not be a detailed analysis but aims to summarise the major threats posed by climate change in the region. This will be followed by a brief section relating to mission history and culture. Fifth, I will explore some opportunities for mission action in recent years, followed by some responses from Pacific Churches. Finally, I will present a seven-point list of faith-based opportunities for engagement on behalf of a sustainable Pacific region.

Theology and Mission

In addressing the theme of "Faith and a Sustainable Pacific", the beginning point must be a recognition that the mission of the Church flows out of the nature and theology of the Church. It is of course set in the context of the environmental crisis and the need to care for creation, but it begins with ecotheology as the inevitable foundation for any approach to the theology and praxis of eco-mission. Langmead (2002, p. 508) expressed the link quite succinctly, "Ecotheology implicitly contains an ecomissiology, just as all theologies give direction for understandings of mission that flow from them". What is critical is to see that ecology is inevitably a part of both theology and mission, and to see the absolute link between ecotheology, eco-mission, and eco-praxis. Forrester's observation is compelling when he asserts that we cannot talk about God or talk to God while setting aside, even temporarily, the ethical or normative question: "What is God calling us to do?" (Forrester 2000, p. 53).

Much could be said about ecotheology, but I will limit myself to several points. First, the grace of God is bound to emerge as a crucial insight if we are to understand God, ourselves, and our place on Earth. Jenkins' phrase is "ecologies of grace", and it demonstrates how grace is an undergirding element in the whole discussion of the way faith relates to the Earth. In the words of James Nash, "The logic of the doctrine of creation does not permit a nature-grace dichotomy.... The whole of nature – the biophysical universe – is ... an expression of grace" (Nash 1991, p. 95).

The Bible begins with the affirmation that "In the beginning God created...." Creation represents an affirmation about the world and us. In all its finitude and limitation, creation is good, even though Duchrow and Liedke (1989) have a point when they argue that the initial reference now should rather be the suffering of creation. This raises a wider identity issue; it is reflected in the *oikos* image of the Church as "the whole household of God", which as Conradie (2007, 2011) notes, has emerged as a significant "theological root metaphor" in ecumenical discussion. We can assume that "the household of God" refers to a unique people called out from the wider community—a position not without biblical support: but when we acknowledge that the whole of creation and all life belongs to God, a different understanding is suggested.

Conradie's question is valid: "if the church is not itself the household of God, what is its place and mission within this household?" (2011, p. 118). Perhaps part of the answer to that question may be in the development of the notion that the Spirit of God seeks to transform a portion of the world into the church so that, as transformed world, the church may live for the future of the world. Such an approach is suggested when the various metaphors are taken together as various facets of a diamond. Or again, while both theologically and sociologically the church is clearly a discreet entity, its life is lived within the wider environs of the earth itself. Conradie concludes that "the church offers a particular vision of the very architecture, building and ownership of the house", and it is "that place within the house where one can find traces that bear witness to the presence of the owner and keeper of the house" (2011, p. 120).

Ultimately, the "house" is the Earth itself and all the life upon it—God's creation—and not merely our nation, or those who are "like us". It is fine to believe that humankind has a special place and responsibility in the economy of God, but we are part of what Sanguin (2007, pp. 170f.) calls the "kin-dom" of God, and we are slowly beginning to grasp something of the importance of that inter-connection.

In biblical terms, the concept of covenant is potentially important. It is, after all, describing a relationship. Bühlmann contends that the primary covenant is in fact what he calls the Covenant of Creation. He argues that "without the creation account it would be unclear just who *is* the God of Abraham, Isaac, and Jacob, the God of Jesus Christ, the God of all humankind" (1982, p. 11). One of the covenantal implications of humankind's creation in the image of God is that we are thereby called to follow God's lead in caring about all life.

Another way of expressing that is in terms of a ministry of reconciliation; if our preaching about reconciliation means anything at all, it must have something to say about healing divisions and even the Earth itself. As the Apostle Paul wrote in 2nd Corinthians, "All this is from God, who reconciled us to himself through Christ, and has given us the ministry of reconciliation". In his book "Intercession" Lukas Vischer wrote:

> Prayer for all the churches who confess Christ is therefore a standing protest against the *status quo* of our divisions. God's gift of reconciliation can and should find full expression. Its aim is that the Church should become a real sign of reconciliation in the world. (1980, p. 3)

If the Church is to be "a real sign of reconciliation in the world", there are some profound questions that all of us might reflect upon. What does it mean to affirm a ministry of reconciliation? What does it look like? How does a ministry of reconciliation relate with the need for a prophetic voice on issues such as a sustainable Pacific? It is never going to be easy, but if the Church is part of the mix, we can hopefully make a positive contribution. A theology of reconciliation is basic to the Christian Faith and has numerous implications.

Partnerships for the Earth

In order for the Church to be "a real sign of reconciliation in the world" in such a fundamental area as the sustainability of life on earth, it will be necessary to seek a series of partnerships which may at times push us beyond our comfort zone. It will take us into the political sphere, working with science, with Christian traditions perhaps quite different from our own, as well as with non-Christian religions or even with no religious affiliation at all.

For some decades now something close to a consensus has been developing in terms of our duty of care towards God's creation, and this process has come about in three main ways. In recent decades there has been a veritable explosion of eco-theological texts globally, and while there may be some differences in emphasis, the overall position has been remarkably similar. A related factor has been that different denominations and confessional bodies have independently reached a position which has been expressed either in the form of resolutions or an official Statement. When taken together those Statements become highly significant in terms of their common vision. The third route has been through official ecumenical channels such as the World Council of Churches and the formation of its Unit, Justice, Peace, and the Integrity of Creation.

If the potential for faith-engaged approaches is significant in Christian terms, it may be even greater when viewed in interfaith terms. In his book *The Church and Climate Change* Ernst Conradie indicates the need for a broad approach: "Remarkably, Christians will only be able to hear God's word in conversation with people from other living faiths at the dialogue table – since climate change, perhaps for the first time in human history, is a challenge that can only be addressed through a collective, indeed a global effort" (2008, p. 101).

Just as Christian environmental Statements express fundamental agreement across virtually all confessional positions and divisions, so there is broad agreement among world faiths. In the year 2000 the United Nations Interfaith Partnership for the Environment published a book called *Earth and Faith: A Book of Reflection for Action.* As the Director of the UNEP explained, this was the result of an effort "to continue the dialogue between the scientific and faith communities" from which, it is hoped, will emerge "a greater commitment to taking responsible actions for the protection of our environment for our common good" (Bassett 2000).

There are two issues in that. The first is a positive partnership between faith and science, especially in the area of ecological issues and challenges. In Moltmann's words, "The sciences have shown us how to understand creation as nature. Now theology must show science how nature is to be understood as creation" (Moltmann 1985, p. 38). Through ecologists and others, science has alerted us to what we are doing to the environment, and what the consequences will be if we keep living in an unsustainable way. Faith and spirituality for its part touches deeper levels of purpose and belonging. We clearly need the contribution of both.

The second issue rising out of the UN book is the way in which the major world faiths converge in their attitude to environmental care. Authorised representatives from the Muslim, Buddhist, Jewish, Hindu, Christian and other faiths each wrote a short first-hand account of their Faith's teaching on environmental care, and the result is striking; in real terms they are in remarkable agreement on this issue, as they all enjoin their followers to a response of care.

The clear outcome of this convergence of belief is that, as the Edinburgh 2010 Conference of the World Council of Churches affirmed, the care of the planet may be a rich area not only for fruitful interfaith dialogue but also for common action. The Director of the UNEP Adrian Amin wrote that "We … view the convergence of spiritual values and their respect for the environment as an inspiration for environmental actions today so that our succeeding generations may all be beneficiaries of a healthy planet and a development that is sustainable" (Bassett 2000, p. 4).

It is not difficult to find examples of the kind of collaboration I am advocating; I myself have been involved in several interfaith fora in recent years, and it has always been a good experience. Then too, a group such as the Australian Religious Response to Climate Change (ARRCC) operates on an interfaith basis, actively seeking to work across religious diversity for the sake of the Earth.

More particularly, in terms of the Pacific in the present time, a Media Release from the Pacific Conference of Churches in September 2019 reported on an interfaith service for the Season of Creation under the heading "Creation Breaks Barriers". "Verses from Islamic and Hindu scripture echoed through Suva's Sacred Heart Cathedral last night as people of all faiths gathered to give thanks and pray for creation". It is surely a sign of hope when about 400 people, members of the Christian, Islamic and Hindu faiths should gather at the Sacred Heart Cathedral in Suva to pray for creation. The congregation heard a call to see God in all creation and take appropriate action in the light of the continuing threat of climate change and irresponsibility towards the environment. The Statement indicates that representatives of the Anglican, Catholic, Methodist and Seventh-day Adventist churches attended the event alongside Islamic and Hindu clerics.

A holistic approach to mission beyond our own congregation or denomination, or even beyond the Christian Faith, also needs to engage politics. But when it comes to Earthcare, adversarial political approaches are not the answer. There will always be differences of viewpoint, but it is imperative that they be held within a cohesive framework. As people of faith, I see it as part of our calling to demonstrate how that may be done, initially through the way in which we interact with each other in the Spirit of Christ. If we are as adversarial as anyone else, then we have no moral authority to presume to set an example for others. My contention is that if communities and political operatives were to address the increasingly serious environmental threats to our future in the search for global ecological health, we would discover an enormous unifying potential. We could help to dismantle the politics of division, and to build bridges instead of walls.

The engagement of Church and State in the search for global ecological health requires a whole of life and whole of community approach. As Christians we cannot do this alone, but rather seek to make our contribution alongside political operatives, social scientists, climatologists, psychologists, economists, educationalists, and others. However, such engagement needs to affirm God's concern for the whole of creation and not just the human world. The Church needs to speak with its own voice.

Some important issues centre on the process of decision-making locally. In dealing with people from another culture, it is essential to know who the decision makers in that society are, and to respect that tradition. It can be easy for non-Pacific visitors, albeit with the best of intentions, to have discussions about a range of topics and make plans for the future, only to find later on that nothing at all is happening because they have been speaking to people who do not have the authority to make decisions about such matters.

Current Pacific Audit

In order to explore faith-based opportunities in the context of multi-faceted climate change, it is necessary to have an idea of the current ecological situation in the Pacific region, even though this is a global crisis that ultimately cannot be resolved by any one section of the globe acting in isolation. Impacts are rarely commensurate with a country's contribution to the problem, but we all must play our part.

At one level it is legitimate to speak of "the Pacific Islands" in a general sense as a regional group of island nations that would include those on the Pacific rim, including Papua New Guinea. Australia and New Zealand also relate to the area, but more indirectly. When considered in this way, it is immediately clear that many of the island nations of the Pacific may appear to some as little more than sparsely populated specks scattered across a vast ocean. Over many decades, such a problematic perception has led some of the major powers to regard the region as ideal for exploitation. Any analysis of this issue would have to include the lasting impacts of war and of nuclear testing in the Pacific, the outcomes of which tend to live on for generations.

At another level, however, despite some areas of commonality, there are significant differences. The geographical and geological contexts are varied, so that the impacts of the global ecological crisis present and are perceived differently in different parts of the Pacific. With that important caveat, and without presuming to cover the crisis in detail, ecological and sustainability threats in the Pacific include a range of interrelated issues such as rising sea levels, the increasing seriousness of natural disasters such as cyclonic activity, issues around waste, impacts on fishing and agriculture, as well as cultural matters (Leal Filho et al. 2020). There may be other consequences such as forced migration (Luetz and Havea 2018). One thing that makes it hard for the peoples of the Pacific is that many of the problems originate in other parts of the world.

The phenomenon of global warming and rising sea levels is well documented; but it is significant that, as Becker et al. have observed, "The western tropical Pacific is considered as one of the most vulnerable regions of the world under present-day and future global warming" (2012, p. 85). Consequently, the research has demonstrated that "Towards the end of the twentieth century, the rate of global sea-level rise accelerated and in Micronesia reached its fastest rate for some 7000-8000 years" (Nunn et al. 2016b, p. 968). They note further that if projections prove correct, "It is probable that lower islands in Micronesia will be rendered uninhabitable in 20-30 years time and their inhabitants will need to relocate". Within that context, Becker et al. note that in the case of Funafuti Island (Tuvalu) "the total rate of rise is found to be about 3 times larger than the global mean sea level rise over 1950-2009" (Becker et al. 2012, p. 85). The IPCC also has a dire prediction if greenhouse gas emissions remain high.

In such a context, climate change and rising sea levels cease to be a future theoretical threat, but a current existential reality that impacts on every aspect of life. Visible inundation leads to images of children wading through the water that has invaded their home. Salt water intrudes into drinking water supplies; it leads to soil salinity and the destruction of essential built infrastructure such as roads or energy systems. As the population increases, and land becomes scarce, the problem is exacerbated by an increasing inability for the soil to grow food crops. Act for Peace (2020), an arm of the NCCA, identifies Vanuatu as "the most disaster-prone country on earth", and that "every year they face the threat of cyclones, earthquakes, tsunamis and volcanic eruptions". Climate Change certainly exacerbates disasters both in the Pacific as in other parts of the world.

A sustainable Pacific must take note of the issue of waste, and that from several points of view. In a personal communication, and based on his experience in the Pacific, Morrison notes that

> a Pacific nation such as Tonga has recognised the problem caused by waste and for several years now has had a Waste Management programme in effect. There are public waste bins along some main roads and a regular pick-up of domestic waste. Tonga has also had for some years a relatively small solar farm operating, but in more recent times has had a much larger solar farm operating in order to lower its dependence on the diesel powered generators which had previously generated the power for the island of Tongatapu.

It is heartening that serious thought and action is being taken in Tonga, as in other parts of the Pacific; but that is only a part of the story.

A major aspect of the problem is the amount of waste in the ocean, especially plastic which originates in many parts of the world, and which is circulated around the globe in a series of gyres. A critical aspect of this issue is that plastic ultimately does not break down, except in the form of almost countless pieces of microplastics. Henderson Island in the Pitcairn group is a good example of the problem. Henderson is a remote and uninhabited coral atoll, and is a UNESCO World Heritage site; but on-going research by Jennifer Lavers of the University of Tasmania has revealed estimates in 2017 of nearly 38 million pieces of plastic weighing a total of 17.6 tonnes on the island, which "accounts for only 1.98 s worth of the global production of plastic" (Lavers and Bond 2017, p. 6055). The research further shows that the problem is exacerbated by the fact that some debris becomes buried just below the surface or is reduced to microscopic particles. Either way, those particles are seriously damaging to wildlife. This waste material thus successfully catches up the Pacific Islands in a global problem over which the people have little influence, and which impacts significantly on their traditional way of life.

A third issue, which in part rises out of the waste problem, relates to impacts on fish stocks and fishing in terms of a traditional lifestyle, in addition to significant disruption of agriculture in some areas. There are various aspects to the crisis; for example, when microscopic plastic waste is eaten by fish or other marine life on which many islanders depend, the risk both to the fish and to human health is magnified. In his chapter on "Climate Change and Food Security in the Pacific Islands", Barnett (2020) provides a useful summary of the way in which a range of inter-related issues are involved, including impacts on tourism and increased cyclonic activity. There is clearly much more that could be said about these matters; but it is also clear that when taken together, the combined and compounding impact of global warming, sea level rises, and ocean waste becomes hugely significant in the lives of Pacific Islands people.

Pacific Mission, History and Culture

While it is not necessary to review the whole history of 18th and 19th century Christian mission in the Pacific region, it is nevertheless an integral part of the current situation, both in a positive and a less than positive sense. From a traditional Western Christian perspective, some may view this history with a sense of pride. However, with the Faith also came the baggage of Western colonialism.

In a personal communication, Morrison observed that "Some Pacific countries have fairly ambivalent attitudes towards the missionary movement of a couple of centuries ago. On the one hand, the earliest missionaries are generally revered because they are the ones who brought the gospel and established Christianity. At the same time, Morrison notes that it is not unusual for some people in the church to express the point of view that 'the missionaries destroyed my culture'. This means that … there can be quite a hiatus between someone's expression of faith and the more modern issues that need to be dealt with".

A complicating factor is that, like many other cultures around the world, there is also the problem of globalisation to be dealt with. In part, that may emerge as a consequence of tourism; but even more, like their peers in many parts of the globe, young people in particular are introduced to a whole new world of technology which can lead to a devaluation of the traditional mores of society and of faith beliefs. Globalisation of course raises a whole range of other associated issues which, while important, need not detain us here.

In addition to the mainstream Christian denominations from the early missionary movement, other faith groups are now very much part of the Pacific scene—Seventh-day Adventists, Latter Day Saints (Mormons), in addition to other religions such as Hinduism. Thus, if people of faith are to grasp the opportunities for effective ecological action, the significance of an ecumenical and interfaith approach becomes critical.

Mission Action

Only now, having dealt with the theological aspects of ecology and mission, and some of the contextual dimensions of the Pacific Islands in terms of the impacts of climate change, can we turn our attention to practical missional opportunities. Realistically, of course, much of that mission will be exercised within our own faith communities. But, as the hymn has it, "The Church's one foundation is Jesus Christ her Lord", and the care of the earth is, by definition, a global whole-of-life responsibility. There are of course many national and regional variations in terms of an environmental profile, but essentially there is one earth, and that means there are even more common elements. This is not a competition, and we do not each need to "re-invent the wheel" in our own denominational community. The global faith community therefore has an important part to play. That fact has been acknowledged, for example through Statements made by global confessional groups.

In July 2014, the World Council of Churches held a significant seminar called *Green Churches: Ecology, Theology and Justice in Practice*. The seminar had places for only 27 participants from 18 countries globally; the South Pacific group included three from the Methodist Church of New Zealand, one from the Pacific Conference of Churches), and Colleen Geyer from the Uniting Church in Australia. It should be noted that other than a representative of the Pacific Conference of Churches (PCC), there were no Pacific Islanders in the group. It is not necessary to try to give a full account of it here; however, there are several points that may be made involving the sharing of resources and enhanced communication. Participants agreed to the development of a Global Green Churches Handbook, which would clearly include some regional variations. A second agreed project was the "development of education resources and courses on eco-theology that will be accessible for church members and theological colleges and other educational institutions" (Geyer, unpublished 2014) as well as resources for Sunday Schools and schools.

More specifically in the area of communication the seminar agreed to the establishment of global and regional networks, with a website for a global green churches network. Colleen Geyer reported from a South Pacific Regional perspective that:

> Together we noted the importance of being in partnership with the rest of the world, including the need for resources for smaller nations; as well as building partnerships between countries in the region for mutual support, encouragement and sharing of resources. We agreed to work together on a number of initiatives over the coming years to strengthen our network and build partnerships, and to contribute to the churches' commitment to the environment and understanding of broader issues requiring action and change.

To that end they agreed on twelve specific initiatives over the coming years, including the development of a regional network, partnerships between Churches in Australia, New Zealand, and the Pacific, the maintenance of contact and an eco-justice narrative. The potential of those goals for mutual support, empowerment and accountability would be difficult to underestimate. This was clearly a wonderful opportunity for ecumenically oriented faith-based engagement in a critical issue for the Pacific Island nations. Yet in a personal note to me in March 2020, Colleen stated that despite some on-going contact with some participants, she was not aware of any progress with the commitments. At the very least, that is disappointing. The propensity for ecumenical gatherings to make grand commitments and then not carry through with them, is a real concern.

If commitments are made, one would expect that they would come with arrangements for a person or group to take the initiative in converting the plan into action; otherwise they are rendered almost meaningless. What is the significance of any progress or lack of progress? Two main points may be made. First, we all have a degree of responsibility to ensure that such a failure to carry through on commitments becomes a thing of the past. We cannot change the past, but we can change the future. With that in mind, the second point is that ecumenical action can still take place to review the 2014 commitments and devise a plan of action.

Pacific Church Responses

The disappointing outcome of the 2014 Green Churches event does not mean that nothing has happened. On the contrary, the Pacific Conference of Churches (PCC) and the PNG Council of Churches (PNGCC) are both being very proactive in terms of the environmental threats in the Pacific region.

At a Conference hosted by the Papua New Guinea Council of Churches in August 2019, James Bhagwan, General Secretary of the PCC, argued that climate initiatives must change. He picked up the theme of "household" and "highlighted common Pacific values of spirituality, family life, traditional economy, culture, mutual care and respect, which celebrates quality of life and values human beings and creation over the production of things". Bhagwan (2019) went on to say:

> There is something seriously wrong with an economic system that produces so much suffering and poverty, exploitation of labour and widespread environmental devastation. The acceptance of things like "planned obsolescence" – intentionally designing things to break down in the interests of making more money because one doesn't have to cover the real costs (pollution, environmental degradation in production methods, etc.) associated with

waste. This leads to the throwaway disposable culture. These kinds of waste and degradation happens because we don't acknowledge and appreciate what we've been gifted…. (p. 5)

But in addition, the problem also traces back to a false notion of economy and what progress looks like. Pacific Island leaders meeting in Nauru in 2018 also "emphasised how the negative externalities of consumption-based economic growth are undermining the health of our oceans and the safety of our climate" (Bhagwan 2019). Or in the words of Meg Taylor DBE, General Secretary of the Forum: "Pacific Leaders presented a united stance on the pressing need for accelerated and ambitious global action on climate change as they joined world leaders recently for the 72nd Session of the United Nations General Assembly" (2019a).

At some point in the not-distant future, there needs to be a serious discussion on neoliberal capitalism, the sometimes-unspoken assumptions that are expected of us, and the implications of an economic model which appears to require an unquestioned and almost faith-like acceptance. For their part, Pacific peoples have justifiably been noted for their generosity. An example of that, which also speaks into the global household theme, emerged in the context of Australia's bushfire emergency. In a Press Release (2020), Act for Peace (an agency of the National Council of Churches in Australia) quoted Vanuatu's Acting Prime Minister, Jotham Napat as saying, "As a neighbour and friend, we offer whatever assistance we can to Australia", despite the fact that "The people of Vanuatu are some of the poorest in the Pacific".

Faith-Based Opportunities

If the locus for theology is the world, it should be clear that the change in behaviour patterns I and many others are seeking is not for the sake of the Church, but indeed for the world. That has local and immediate implications, since there is no local situation that is devoid of global implications. As Bosch argues, "missiology means globalisation", and that "in order to achieve globalisation, it needs specificity, concretisation" (Bosch 1991, p. 496).

The importance of a faith perspective in the Pacific region is underscored in research on "Spirituality and attitudes towards Nature in the Pacific Islands" by Nunn et al. (2016a). Based on a sample of 1226 students from the University of the South Pacific, the study established that the students shared a "high degree of concern for the global environment and a sizeable majority considered their future would be affected by climate change" (2016a, p. 491). More than that, Nunn et al. concluded not only that faith-based communications are likely to be more effective than secular ones, but also that the quality of the information needs to be upgraded and expressed in a culturally appropriate manner. This might offer both encouragement and a challenge to faith groups wishing to be involved.

At a practical level, many of the faith-based opportunities for engagement on behalf of a sustainable Pacific region will involve revisiting, and revising if necessary, the regional goals set at the Green Churches event in 2014; but as I have suggested, goals without action are relatively meaningless. Yet those opportunities will not appear in a vacuum; rather, they will emerge as a response to the theology that has shaped our life as the Body of Christ. In addition, it is difficult to over-state the importance of our attitude either to help or hinder the process of climate change adaptation in the Pacific. This is not just a matter for Pacific people and nations; the impacts of climate change and other sustainability issues involve all of us. Ultimately it is for the Pacific Churches and nations to determine the priorities within that region, but my contention is that in addition to points already made in this essay, there are at least seven main opportunities for Christian eco-mission.

First, we each need to deal with our ecological issues based on holistic ecotheology. The crisis of sustainability and climate change tends to live at the margins of awareness for too many of us, so that often we do not perceive the care of creation as part of our mission charter. It is important that Churches globally are deeply aware of the undergirding theology and that they have a viable eco-mission aspect to their mission goals. It may seem altogether too obvious, but since to a greater or lesser extent we are all part of the problem, a commitment increasingly to be part of the solution is a necessary starting point.

The second opportunity is actively to encourage the development of partnerships between congregations in Australia/New Zealand with Pacific Islander congregations. This has the potential to encourage the growth of personal relationships from across the Pacific within a Christian framework, and that alone has enormous potential for the encouragement of trust and friendship across the whole region. Among other things, it will require us to learn about the culture of Pacific nations and enable a growth in understanding of the real situation in the Pacific on the part of all participants. An important part of that process may be the encouragement of group visits or pilgrimages, such as the regular exchanges in recent years between the Caloundra Uniting Church congregation in Queensland and Churches in Tonga.

The third opportunity is the establishment of a Pacific regional Green Churches Network. In terms of the 2014 proposals, this will entail the maintaining of "contacts across the region for accountability and mutual support, including regional gatherings". Over time, that may include meeting "as a Pacific 'congress' for encouragement, support and accountability". But two additional points may be made. First, someone must be responsible for ensuring that action is taken. And second, the participating congregations should not be limited to the established ecumenical movement.

Fourth, an eco-mission agenda will necessarily include a viable eco justice narrative. This will include an appreciation of the real concerns of Pacific people and a preparedness to push for the rights of Pacific people.

Fifth, faith-based opportunities for climate change adaptation in the Pacific will call us to act within the context of the Church as a whole, rather than simply our own denomination. But in addition, the opportunities offered by an interfaith approach are significant; the issues are too important for an insular approach. In addition to the potential for enhanced effectiveness, it will also be a demonstration to the whole community of a Christian spirit of reconciliation.

Sixth, and very importantly, seek a viable partnership between Church and State as we address the issues of climate change. However, it should be clear that while this opportunity may take us into areas of possible conflict, we are not talking about partisan politics. There is ample evidence that Pacific leaders are taking a strong and passionate stand on what for them is a life and death issue, namely the environmental threats to the region; that is well-documented following the Pacific Islands Forum in 2019. While the position of Pacific leaders is clear, it is disappointing to note that countries like Australia or the United States have not been fully supportive of stronger action. This complex opportunity therefore invites us not only to pray for the leaders and to support them where it is appropriate, but also to raise our voices in opposition when that is called for.

Seventh, the giving of practical or emergency aid will always be a constant dimension, wherever we are in the world, and various agencies are already doing a significant job in that area. But there are some dangers, such as when aid is given from the perspective of donor assumptions or as a salve for our conscience. The humility of Christ suggests a better approach, namely a partnership that enables mutual learning and understanding, and ministering to one another in times of crisis, as when Vanuatu gave aid to Australia at the height of the fire crisis. But giving is not a substitute for a more sustained effort to deal with the underlying issues in a spirit of partnership.

Concluding Reflections

Earlier in this chapter I referred to Forrester's point that we cannot talk about God without asking what he called the normative question, "What is God calling us to do"? I have sought to show that not only does Christian mission begin with theology, but it inevitably includes an ecological dimension that becomes ever more crucial with the passage of time. Furthermore, in reflecting on the Pacific region, I have identified some important opportunities for mission with implications for the whole creation. The starting point was to establish the theological base for what we seek to do. But beyond that, it is not the function of non-Pacific nations or churches to tell Pacific Islanders what to do. Rather, as I have sought to show, their purpose is to stand with Pacific peoples in a practical faith-based partnership for a sustainable future

The Gospel of Matthew records Jesus' well-known blessing of the children: "Let the children come to me, and do not stop them; for it is to such as these that the kingdom of heaven belongs" (Matt 19:14). That blessing of the children and of all the "little ones" of the world adds special meaning to a comment by Meg Taylor DBE (2019b) in a report on the Pacific Leaders Forum in 2019. She wrote:

> It was indeed an emotional moment when Pacific leaders and delegates were greeted by small children of Tuvalu – submerged in water surrounding a model of their sinking islands with their call to "Save Tuvalu, save the world", upon their arrival to the 50th Pacific Islands Forum held in Funafuti, Tuvalu, from 13 to 16 August (2019). Tuvalu's message could not be clearer or more important – climate change is real and the future of their children and

all the children of our blue Pacific is uncertain – unless urgent, ambitious action is taken to limit global warming to below 1.5C.

At the time of writing, the COVID-19 pandemic is devastating the global community, and it is far from clear how or when it will end, what the post-pandemic world will look like or indeed whether the Pacific Island region will continue to be spared (Leal Filho et al. 2020). But it is worth noting that some national and local leaders are emphasising that "we are all in this together"; certainly, it can bring out the worst in some folk, but there is arguably an even stronger focus on care for one another. The feeling of powerlessness through isolation is a problem. But some are trying to encourage the community with phrases like "We can do this"! Within the framework of a sustainable Pacific, a similar mantra can apply if as Churches in the region we can apply ourselves to the needs and opportunities presented.

References

Act for Peace (NCCA) (2020) (Press Release). Weathering the climate emergency together. www.actforpeace.org.au

Ayre C (2013) Earth, faith and mission: the theology and practice of earthcare. Morning Star Publishing, Northcote

Barnett J (2020) Climate change and food security in the Pacific Islands. In: Connell J, Lowitt K (eds) Food security in small Island States. Springer, Singapore

Bassett L (ed) (2000) Earth and faith: a book of reflection for action. The United Nations Environment Programme, New York

Becker M, Meyssignac C, Letetrel W, Cazenave T (2012) Sea level variations at tropical Pacific islands since 1950. Glob Planet Chang 80–81:85–98. https://doi.org/10.1016/j.gloplacha.2011.09.004

Bhagwan J (2019) The effect of modern "development" trends on the Pacific. https://www.pacificconferenceofchurches.org/f/climate-initiatives-must-change

Bosch DJ (1991) Transforming mission: paradigm shifts in theology of mission. Orbis Books, Maryknoll, NY

Bühlmann W (1982) The chosen peoples. St Paul Publications, Middlegreen, Slough, UK

Conradie E (2007) The whole household of God (oikos): some ecclesiological perspectives (Part 1 and 2). Scriptura 94:1–28

Conradie E (2008) The church and climate change. Cluster Publications, Pietermaritzburg, South Africa

Conradie E (2011) Christianity and earthkeeping: in search of an inspiring vision. Sun Press, Stellenbosch, South Africa

Duchrow U, Liedke G (1989) Shalom: biblical perspectives on creation, justice, and peace. WCC Publications, Geneva

Forrester DB (2000) Truthful action: explorations in practical theology. T & T Clark, Edinburgh

Geyer C (2014) Report on Green churches: ecology, theology and justice in practice. Unpublished

Langmead R (2002) Ecomissiology. Missiology: Int Rev XXX(4) Oct 2002

Lavers JL, Bond AL (2017) Exceptional and rapid accumulation of anthropogenic debris on one of the world's most pristine islands. PNAS 114(23):6052–6055. www.pnas.org/cgi/doi/10.1073/pnas.1619818114

Leal Filho W, Ha'apio MO, Luetz JM, Li C (2020) Climate change adaptation as a development challenge to small Island states: a case study from the Solomon Islands. Environ Sci Policy 107:179–187. https://doi.org/10.1016/j.envsci.2020.03.008

Luetz J, Havea PH (2018) "We're not refugees, we'll stay here until we die!"—Climate change adaptation and migration experiences gathered from the Tulun and Nissan Atolls of Bougainville, Papua New Guinea. In: Leal Filho W (ed) Climate change impacts and adaptation strategies for coastal communities. Springer Nature, Switzerland, pp 3–29. https://doi.org/10.1007/978-3-319-70703-7_1

Martin PCM, Nunn PD, Leon J, Tindale N (2018) Responding to multiple climate-linked stressors in a remote island context: the example of Yadua Island, Fiji. Clim Risk Manag 21:7–15. https://doi.org/10.1016/j.crm.2018.04.003

Moltmann J (1985) God in creation: an ecological doctrine of creation. SCM Press Ltd., London

Morrison A (2020) Observations. Personal correspondence

Nash J (1991) Loving nature: ecological integrity and Christian responsibility. Abingdon, Nashville

NCCA (2019) (Press Release). National Councils of Churches in the Pacific meet. https://www.ncca.org.au/newsletters/june2019#2

Nunn PD, Mulgrew K, Scott-Parker B, Hine DW, Marks ADG, Mahar D, Maebuta J (2016a). Spirituality and attitudes towards nature in the Pacific Islands: insights for enabling climate-change adaptation. Clim Chang, 136, 477–493. https://doi.org/10.1007/s10584-016-1646-9

Nunn PD, Runman J, Falanruw M, Kumar R (2016b) Culturally grounded responses to coastal change on islands in the Federated States of Micronesia, northwest Pacific Ocean. RegNal Environ Chang 17(4). ISSN 1436-3798. Springer-Verlag, Berlin, Heidelberg

Pacific Conference of Churches (2019) (Press Release) Creation breaks barriers. https://www.pacificconferenceofchurches.org/f/creation-breaks-barriers

Sanguin B (2007) Darwin, divinity, and the dance of the cosmos: an ecological Christianity. Wood Lake Publishing, Kelowna, BC, Canada

Taylor M (DBE) (2019a) Pacific Leaders reaffirm their leadership for Climate Change and Resilience. https://www.forumsec.org/2017/09/08/pacific-leaders-reaffirm-leadership-climate-change-resilience/

Taylor, M (DBE) (2019b). Pacific leaders set new bar by collectively declaring climate crisis. https://www.climatechangenews.com/2019/08/28

Vischer L (1980) Intercession. WCC, Geneva

Clive W. Ayre is an Honorary Research Fellow at the University of Queensland, and a retired Minister of the Word in the Uniting Church in Australia. He has served in a variety of congregations, mainly in Queensland, and has been significantly involved in the ecumenical movement. His published work includes Earth, Faith and Mission: The Theology and Practice of Earthcare, based on his PhD thesis. His principal interests relate to practical eco-theology and mission, ecumenism, and reconciliation. Clive was born and educated in Brisbane, Australia and has been married to Gail since 1968; they have three adult children and ten grandchildren.